Taxation of
Nonrenewable
Resources

Taxation of Nonrenewable Resources

Albert M. Church
Lincoln Institute of
 Land Policy
University of New Mexico

LexingtonBooks
D.C. Heath and Company
Lexington, Massachusetts
Toronto

Library of Congress Cataloging in Publication Data

Church, Albert M.
  Taxation of nonrenewable resources.

  Bibliography: p.
  Includes index.
  1. Natural resources—Taxation—United States.
I. Title.
HC103.7.C48   336.2'783337'0973   80-8784
ISBN 0-669-04367-2   AACR2

Published simultaneously in Canada

Printed in the United States of America

International Standard Book Number: 0-669-04367-2

Library of Congress Catalog Card Number: 80-8784

*To Michael and Eric
and to my parents*

Contents

List of Figures

List of Tables

Acknowledgments

First I must express my appreciation for financial support and patience to the Lincoln Institute of Land Policy and to Arlo Woolery and Charles C. Cook specifically. Without their support this effort would not have been possible. Similar support came from the National Science Foundation (Grant #AER 77-18684) and my project officer, Dr. Lynn Pollnow. The standard disclaimers hold that neither institution endorses the methodology or conclusions of the research reported here.

I am also deeply indebted to Joel P. Clark and Patricia T. Foley, Department of Materials Science and Engineering, Massachusetts Institute of Technology, and to Martin B. Zimmerman, The Sloan School of Management, Massachusetts Institute of Technology, who developed the copper and coal simulation models. This research permitted the calculation of effective tax-rate estimates and simulation of alternative tax regimes. I am grateful for the liberty to summarize their work in chapter 6.

A large number of other people contributed to this effort. Channing Kurry made detailed suggestions and critical comments on the manuscript, and his encouragement came at a much needed time. Edna Loeman helped to formulate the model of pragmatic tax policymaking. Recommendations and insights into policymaking and administration came from individuals too numerous to mention. Included are the Western Association of Tax Administrators, the Western Tax Association, other academic forums (including some of my classes) where portions of the contents of this book were presented, and tax officials in many states. Perhaps it is regrettable that these latter individuals are classified as bureaucrats and legislators. The connotations of these words are in no way negative, for I found them to be cooperative, frank, and dedicated to their jobs. The advisory panel to the NSF grant made helpful suggestions.

Finally, special thanks go to Sharon Shea and Sharon Woolery for putting up with sloppy writing and editing and for producing a crisp, clean final draft. Their patience and support were extraordinary.

Introduction

This book is about the taxation of nonrenewable natural resources. While it is comprehensive, it is not exhaustive. The reader will notice that the scope of vision changes from a view of a wide-open landscape to one of a narrow and detailed examination of selected facets of a single element. This variation is caused by my experience, interests, and limitations. Because of their own interests and requirements, readers will want to linger at some points and skip over others. This procedure is both possible and recommended, for the chapters and sections are relatively independent and the assumed background knowledge of the reader is not uniform. Those lacking background in economics should be able to comprehend most of the material. However, certain sections rely heavily on economic theory and jargon and therefore will be of interest to specialists mainly.

The subject of natural-resource taxation is an important one, and it is indisputable that it will become increasingly important. The reason for its increasing importance is the rapidly escalating prices of certain nonrenewable resources and the significant share of resource wealth being captured and redistributed by means of direct ownership (via royalties and bonus payments), direct regulation (price regulation, pollution laws, and so on), and indirect socialization via taxes. However, taxes, regulation, and ownership do more than extract income from one set of individuals and redistribute it to others via either transfer payments or expenditures that benefit certain individuals. These government policies also change how individuals make economic decisions. This is desirable if it results in natural resources' being used more efficiently. However, policies that create distortions and inefficient resource allocation are undesirable. While these statements apply to all services and commodities that are valued either positively or negatively in our society, nonrenewable natural resources introduce an added dimension— they are finite and exhaustible. Once used up, they are irreplaceable and are unavailable to future generations. The limits-to-growth literature of the 1960s stressed how the exhaustibility of natural resources and the sensitivity of the natural environment are affected by temporal production and consumption decisions and thus the importance of the intergenerational legacy. Events in the Middle East have established the vulnerability caused by our reliance on imported oil.

Tax policy of federal and state governments toward natural resources is becoming economically and politically significant, and these policies are at loggerheads. Price deregulation, increasing resource scarcity, and the Organization of Petroleum Exporting Countries (OPEC) have increased market prices and industry profits. State and federal governments have enacted tax legislation that determines who bears the burden of more costly natural resources and how and where those resources are developed. The level of conflict among the natural-resource industry, consumers in the industrial Midwest and East, and residents of natural resource-rich states is growing. For example, Alaska is expected to collect an additional $37 billion in tax revenues from 1980 to 1990 as a result of oil-price deregulation and the exemption of state severance taxes and royalties from the windfall-profits tax; and Montana's 30 percent coal severance tax has caused court challenges by consuming regions under the commerce clause of the U.S. Constitution. A 1979–1980 exchange of letters and analyses between the Goverment Accounting Office (GAO) and the Department of Energy (DOE) makes it apparent that the federal government is failing to conduct tax-policy analysis. This book reviews the current state of knowledge regarding natural-resource taxation and extends that frontier ever so slightly.

The basic questions addressed in this book are: How do taxes affect the distribution of income among owners of natural resources and extractive firms, other inputs into the production process, and consumers? How do taxes distort the use of natural resources away from their most efficient uses? Notice that the questions are phrased as "how do" taxes affect resource and income distribution, and not "how should" these taxes be applied. The latter question can never be answered by economists because there are no acceptable criteria that can be used to evaluate how incomes should be distributed. However, economists have formulated criteria to define economic efficiency, and these may be employed to evaluate whether government policies increase or decrease efficiency. These criteria are developed here, as are the ways in which tax policy affects incomes and resource use.

Natural-resource taxation is approached pragmatically, theoretically, and empirically. Chapter 1 reviews the theoretical and pragmatic goals of tax policy, and an empirical investigation of state coal taxes is undertaken to determine the preferences of state tax policymakers and their constituents. The literature concerning the economics of nonrenewable natural resources is surveyed, with particular attention to information concerning the effects of severance, property, and income and other taxes in chapters 2 and 3. Various tax bases are ranked according to a set of multiple tax objectives. A theoretical model of tax incidence is developed to analyze questions of income distribution, and an empirical analysis of who bears the tax burden is made of the corporate income tax and federal coal leasing in chapter 4. The nominal and effective taxes imposed on the extractive industries, with special emphasis on copper and coal, are described and cataloged in chapter 5. Empirical models of the domestic coal and copper industries are developed, and simulations of several tax regimes are analyzed for the effects on the distribution of income among labor, firm owners, and consumers as well as the effects on economic efficiency in chapter 6.

1 Tax Policy Goals

Nations and their political subdivisions of states, provinces, and municipalities are in the unceasing process of making, administering, and enforcing tax policy. We read reports, deliberations, pronouncements, and other propaganda released by tax policymakers and their staffs, lobbyists, and organized taxpayer groups about proposed and operational taxes. Some of this information is accurate, some is somewhat biased, and some of it is purposeful obscurantism. This information is sometimes in the form of criticism, sometimes justification, and sometimes advocacy for recommended policy changes. Conflicting viewpoints and unstated and imprecise statements about what tax policy goals "should" be make it difficult to evaluate alternative proposals. Both theoretic and pragmatic tax policy goals, as seen through the eyes of an economist, are reviewed in this chapter.

There is little doubt that the 1970s could be labeled as the period when historical trends in energy use and expectations about the future availability and cost of nonrenewable resources were revised. Although the specter of natural-resource exhaustion has been raised on many occasions, until the 1970s history had shown that the real costs of both fuels (energy) and nonfuels had been falling and their availability had been rising. These trends were essential to the industrial revolution. Awareness of these trends diminished the credibility of reported impending constraints on growth. The report by the Club of Rome entitled *The Limits to Growth* (Meadows et al. 1972), which portrayed an impending exhaustion of nonrenewable resources, failure to produce sufficient renewable resources, and the depletion of earthly life as we know it because of toxic levels of pollution and human-induced changes in the earth's environment, was widely reported but heavily discounted.

The successful OPEC-organized oil embargo, gasoline lines, and the multifold increase in price of imported oil in 1973–1974 and in 1979 were hard evidence that the Club of Rome might be right. Prices for spot and future delivery of virtually all natural resources skyrocketed in 1973. However, it takes more than one sudden shock to change people's habits and expectations. Shortly after the gas lines disappeared, the demand for heavy, gas-guzzler automobiles could not be met by the Detroit production lines, and the spot and future prices of natural resources, save oil, plummeted to "near-normal" levels.

However, by the mid-1970s legislators in the energy-rich Western states in the United States and western Provinces in Canada, who observed profits of resource firms and read research reports on energy supply and demand, became aware of both the economic activity which resource development was bringing and the associated social costs fostered by the development. A few states and provinces began increasing taxes on the extraction of coal, oil, natural gas, and uranium; and this cascaded into widespread tax increases by the end of the 1970s. A 1976 report issued by Rand Corporation consultants stated that "...the emerging pattern of state coal tax policy in the Northern Great Plains is one of OPEC-like revenue maximization" (Nehrig and Zycher 1976, p. 148).

By the end of the 1970s, the festering conflicts between energy-rich states and energy-consuming states had surfaced to open combat, with the federal government acting schizophrenically by pursuing multiple goals. Having multiple goals and being buffeted by unexpected world events (such as the 1979 Iranian revolution), the federal government has been notably unsuccessful in achieving a coherent, practical policy toward natural-resource development, particularly energy.

Prior to the shocks of 1973–1974, the primary goal of federal policy was to keep nonrenewable resource prices low. A complex patchwork of taxes and direct regulation had achieved this end. Various tax preferences, which grew dramatically through the twentieth century, benefited these industries. One effect was to make the after-tax rate of return more attractive than alternatives and thus attract investment dollars. The result was greater extractive capacity; low prices; efficient and inexpensive transportation of energy, goods, and people; and a conversion to a high-technology, energy-intensive, automobile-oriented society. The acknowledged major objective of the Federal Power Commission (FPC) was to keep the price of natural gas low, and this encouraged a rapid growth of pipelines, electrical power generation, and residential and industrial consumption.

Energy consumers benefited temporarily from stimulative tax and regulatory policies. The most readily accessible and inexpensive resources were extracted first. However, voices of economists and others pointed out that when tax and regulatory policies keep prices below what they would have been in their absence, resources are extracted more rapidly than is socially desirable and thus consumption is artificially encouraged.

During this period, the Western energy-rich states were sensitive to what some have called their colonial and exploited status. The Texas Railroad Commission and other state conservation regulations controlled pumping rates and sought to increase prices. One federal response was to extend enormous tax advantages to imported oil and to phase out the lucrative percentage depletion allowance which benefits domestic producers. Government intervention in and conflicts over resource development and intergovernmental, consumer, and producer conflicts over who should benefit and who should pay became clearly defined.

The shocks of 1973–1974 deepened the bias in favor of the consumer and intensified the battle over resource income. Oil prices were regulated at the production, distribution, and retailing phases. Natural gas was maintained at an artifically low price. Pollution regulations made nuclear and oil-fired electrical power generation more attractive than coal, particularly low-sulfur Western coal. The combined effects of regulation and taxation made high-cost imported oil more profitable than searching for new domestic sources and extracting higher-cost reserves by using various enhanced recovery techniques. Domestic energy production stagnated.

Although a coherent and implementable national energy plan was avoided, a realization was born that conventional domestic energy production had to be encouraged by means of discovering and recovering high-cost oil and gas to replace expensive, unreliable foreign sources; an increasing reliance would have to be placed on coal, nuclear power, and renewables (solar, geothermal, and so on); and energy would have to be conserved and used more efficiently (in the 1970s the United States consumed 30 percent of the world's oil output). These goals required a revamping of tax and regulatory policies. Oil and gas were deregulated (complete decontrol of oil prices in 1981 and natural-gas decontrol in 1985), so that users would face the true social cost of consuming the resource. Some of the tax preferences for conventional resource extraction were removed (for example, the percentage depletion for oil), and tax incentives were installed to encourage conservation and the use of renewable energy. A preferable policy would be to remove all tax incentives and tax disadvantages which impinge on the transition to nonconventional and renewable-energy sources. However, it is politically easier to grant new tax concessions than to remove old ones.

In 1979 the Iranian revolution and oil cutback, OPEC price increases in excess of 100 percent and deterioration of their price discipline, and Russian expansionism lent additional credibility and urgency to these evolving and required changes. However, the policy changes of deregulation and tax adjustments have created an enormous political and social problem. As domestic energy prices increase, profits to owners of previously discovered reserves of oil, natural gas, and coal with low recovery costs increase along with the prices of new high-cost OPEC oil, Alaskan North Slope oil, and oil recovered by means of secondary and tertiary techniques. These immense riches are both a political liability to those who have advocated deregulation and equal tax treatment and an opportunity for a new source of government revenue. The intensity of the conflict over these "blue-eyed Arab" rich is continuously escalating. The energy industry maintains that it requires profits for investment in exploration, development, and production; yet some energy firms embarrass their colleagues by acquiring nonenergy enterprises for inflated sums (for example, Exxon's acquisition of Reliance Electric Co.). Stocks in domestic energy producers rally as profits increase, and expectations are created among investors of future wealth. Energy-related industries, particularly those which possess monopoly power such as railroads, barge carriers, pipelines, and electric power producers, aggressively press for rate increases (for example, see Zimmerman 1979*b*).

As a condition for the deregulation of domestic oil production, Congress enacted the windfall-profits tax. And not far behind, governments of energy-rich states were enacting new tax legislation, and governments of energy-consuming states were resisting the former tax increases but pushing for energy-related taxes of their own. There is little doubt that the 1980s will be a decade in which owners, producers, consumers, and federal and state governments will continue to battle for control of resource wealth, and that conflict will escalate.

The size of this wealth and, by implication, the burden on consumers are staggering. The U.S. Treasury has estimated that the windfall-profits tax, which began in March 1980 and expires in 1990, will raise $228 billion. Royalties going to state governments and Indians are exempt from the tax. The U.S. Treasury estimates that this tax, when coupled with deregulation, will result in $127.7 billion in revenues to oil and gas states, as shown in table 1–1.

The Prudhoe Bay discovery in 1968, the trans-Alaskan pipeline in 1977, oil-price deregulation (1980), and state ownership of resources have brought an embarrassment of riches to Alaska. July 1981 brought a $3 billion state surplus, and $53 billion in surpluses is expected over the remainder of the decade as a result of the 12.5 percent tax on oil's wellhead price, a 12.25 percent production tax, and the state income tax on oil companies. Royalties from production on Indian and state lands are additional flows not included

Table 1–1
State Revenues Resulting from the Phased Decontrol of Domestic Oil, 1980–1990

Area	Total Revenue (Billion $)	Percentage
Texas	33.2	26.0
Alaska	37.3	29.0
Louisiana	13.8	11.0
California	21.8	17.0
Oklahoma	3.1	2.0
Wyoming	3.5	0.3
New Mexico	1.5	1.0
Kansas	0.9	0.6
All other[a]	12.6	10.0
Total	127.7	100.0

Source: U.S. Department of Treasury, unpublished.

[a]No breakdown by state is available.

in these numbers. The *Anchorage Times* ran a regular feature, "Helping the Governor Spend a Billion," where readers' suggestions were printed. Creative legislators came up with $16 billion in appropriations bills for 1981, although the 1980 state operating budget was only $1.3 billion. Subsidized home loans, immediately paying off the state's obligation to Alaska natives, a permanent fund which would be invested in government and corporate financial securities whose income will replace oil revenues (Prudhoe Bay production is expected to decline starting in 1987), subsidized loans for job-creating businesses, and cash grants to residents in lieu of tax reductions are some of the ways the state will spend its riches.

In 1975 Montana increased the maximum severance tax on coal to 30 percent. The tax was challenged in the state courts, and it is certain to be appealed to the U.S. Supreme Court. Former Secretary of State William P. Rogers argued on behalf of a group of Midwestern utilities. Detroit Edison Company expects to pay as much as $1.36 billion to the state by 2002, although the state's coal-tax revenue in 1980 amounted to $42 million. They maintain the tax is confiscatory, unfair, and unconstitutional. The state is convinced the tax is constitutional and is prepared to fight. One argument is that coal development requires government-supplied services and state residents deserve compensation for the disruption and pollution. Wyoming is contemplating a rise of the state's 17 percent severance tax ($0.89 per ton in 1980) to 25 percent for the same reasons. North Dakota expects a $50 million deficit by 1983 with severance taxes at their 1980 rates. However, critics and sometimes proponents of these taxes make it clear that resource taxes are a way of exporting their tax burdens to consumers in other states.

The 1976 Rand report (Nehrig and Zycher 1976) was prophetic in its statement that consuming states, electric utilities, and coal producers might support a federal limit on severance taxes. In 1980, forty cosponsors introduced a bill in the House of Representatives to limit state coal severance taxes to 12.5 percent. One Montana representative said the bill could ignite an economic war among states that would "make the Sagebrush Rebellion look like a garden party." Consuming states are getting into the act on their own by taxing the refining and distribution of oil. Connecticut instituted a tax on wholesale oil distribution of 2 percent of gross receipts in 1980. Its expected revenue is $60 million per year. New York, Illinois, Kentucky, Louisiana, Massachusetts, Maryland, North Dakota, and Pennsylvania considered similar special tax measures. In spring 1980, California voters defeated Proposition 11, a proposal to enact a 10 percent surcharge on the income of large oil companies in the state. This would have raised between $300 million and $420 million in the first year.

The ultimate constraint on these tax and regulatory schemes hatched by producing and consuming states is Article 1, section 8, of the U.S. Constitution. This is the commerce clause, which grants Congress the power to regulate commerce among the states. However, precisely how the commerce clause constrains states' taxing powers is not specified in the original document. This has been called the "great silence." The courts have ruled consistently that state legislation which impedes interstate commerce and is discriminatory violates this clause. However the Southerland Supreme Court ruled in the 1920s that Pennsylvania's 25 percent severance tax on coal (90 percent of which was exported) was constitutional because the act of severing coal did not involve goods in interstate commerce. Consequent to this test case, the court has consistently ruled in favor of state severance as well as privilege and use taxes as long as such taxes (1) impose no multiple burdens and are fairly apportioned, (2) do not discriminate between interstate and intrastate commerce, and (3) are reasonably related to the services provided by the taxing state.

Montana's defense of its 30 percent severance tax rests directly on the severance-tax precedents. The utility company plaintiffs maintain that the tax violates the commerce clause because of its burdensome level and that expected revenues far exceed the costs of any services provided by the state and local governments.

Since most of the public debate over any single tax measure centers on rather specific, fine points, it becomes unclear what tax policy objectives each interest group is striving for. The publicly held position is usually that certain aspects of proposed or existing tax legislation will be damaging or favorable to the economy and it is only incidental that it will be detrimental or beneficial to their own incomes as well. Thus one meaningful way to separate tax policy goals is by analyzing what is "best" for the entire economy and how alternative tax structures affect incomes and the economic well-being of different interest groups.

Most people will certainly agree that special interest groups have little trouble ascertaining which aspects of tax policy are beneficial to them and which are not. When special interest groups speak for consumers, workers, the affected industry, or any other agglomeration of people, their motives must be suspect. How, then, does a policymaker go about specifying overall tax policy goals and deriving optimum tax policy? The answer is that it is obviously a complex task, and maybe no single optimum optimorium exists. If this is true, can we trust supposedly unbiased experts who state that an optimum tax policy does exist? The most difficult questions are addressed in this chapter: What is the intent of tax policymakers? What "should" it be? And how can their behavior be analyzed?

"Optimal" Public Policy

A key to answering these questions obviously centers on how *optimal* or *best* is defined. What is optimal for the owners of Exxon with respect to taxes on oil, gas, and other fuel production and distribution may not be optimal as defined by consumers. Economists have sought to define a universal criterion for *optimum* that is independent of the narrow self-interests of any individual segment in society. The heart of the criterion is the single goal—economic efficiency. This is the ability of society to satisfy the preferences of its members to the greatest extent feasible, given scarce resources and the limited capability for an economy to produce goods and services. The efficiency criterion is attributed to the nineteenth-century Italian economist Pareto. Simply stated, the *Pareto-optimum criterion* holds that the limited resources of a society are being utilized optimally when the only way to make one person better off is to make one or more other persons worse off. Conversely, a *suboptimal state* exists when resources may be rearranged in a manner which permits improving the welfare of one or more individuals without diminishing any individual's welfare. This situation implies that a reallocation of resources can produce a net benefit to society, and consequently the suboptimal state is inefficient.

As abstract as this criterion appears to be, it has proved useful to theoretical economists. From the time of Adam Smith, who published his *Wealth of Nations* in the year of U.S. independence, economists have attempted to prove that an unregulated market allocates resources efficiently (in Pareto's sense) by means of the price mechanism. It was not until the twentieth century that modern mathematical techniques permitted economists to formally prove the correspondence between the competitive market allocation of resources and Pareto optimality (economic efficiency). The practical importance of Smith's assertion of market efficiency and the history of thought leading to its mathematical proof are enormous. The doctrine has served as an alternative to social Darwinism and other cruder justifications for minimal government regulation and intervention in capitalistic nations. It is repeatedly put forth by political conservatives and chamber of commerce businesspeople as the justification for fiscal conservatism and small government.

A second application of the Pareto-optimality criterion is in judging the soundness of public policy. Often, to the consternation of these same political conservatives, economists from Smith to the present have described situations where the competitive markets' allocation of resources may fail to achieve Pareto optimality. These are so-called market failures. Smith stated that government is necessary for national security, to maintain internal peace and to protect private property and the right of contract as administered through a system of laws and courts. Also, he noted that whenever a group of businesspeople meet, they will attempt to conspire to set prices and behave jointly as a cartel or monopoly. It is easily shown that a monopoly will produce less and charge a higher price than an industry organized on a competitive basis. If the monopoly or cartel could be made to behave as a competitive industry by means of public policy, then industry output and consumption would increase and prices would fall. This is clearly one instance where government intervention might enable society to move from a Pareto-suboptimal allocation to a more efficient utilization of its scarce resources. In fact, conceivably this result could be achieved by means of a tax and subsidy scheme. It could be designed to ensure that some members of society are better off and none are worse off—a Pareto-superior change.

The solution to the monopoly market-failure problem is a classic one which is included in every introductory textbook of economics. In the monopoly case, the Pareto-optimality criterion applied to public policy would conclude, "Go ahead and implement the tax-subsidy scheme." The empirical problem, of course, is that it is not always so easy to identify monopolies and cartels or to control them. The most powerful cartels such as OPEC are not so easily controlled as in the textbook example of a single domestic monopoly (Griffin and Steele 1980). Furthermore, as the OPEC example clearly illustrates, a monopoly or cartel earns a greater profit than a competitive industry. It can use a portion of these profits to protect and defend its monopoly position. And if the effects of higher international monopoly prices can be primarily exported to consumers, whose welfare does not primarily concern the country of origin, then a cartel may increase national welfare and serve national policy.

International trade is another classical example of unregulated markets achieving Pareto optimality which has been examined by economists. It is easily shown that free trade unhindered by tariffs, quotas, or other restrictions will result in a Pareto-optimal allocation of resources for the countries engaged in trade. This implication has been preached by economists since the early nineteenth century. However, in large part these calls have gone unheeded since protectionism, mercantilism, and imperialism have been a cornerstone of many nations' trade policies up to this day. The birth of the European Common Market and the General Agreement on Tariffs and Trade (GATT) negotiations and agreements beginning under President Kennedy are testimony that the economists' scribblings have been read in some quarters and acted on. The finding that free trade lends to Pareto optimality rests on a foundation which encompasses the welfare of the citizens of all trading countries. If one were to take a narrower, parochial point of view in which the welfare of a single country is considered, then a policy of tariffs and trade restrictions may produce a net benefit to that society. This contradiction reveals the fatal weakness of the Pareto-optimality criterion in judging public policy: it makes no distinctions about the distribution of income and wealth; in other words, it fails to account for who benefits and who is hurt by a public policy.

The only public policies which pass the Pareto test, that is which correct market failures, are those which end up benefiting one or more individuals and hurting no one. There are regrettably few changes in resource allocation that can be achieved through taxation, regulation, or any other government policy which in practice can "cut Pareto's mustard." Even though Pareto-superior policies can be imagined, they are virtually impossible to implement. This is because taxation, expenditure, and regulation are broad policy tools which are incapable of being subtle enough in actual implementation to both reallocate resources to increase welfare and do it in such a way that nobody is made worse off. Almost without exception, government policies make some people better off, which is fine by Pareto, but also make some people worse off, which is not acceptable for the Pareto test. What is even more interesting is that who ultimately is made worse off and better off by government intervention is not always immediately clear, particularly when policy is being formulated. This places policymakers in a double dilemma. No policy can make everyone happy, and it may be difficult to predict who ends up where. Of course, this uncertainty also proves useful to the politician, bureaucrat, and special interest group, for it is possible to make all sorts of claims and engage in purposeful obfuscation.

Pareto Optimality and the Distribution of Income and Wealth

Frustration over the inability of virtually all feasible policies to pass the strict Pareto test of making no one worse off has led to various suggestions for a modified criterion. It is intuitively appealing to amend the criterion to permit changes in the distribution of individual welfare, income, and wealth (often employed as a measure of welfare) so that an overall or net benefit results. The necessary modification is called a *compensation test*. If the gainers in a proposed change in policy are able to compensate the losers so that the losers will voluntarily agree to the change, then the compensation criterion holds that the change does cut the mustard. Debate has arisen over the logical consistency of the test and whether compensation actually needs to be made or whether it is sufficient to be potentially feasible.

The consensus is that the compensation criterion is itself potentially inconsistent and that the issue of actual payment for compensation is a societal value judgment. This occurs because the strict Pareto criterion can be achieved with any conceivable distribution of income. Of course, even if the compensation test were acceptable theoretically, it could never be implemented because of information and measurement problems. Regretfully, it must be concluded that theoretical welfare economics provides little practical guidance in establishing optimal tax policy. Situations in which tax policy may theoretically shift society from a Pareto-suboptimal state to a Pareto-optimal one are discussed in chapter 2, but these are limited in number. No existing tax levied on industries extracting and processing nonrenewable resources nor any proposed tax could be justified on narrow Pareto grounds. It must be concluded that although the strict Pareto criteria have their uses in theoretical discussions of tax policy, they offer no useful guidelines for setting pragmatic policy.

There is a simple test which can support this assertion. Given the existing tax regime and economic conditions, the distribution of income and wealth is determined. In order to pass the strict Pareto test, a proposed change in tax policy would have to improve at least one person's lot without diminishing any other person's income or wealth even to the most minute degree. I defy the reader to visualize any implementable change in taxes which could pass this test in the real world. Pragmatic tax policy making requires that changes in real income engendered by tax and expenditure policy be accepted by a majority of legislators and ultimately, by implication, by a majority of voters. The Pareto criterion for economic efficiency implies a unanimous voluntary agreement. Unanimity except in small groups is inconceivable.

A majority criterion opens up the grounds for making tax policy in the real world even though it may fail a strict Pareto test or even to meet the compensation criterion. Probably the most important criterion in the tax and expenditure decisions is how incomes of taxpayers and of those receiving the benefits of government are perceived to be affected. Individuals and special interest groups are primarily interested in using the public sector to enhance their welfare. They have far less concern for how others are affected and place virtually no importance on the goal of economic efficiency.

Pragmatic Tax Policy

The major concern of national and state legislators is that government budgeting and regulation of the private sector will benefit their supporting constituents as defined by interest, philosophy, and geographic region. Elected officials who fail to convince a majority of the voters in their jurisdiction that government is helping them will fail to return to their posts after election. What is meant by "helping" is that voters must feel that when all government programs, budgeting, and regulations are lumped together, the voters are receiving more value from services than the costs of their tax dollars. Individual voters thus look at their own interests and the interests of others in society they believe are worthy of support. The notions of efficient use of resources and Pareto-optimal public-sector policies carry little weight in determining what candidate, party, or issues to support. Those running for office realize that public opinion and perceived changes in the distribution of income must be entertained. We analyze the effects of alternative tax structures on the net incomes of various groups in chapters 4 and 6 from a theoretic and an empirical basis. Here the incidence of taxes is painted with a broader brush.

Although there are those who would contest some or many of the statements made here in asserting that elected officials are primarily interested in the income-distribution effects of public policy because their constituents are primarily concerned about their own preferences, most will agree that this position is a tenable one. Accept this position for a bit, and let us follow its implications. If different interest blocs in society favor policies which enhance their absolute and relative positions in the structure of income and wealth, then, of necessity, conflicts will result. The art of politics is the art of balancing conflicts and making compromises. This is easily stated but obviously difficult to quantify and make operational, as every politician, successful or not, knows. We attempt to quantify these conflicts among various interest groups into what is a simplified, but hopefully insightful, specification of the pragmatic tax policy goals in an individual state. Because voter constituencies are different in each state, we expect variation in tax policies which reflect the majority position. We ignore federal tax policy for the moment.

One nonrenewable-resource tax policy goal which has been suggested as providing insight into the involved political process is that of maximizing state tax revenues. The philosophical root of this objective can be traced to the concept that the sovereign owns the mineral rights to all the land of the realm and must be compensated for their use. This doctrine is followed to this day in British Commonwealth countries. For example, in Canada, mining on public and private lands is assessed both royalties and taxes as determined by the provincial and federal governments. A legal justification of the state severance tax levied on the extraction of nonrenewable resources in the United States is based on the same line of reasoning (Bingaman 1970). An extension of this argument is that the state should maximize tax revenues from nonrenewable resources since the state is the trustee of present and future generations and once extracted, the resource is gone forever. The implications of a policy of a state maximizing resource-tax revenues is analyzed by Shelton and Morgan (1977), Church (1978b), Gaffney (1977a), Gillis (1977), Hogan and Shelton (1973), Link (1978), McLure (1967, 1969, 1970a, 1975, 1978), Sandler and Shelton (1972), and Sorenson and Greenfield (1977). The income distributive effect of tax policy is what we hypothesize that tax policymakers are most interested in. Severance tax revenues benefit taxpayers by making other tax rates lower than they might have been had the resource tax not been in effect. Expenditures, funded by the severance tax, may also provide government services which benefit the citizens in the tax jurisdiction.

The exploitation of a nonrenewable resource, particularly in cases where rapid economic development is taking place, results in accelerating demands for public services such as schools, utility systems, highways, police protection, and so forth. Furthermore, the resource frequently is found in remote locations. The development and construction phase requires more workers and support personnel than the extraction and operation phase, as in the case of coal-fired, mine-mouth electrical power plants and proposed coal gasification plants. Consequently, employment and population changes are lumpy, and in all cases when resource exhaustion takes place, employment and population of the region will diminish. The typical life cycle of a resource-based boomtown is one of initial rapid employment and population growth during the construction and development phase. Nearby towns are unable to supply basic public services to the new arrivals. Initially a sufficient tax base does not exist to support capital improvements such as streets, schools, and so forth. Resource extracting and processing improvements are not complete, and no exploitation or production has started, so property and production (severance) tax revenues do not flow initially. Thus insufficient lead time and capital exist to construct the necessary schools, water-supply systems, roads, and other social overhead. During the exploitation and operation phase, employment diminishes. This may result in unused improvements if government agencies have overbuilt to accommodate the recently passed peak construction population. When this occurs (past examples indicate this takes anywhere from ten to thirty years), a reduced population or ghost town may result (Cummings and Mehr 1977; Cummings and Schulze 1978; Gilmore et al. 1976). Not only may this development life cycle result in utilizing public services and capital inefficiently, but also the human costs may be far greater. This growth pattern necessitates large-scale migration as construction personnel are replaced with operational personnel, who must in turn depart when the exploitable resource is fully depleted. Rapid growth, turnover, and departure cause severe disruptions to those directly involved and to all other residents who are indirectly affected by the social instability and associated fluctuations in property values. The economic and human costs of the natural-resource-based boomtown phenomenon are now well documented.

The development, extraction, and eventual depletion of a natural resource engender environmental damage and disruption as well as social and economic dislocations and inefficiencies. These environmental damages remain long after the departure of the extracting industry and its employees. In fact, some damage may be irreversible and thus place a burden on all future generations in addition to the costs borne by those earning income from and consuming the fruits of resource development (Kneese and Schultz 1975). Extraction-based taxes may be used to compensate present and future generations for these social, economic, and environmental costs (Schulze 1974).

One motive behind Montana legislators enacting a severance tax with a maximum 30 percent rate on coal extraction in 1975 was compensation

(Dumars, Brown, and Browde 1979). Of the tax revenues 25 percent went into a trust fund to be used to provide future employment in mineral-depleted areas, and this was increased to 50 percent in 1979. A coalition of ranchers, who wished their lands and lifestyles protected; environmentalists, who deplored the strip-mining techniques of the coal companies; and general citizenry, who feared that general taxes would have to be increased to provide public services for coal-development regions were responsible for passing this tax and other restrictive legislation. Although the 1970s witnessed rapid development of Montana's strip-mined coal, mine workers and coal company managers who argued for unhindered development were not a strong enough political force to prevent the enactment of the tax.

Tax Exporting and Regional Competition over Income and Wealth

Another more selfish motive lurks behind increases in natural-resource taxes—an effort to redistribute income from resource-consuming regions to resource-producing areas. Certain geographic areas possess the good fortune of having scarce natural resources underfoot. When these resources are discovered and determined to be less costly to extract, process, and transport than in alternative regions, a tax bonanza and mechanism to shift incomes from consuming to producing areas may well be at hand. The OPEC cartel discovered its monopoly powers during the 1973 oil embargo and, in succeeding rounds of price increases, has utilized this power. Although the cost of lifting and transporting oil in Saudi Arabia is estimated to be as little as $0.25 per barrel, the export price that, for various economical and political reasons, is made up of taxes, royalties, and a base price, which all accrue to the Saudi government, has risen from less than $2 per barrel to many times this amount since the embargo. An economic effect of oil exporters possessing a near-monopoly has been the transfer of hundreds of billions of dollars from oil-consuming nations to oil-exporting countries.

The oil sheikhs are not the only ones who comprehend the potential of exporting a tax burden. Virtually every state legislature in the United States at one time or another has levied selective taxes with the intent of exporting its burden. This is particularly true of taxes on the extraction and processing of natural resources when a state discovers that it is blessed with a resource which reaps great profits because of its location and low recovery costs. Rather than allowing all those profits to end up in the hands of the owners of the resource and owners of the extractive firms, states have cashed in on the bonanza by socializing a portion of the profits through their taxing authority.

For example, at one time Texas was the major source of oil for the United States and held a partial-monopoly position because of its enormous reserves, pumping capacity, and low production costs. Rather than allowing unregulated competition among producers which would establish output at the point where the most expensive well just made an average return on invested capital, the state intervened. In the name of conservation, but also with the motive of increasing the returns to all producers and to the state which owned and leased great tracts of Texas, regulations were enacted that restricted output. One result was to drive up the price above its competitive level and to produce monopoly profits. The Texas Railroad Commission (TRC) was empowered to establish pumping rates as a percentage of the maximum flow rate for each well. The TRC presumably was acting to "protect" and "conserve" this nonrenewable resource. However, several times each year the TRC surveyed production in other states, nationwide consumption, and prices. It then established pumping rates at a level which reaped monopoly profits to Texan producers and to the state treasury. The TRC became so expert at this activity (and was able to affect the U.S. oil price because of the large share Texas had in total U.S. capacity) that other producing states imitated them with regulatory agencies, which generally followed the lead of the Texas Railroad Commission. The losers were consumers who paid a higher price for oil-based products than they would have if each producer pumped at an unregulated rate. The great public and private fortunes which were amassed as a result led to political power. This power was used to enact federal tax provisions in the 1920s and 1930s which discriminated in favor of the oil and gas industry. It also resulted in establishing import quotas and other restrictions, developed in the guise of protecting and developing a national supply in the interest of national defense.

Consumers and legislators from nonproducing states eventually became aware that prices were being regulated indirectly by means of production restraints and tax advantages. Economists explained that the special tax considerations led to higher profits, and these arguments enjoyed wider and wider coverage in the 1950s and 1960s. Consuming states were not prepared to be gouged by the same geographic block when natural gas came into widespread use after World War II as a result of improved pipeline technology. The strategy to prevent a similar situation as was found in oil was to establish a federal authority to directly control the price of natural gas transported in interstate pipelines—the Federal Power Commission (FPC). Since the Midwestern and Eastern consuming states contain many times the number of voters found in Texas, Oklahoma, Louisiana, and other producing states, gas was purposefully maintained at a low price. The effect was that the consuming regions gained an income advantage over producing regions. The monopoly or cartel power was created by the federal regulatory agency to maintain prices below a competitive market rate or below a potential cartel rate which might possibly be established by producing states in a manner similar to the Texas Railroad Commission.

Events of the 1970s have demonstrated that when the price of natural gas is held below the true competitive market price, consumption demand outstrips production capacity and shortages develop. In a sense, *shortage* is the incorrect word for this phenomenon. Economists prefer to call it *excess demand*, because prices are below the level where quantity supplied and quantity demanded are equal. They argue that if prices were allowed to increase, the growth of consumption would be slowed voluntarily and production would increase as it proved possible to recover higher-cost gas. Economists' calls for a market-determined price and allocation of natural gas were not voices in the wilderness. The advent of gas shortages engendered competition among consumer states for gas allocations and diminished their political power. Producing states began to sell increasing amounts of unregulated intrastate gas at higher unregulated market prices internally and used the resource to lure new industry. This increased consuming states' resolve to bring about a change in interstate gas regulatory policy. Congress approved a phased decontrol of natural-gas prices, which will be completed in 1985.

After the 1973–1974 international oil-price escalation, Congress legislated a two-tiered domestic price structure—approximately a $5 per barrel price for wells which were in production at that time, so-called old oil, and one somewhat below the prevailing international price for "new oil." The Department of Energy was a progeny of the domestic energy plan and was granted wide regulatory authority. A major goal of the agency's regulation on domestic oil production and refining was to maintain artificially low oil prices. To accomplish this, allocations were established for every link of the production and distribution chain. The resulting excess demand, as evidenced in gasoline lines, demonstrations by independent truckers and farmers, and other disturbances, produced simultaneous calls for more regulation, a rationing plan, oil-price deregulation, and windfall-profits taxes. It is apparent that producing and consuming regions and a complex array of special interest groups are engaged in a complex, ongoing battle over the profits of a depleting resource which are potentially available as the market price continues its way up a veritable Mount Everest. The complete story of the tax and regulatory thrusts and parries in the battle over gas and oil profits is a very complex and long one which is only briefly recounted here.

This synopsis is sufficient to show that taxes are one weapon used in this war and that the greater the stakes become, the more violently the war is conducted. If Exxon and the other oil processing and distribution giants were making profits in the millions instead of in the billions and energy costs were less than a few percent of consumer budgets, then the war would never have escalated to its present intensity. Another example of this is the firefight taking place in state and federal courts over the taxation of electricity produced in New Mexico, and a potential full-scale war is heating up over revenues from the exploitation of the rich and vast strippable coal reserves of the Northern Great Plains.

In 1977 the New Mexico legislature enacted a tax on the production of electricity. A majority of electricity produced by the Four Corners coal-fired plant is exported to Arizona and southern California. Although it would be unconstitutional for the tax to apply only to exports, a special provision enabled the electricity tax to be taken as a credit against the gross-receipts tax in New Mexico. This in effect placed a differential burden on exported power. The importing utilities challenged the tax on several grounds and took their case to the U.S. Supreme Court, which struck down the tax. A provision of the 1976 tax reform act was written in such a way as to single out the New Mexico tax for extinction. The matter is not entirely a dead issue since the court's ruling did not preclude enacting an electricity generation tax in another form (Chung, Church, and Kurry 1980).

Coal reserves underlying Montana, Wyoming, and North Dakota are attractive targets for resource-development groups because of their potential wealth. These coal seams are some of the deepest in the world, and little overburden needs to be removed to expose them. The coal has a low sulfur content compared to Midwestern and Eastern coal, which permits it to be used with fewer pollution-control devices than other regions' coal. These characteristics mean that Northern Great Plains coal is among the least expensive to extract and burn to be found anywhere. Electric utilities in an arc running from the Midwest to Texas are purchasing and transporting this resource, making it the fastest-growing coal supply region in the United States. The Carter administration's energy plans from 1977 through 1979 were predicated on the growth of coal's share in total energy production within the United States. In late 1977, analysts predicted that by late 1985 an increase of 200,000 to 532,000 coal mines and, by the year 2000, investment of $45.5 billion in mines, $5.2 billion by private railroads and $4.9 billion by Conrail, the federally subsidized rail program, would be required to achieve the Carter administration's objectives. Although doubt continues that these forecasts will be achieved, there is no doubt that investment in mining in the Northern Great Plains has grown rapidly—so rapidly, in fact, that rail-car shortages and deterioration of railbeds became serious problems in 1978–1980. Costs have also escalated, engendered by provisions of the Surface Mining Control and Reclamation Act of 1977, federal taxes collected to compensate miners for black-lung disease and institute more stringent health and safety regulations.

The activity has not gone unnoticed in the host states. Although North Dakota enacted a new coal severance tax in 1977 and Wyoming levies substantial taxes on coal extraction, it was the maximum 30 percent severance tax on coal extracted in Montana that gained the bulk of attention. Commonwealth Edison of Chicago, Detroit Edison, and Texas-based Lower Colorado Authority all have contracts with Decker Coal in Montana. Annually 10.5 million tons of coal is shipped to the Midwestern utilities, and 3.5 million tons goes to the Texas utility. The contracts specify that tax increases are passed on to the utility producers, who pass them through to their consumers by means of rate adjustment and through automatic fuel-adjustment clauses. L. Cohan, Vice President and General Counsel for Detroit Edison, stated, " ... this tax will cost our customers $1 billion over a 25 year period. We contend that this is unfair and illegal." The major contention in the utilities' lawsuit against Montana is that the tax violates Article I, section 8, of the U.S. Constitution which empowers Congress "to regulate commerce."

Montana Governor Thomas Judge was not about to passively take the consuming utilities' position and called the suit "attempted economic blackmail by giant Eastern and Midwestern utilities for whom the welfare of the people of Montana is nonexistent." He stated further: "We do not believe that Montana alone should pay for coal impacts. We recognize the responsibility to provide energy, but we're not about to see our state torn up and our lifestyle destroyed." Similar remarks came from North Dakota Governor Arthur A. Link at a tax conference held in 1978 (Link 1978). The plaintiffs in the Montana case maintained: "There is no rational basis for the services performed by the state to warrant too high a tax." However, Governor Judge rebuts: "The philosophy of the tax is not just to offset impact of development. Once the resource is gone, it is gone forever. The state has a right to impose a tax to benefit future generations."

By implication, one intent of the tax is not only to pay for roads and capital costs in coal boomtowns, but also to squeeze as much tax revenue from the source as practicable. Virtually all industry observers agree that the tax is the major reason that new project development has slowed to a standstill in Montana and shifted to Wyoming. This has produced little political backlash among Montana voters because the tax has put no one out of work and most ranchers are apparently willing to forgo profits from the sale of mineral rights so that their lifestyles are protected. At this point, the war is in its formative stage. Other participants are fighting over potential wealth. The Burlington Northern has increased its shipping rates for unit coal trains several times (Zimmerman 1979b). It has also refused to grant a consortium rights of way for a proposed coal-slurry pipeline and fought legislation to grant to such lines the right of eminent domain. The process of recurring but independent resource-tax increases among coal-rich Northern Plains states was labeled "tax leadership" by this author in 1978 (Church 1978b).

Similar behavior by coal-rich Midwestern and Appalachian states is nearly unthinkable. In contrast to the Western states, the coal industry is a long-established one, which means that it has a large, organized constituency. The coal operators' organizations and the United Mine Workers have vested interests in the continued operation of existing mines and development of new sites, for without continued development, miners would be out of work and equipment would be idle. These interests are represented in state legislatures and reflected in state tax structures. When all taxes on inputs, output, property, and corporate incomes are considered, tax rates in Midwestern and Eastern states as a percentage of mine-mouth value are significantly lower than those found in the West. These objectives of job and industry long-run security explain why President Carter's energy plans calling for vastly increased coal output are welcome news to regions where coal is a long-established industry. Western strip-mining technology means that significantly fewer miners are needed per ton of coal, and the industry is a relatively new one, lacking well-developed political representation. Without extensive and organized opposition, tax increases are easier to pass, as 1977–1980 activity in New Mexico, Colorado, Montana, North Dakota, Wisconsin, and Alaska attests.

Specifying Conflicting Pragmatic Tax Policy Goals

When one is trying to describe tax policy, the goal of Pareto optimality is a vacuous one—both because an understanding of it is not widespread and because it offers little guidance to those making actual policy decisions. Legislators realize that virtually every decision made by them is detrimental to some constituents and beneficial to others. In this case, the relative benefits and costs are weighed, and the position which best supports political success—over both the short and the long term—is selected. In order to describe tax policy making, we chose the conflicting pragmatic tax policy goals of raising tax revenue versus protecting and stimulating an industry and its labor force as representing the dilemma which tax policymakers must face collectively and individually.

In order to depict the wide range of choice available to tax policymakers, objectives encompassing two polar extremes are defined and the manner in which tax policies in coal-producing states fit within these extremes are investigated. The first objective is maximization of resource-based revenues. This is of primary concern when policymakers believe that all or a workable majority of state residents are entitled to some benefit from the extraction of a nonrenewable resource. State residents should be compensated for the unavoidable social, economic, and environmental costs accompanying this development, and the state may export the tax and thereby transfer tax burdens to out-of-state consumers. However, as is shown shortly, as tax rates increase, the incentive to develop new resources diminishes since taxes diminish demand and increase production cost. Sufficiently high resource taxes will discourage the lengthy and costly process of resource discovery and development. Over time the resource industries are atrophied if taxes increase costs sufficiently to make development in other taxing jurisdictions more attractive. If alternative resoures become competitive, then economic growth will be slower and the specialized workers (miners, drilling crews, and so forth) may have to relocate in order to find employment. Incomes of others in the state may diminish because of indirect effects of diminished economic activity.

The second pragmatic tax policy objective we specify is to develop the state's natural resource and the incomes of those directly and indirectly involved in its development. The vague goal of encouraging resource development is made operational by measuring it by the total wage bill of the mineral extraction and processing industry. Any resource tax which reduces output or employment would be minimized if one goal of maximizing the industries' wage bill were followed single-mindedly.

Quantifying Pragmatic Tax Policy Goals

Two measurable goals have been specified as of primary interest to pragmatic tax policymakers—maximizing tax revenues from the resource and maximizing the resource industry's wage bill. Although this makes for a highly simplified view of the world, it is necessary so that the goals themselves are both quantifiable and plausible. While the two variables chosen to measure the "tax the hell out of them" school of thought versus "encourage development regardless of social and environmental costs" thinking may not describe the subtlety and complexity of pragmatic tax policy making, data are available to estimate each variable to see how states may actually balance the tradeoffs inherent in these two conflicting philosophies. In fact, we enter the real world by observing the tax rates on coal extraction of producing states and from these infer where each stands in balancing the tradeoff between development and tax revenues in establishing tax policies.

We hypothesize that all states seek to maximize a weighted sum of the two policy objectives—the wage bill plus direct resource-tax revenues. The weight α_i indicates the relative importance of each tax objective. The weights on the tax-objectives function θ_i are determined by the preferences of voters and elected officials in each state (each state is denoted by the subscript i). This objective is summarized in algebraic form as

$$\theta_i = \alpha_i w_i L_i + (1 - \alpha_i)(Q_i T_i)$$

where θ_i = measurement of objective of overall tax policy

α_i = policy weight on wage bill

$(1 - \alpha_i)$ = policy weight on tax revenues

L_i = total industry employment

w_i = average industry wage rate

Q_i = total annual industry output

T_i = tax rate per unit of output

When $\alpha_i = 1.0$, the sole objective of state tax policy is to stimulate industry development. When it is any level between 1.0 and 0.0, an implicit tradeoff judgment is made between development and raising tax revenues. We first analyze how state policymakers might go about determining optimal tax policy if voters and elected officials had determined what they preferred the policy weight to be. We then observe effective tax rates in the sixteen major coal-producing states and infer the level of α_i for each state.

If a state bureaucrat were presented with the policy objective function θ_i and the appropriate policy tradeoff weight α_i, he or she would be able to specify the optimal tax rate T_i which would maximize the objective function. However, at a minimum, the bureaucrat would require information about the technology of coal production and the demand for coal for the state. Let us look at these statistics. The demand for coal is not a static number or one which simply rises over time. Rather, the quantity demanded is a function of many factors, the most important of which are the location of developed coal reserves relative to potential consumption centers, the technology of electricity production and steel production, and the delivered price of the coal. Since steam coal represents the vast bulk of U.S. production, we look at it alone and ignore the far smaller market for metallurgical-quality coals. A most important factor to realize from the demand point of view is that political boundaries do not correspond to isolated producing and consuming regions. Thus a single state's demand for coal is sensitive to what is going on in competing states.

We play the role of government bureaucrat and estimate the demand relationship for each state in a manner somewhat similar to that which a sophisticated analyst employed by a state taxation department might follow. We suppose that this hypothetical individual is aware of a fundamental proposition of elementary economics—the quantity demanded is inversely related to price. Most bureaucrats and legislators ignore this proposition, since a common implicit assumption in establishing tax policy appears to be that even if the tax increases the cost of production, there will be virtually no change in quantities demanded and produced. There are unique situations where this assumption is warranted, such as in the tax on heroin or Picasso paintings because such commodities will be bought at virtually any price. In the case of coal demand in any single state, it is unwarranted for there are alternative sources of supply (coal can be extracted in other states) and energy (for example, nuclear, residual oil, solar, and natural gas are all technologically possible methods to generate electricity). The demand for coal within the confines of any single state is relatively price-sensitive, or, as economists say, price-elastic. *Price elasticity* is defined as the percentage change in quantity demanded divided by a percentage change in price. The price elasticity for coal is negative and varies from state to state because of locational advantages (transportation costs make up a large portion of the delivered price of coal) and the quality of coal (which determines extraction costs).

How would our bureaucrat quantify the demand function for coal? The first action would be to check research conducted to date, but the search would reveal no single demand function for individual states although complex models describing the entire coal market are available (see chapter 6). A good deal of related research is available, and this would undoubtedly provide the analyst with some ideas and methods. We hypothesize that the bureaucrat gathers data on the price of coal and quantity produced and sold from each state as well as a few other important factors from 1967 through 1976. These data could be simply plotted on a piece of graph paper where coal price is depicted on one axis and quantity on the other and the points are analyzed. One way to estimate the demand relationship is simply to "intuitively" draw a "best-fitting" line. Another possibility is that somewhere on the staff of the hypothetical bureaucrat lurks a person with some training in statistics. This person should be able to relate quantity with price and a number of other factors by means of a statistical technique known as linear regression analysis. We pursued this method as a well-trained analyst might.

Data on the coal industry were collected from *County Business Patterns, 1967–1976* (U.S. Department of Commerce 1978) for wages and employment and U.S. Bureau of Mines (1976) for coal output, prices, and productivity. Certain data unpublished for reasons of confidentiality were gathered from alternative sources. Other possible factors affecting demand in the 1967–1976 period were the price of the most important substitute energy, residual fuel oils and natural gas, and environmental regulations which affected the use of coal. An index of the wholesale price of oil came from *Petroleum Independent* (May–June 1977, p. 63) and the *Minerals Yearbook* (U.S. Bureau of Mines). Changes in the Clean Air Act of 1970 (as amended) were statistically accounted for as well. Along with the time trend, oil prices and environmental regulations are hypothesized as shifting a linear demand function for coal. They key assumption behind this approach is that production costs, reserves, prices, and taxes remain constant in all other

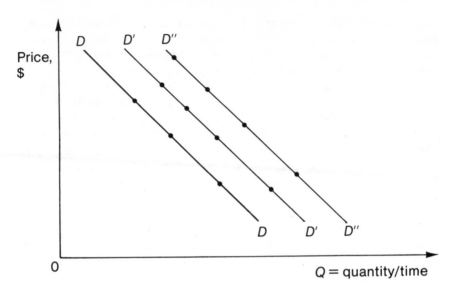

Figure 1–1. Demand for Coal

states. While this is certainly not true over the long haul, it is reasonable to assume they change little over a period of two to four years. What is more important is that tax and other policies in other states cannot be predicted, and to assume constancy is the most plausible assumption. Observations of price and output are depicted in figure 1–1 as dots (where price is on the vertical axis and quantity is on the horizontal axis). A demand function is shown as sloping downward and to the right and shifting (D, D', D'') as demand-related factors other than price change. The reason for the dispersion of the observations on any single demand function shown in figure 1–1 is that the costs of production are changing over time, and this leads to various prices and quantities of output on each demand function.

It is readily apparent that the "economic model" depicted in figure 1–1 is able to account for many different patterns of prices and output (dots). A demand function for each producing state is estimated statistically. The results are not reported here, for they are not of direct interest. We do report the price elasticity in table 1–2 as estimated from a constant-elasticity version of the demand function just described. The linear demand function used here generates a variable demand elasticity. As expected, the demand curve does slope downward and to the right for each state, and there is a good deal of variation in it among states as the reported elasticities attest (column 3, table 1–2). Demand is but one step in determining an "optimal" tax rate for our bureaucrat to recommend to the governor or legislative committee. The next step is to find how technology, the cost of production, and tax rates

Table 1-2
State Marginal-Cost and Demand Estimates and Derived Revealed Preference Weights on the Wage Bill

State	Hourly Wage Rate, 1976 ($)	Output per Worker-Day (Tons)	Estimated of Demand	Effective Tax Rate, 1977						Weight on Wage Bill in Objective Function Utilizing a Tax Rate		
				Underground		Surface		Weighted		Under-ground	Surface	Weighted
				Percent-age	Tax per Ton ($)	Percent-age	Tax per Ton ($)	Percent-age	Tax per Ton ($)			
Alabama	7.79	10.7	-0.05	3.92	0.32	5.65	0.47	5.06	0.42	0.67	0.82	0.77
Arizona	7.45	69.1	-2.80	—	—	27.89	1.30	—	1.30	—	0.41	~
Colorado	7.57	24.7	-0.98	5.48	0.84	8.90	1.36	7.71	1.18	0.11	0.37	0.29
Illinois	8.07	16.6	-0.40	7.28	1.12	11.33	1.75	9.17	1.41	0.87	0.86	0.87
Indiana	8.07	26.1	-0.72	1.14	0.14	2.57	0.32	2.55	0.31	0.79	0.86	0.85
Kentucky	8.07	17.3	-0.11	7.46	1.48	10.25	2.03	10.04	1.99	—	0.10	—
Montana	7.58	120.7	-0.42	—	—	45.99	2.25	—	2.25	—	0.22	—
New Mexico	7.45	23.0	-3.09	—	—	6.27	0.44	—	0.44	—	0.72	—
North Dakota	7.58	87.0	-0.50	—	—	15.28	0.57	—	0.57	—	0.66	—
Ohio	8.07	17.4	-0.12	2.57	0.43	4.31	0.72	3.69	0.61	0.89	0.86	0.87
Pennsylvania	7.94	11.5	-0.14	3.18	0.81	4.70	1.19	3.92	0.99	—	0.02	—
Tennessee	7.79	11.0	-2.53	5.51	0.90	9.04	1.47	7.36	1.20	0.66	0.70	0.68
Utah	7.57	13.8	-0.22	—	—	2.56	0.59	—	0.59	—	0.24	—
Virginia	7.94	9.9	-0.20	2.44	0.59	2.73	0.66	2.54	0.61	0.42	0.70	0.52
West Virginia	7.94	8.4	-0.81	5.02	1.51	6.23	1.88	5.26	1.58	0.78	0.82	0.79
Wyoming	7.58	57.0	-0.20	—	—	19.85	1.39	—	1.39	—	0.37	—

determine where each observation (dot) falls on each demand curve. Then the tax rate optimizing the weighted objective can be calculated.

The technology of coal extraction and processing is described here in a very simple manner by assuming a fixed and constant technology. Although this approach is unrealistic over the long haul, it is plausible over the short term. This also happens to be the length of time so dear to politicians, the two to four years between elections. Our hypothetical bureaucrat assumes that the necessary labor to produce one ton of coal is constant, and this number (workers-hours per ton of coal, denoted as k) describes technology in each state:

$$\text{Worker-hours per ton of coal} = k = \frac{L}{Q}$$

Note that for the sake of clarity the subscripts denoting each state (i) are no longer shown. However the data unique to each state are used in the empirical version.

Since union bargaining and labor practices are complex, the state analyst simply takes the going wage rate w as a given. It may change over time, but it is difficult to forecast what the new level will be. Thus the labor cost per ton of coal is the hourly wage rage w times the worker-hours necessary to extract one ton of coal k, or

$$\text{Unit labor costs} = wk$$

There are published data which may be used to estimate these unit labor costs (cited previously.) No data are available for the capital costs per ton of coal in each site in each state. These are proprietary information and tightly guarded by each company. However, in the short run, most costs except for labor and some operating costs are primarily fixed and are invariant with respect to output. For this reason and because of unavailable data, labor costs are assumed to constitute all per-unit costs of output. In a competitive industry, and given the large number of coal producers and consumers nationwide, the industry may be described as being competitive overall but with some regional monopoly power. The price of coal per ton P will equal unit labor costs (this is equal to marginal costs because of fixed technology) plus the tax per ton of coal T. For example, if price temporarily exceeds production costs and taxes, producers can increase profits by producing more coal, and given a fixed demand curve, price declines. This will continue until the point where price equals unit labor costs plus the tax per ton. If price falls below this level, then production is curtailed because profits are negative, and price rises upward along a demand curve (D' in figure 1–1) until equilibrium is achieved, where

$$P = wk + T$$

It is now possible for our hypothetical bureaucrat to pull all this information together so as to compute an optimal tax-rate recommendation for legislators and other state officials after they advise her or him of the "proper" weight to place on the importance of development, measured by the wage bill WL vis-à-vis tax revenues TQ. If maximum development is desired ($\alpha = 1.0$), the tax rate should be zero. If the maximum tax revenue is desired ($\alpha = 0.0$), then a tax rate can be calculated from the information at hand as well as for a tax rate which corresponds to any given policy weight ($0.0 \leq \alpha \leq 1.0$). The objective function, after substitution and given a linear demand ($Q = a + bP$), is

$$\theta = \alpha wk[a + b(wk + T)] + (1 - \alpha)[a + b(wk + T)](T)$$

The optimal tax rate (T^*) can be shown to be

$$T^* = \frac{a + bwK - \alpha a}{2.0b(\alpha - 1.0)}$$

The entire spectrum of options for various weights (α) is shown in figure 1–2, where tax revenues TQ are on the vertical axis and the wage bill wL is on the horizontal axis. The parabola shown in figure 1–2 depicts all feasible outcomes for the wage bill and tax revenues generated by all possible tax rates.

Clearly, all the combinations of the two factors in the objective function to the left of the dotted line are undesirable, for the state nets less of both. Once production costs, technology, and demand are known, the feasible alternatives are known and the optimal point to the right of the dotted line depends entirely on the policy weight α. It can be shown mathematically that the tangency of a straight line, with a slope equal to $\alpha / (1 - \alpha)$, with the parabola of feasible combinations indicates both the "optimal" tax rate and the levels for the two tax goals in the objective function.

Of course, the problem for our hypothetical bureaucrat is what tax rate to recommend to the executive and legislative branches of government. This is divided into a technical problem, that of accurately forecasting the feasible alternatives available (the parabola in figure 1–2), and a political one, that of looking into the minds of the body politic (special interest groups and the elected public officials who simultaneously lead and are led) and quantifying the preference tradeoff of economic development versus tax revenues by the weight α. The technical problem is one that any competent bureaucrat could handle either with the resources available within the department or by means of outside consultants who are only too happy to supply answers, at a price. The major risk in estimating the feasible alternatives is that the bureaucrat

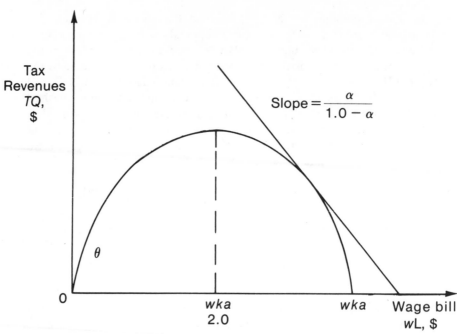

Figure 1–2. Policy Tradeoff

or consultants have committed some grievous error or oversight which the hired guns of the special interest groups detect and exploit to their advantage and his or her embarrassment. It is always preferable to blame someone else for a technical error (or for that matter any error), and this predisposition explains the popularity of consultants. It is possible for risk-averse bureaucrats to spread the uncertainty of a technical error; this is why redundant reports and studies are sometimes made on projects requiring critical decisions.

The problem of estimating preferences and how these are manifested politically is difficult to load on some other unsuspecting person's back. Estimates of which kind of tax legislation will actually pass are difficult to make. The bureaucrat who is responsible for making specific legislative recommendations and who has met the test of survivability knows as much about politics as any elected politician. Examples of failure readily come to mind, such as Dr. James Schlesinger, former head of the Department of Energy, while those who endure are usually not known. Elected officials become jealous of bureaucrats who garner too much public attention. So after very few mistakes the publicity-prone bureaucrat is out of a job. The exceptions to this rule, such as J. Edgar Hoover, former director of the FBI, are able to compete in the publicity arena and survive only when they possess a veritable Fort Knox of privileged and specialized information. All others are vulnerable, and the majority of enduring bureaucrats are those who gather and dispense worthwhile information, make politically astute recommendations, and stay crouched deeply in their previously prepared trenches when the major firefights take place. The bureaucrat who stands up and engages in battle may serve a useful function for others, but usually does not last too long. Barry Bosworth of President Carter's Council on Price Stability and James Schlesinger and John O'Leary of Carter's Department of Energy can be classified as samurai warriors and flack catchers who diverted a certain amount of criticism away from the beleaguered President. Take a look at their longevity, for one can only stop flack for so long before one is useless. The truly successful survivors are those who do their technical homework and who are sensitive politicians who can stay out of the limelight. Their recommendations are both technically well supported and politically realistic.

There is another interesting observation to be made. Elected officials often do not wish to make firm determinations about what policy should be, for they are simultaneously leaders and followers and vote traders who must stay flexible. In the context of this tax policy model, this means that the bureaucrat is not able to get away with recommending a range of feasible alternative policies, but is forced to make a specific recommendation over which the politicians can haggle. This suggested starting point provides the base for vote trading and the other instruments of political interaction. Compromise is the art of politics, but negotiations have to start from an initial position, and this often comes from the supporting bureaucrats.

The political aspect of decision making is so complex, subtle, and issue-dependent that we can at best describe it in broad terms, but are unable to make it operational in this model. Consequently, the model of optimal taxation presented here is not able to predict actual decision making. It is possible, however, to use this model to infer something about the coal-producing states' preferences.

Estimating States' Tax Preferences

The data sources just described supply information about the technology of coal production and states' demand functions. Before this information is exploited to estimate the policy preferences of the coal-producing states, a thumbnail sketch of the industry is required. The two basic technologies prevalent for coal extraction are underground mining and surface mining (an informative and more detailed analysis can be found in Zimmerman 1977, 1979*a*). The major geologic characteristic determining the cost of underground mining is the thickness of the coal seam, which may vary from a few inches (this is not economically exploitable) to many feet, and the depth of the seam from the surface. In surface mining, the thickness of the seam itself and the amount of surface material which must be removed to yield the coal are the significant factors. Exposing the seam and stripping it are done by a highly mechanized operation employing incredibly large power shovels, off-the-road trucks, and scrappers. Thus strip mining is highly capital-intensive. Western strip-mining sites are developed as large-scale, high-labor productivity mines, particularly in the Northern Great Plains where seams are close to the surface and among the thickest in the world. Underground mining is relatively labor-intensive, and productivity gains have been nonexistent and even negative in recent years compared to strong improvements in Western stripping operations.

Although the majority of miners are unionized, the Eastern and Western United Mine Workers are somewhat autonomous, and many Western miners belong to other unions or are nonunion. This has resulted in regional wage patterns, as is revealed in the first column in table 1–2. The second column indicates the tons per worker-day of coal output on the average for underground and surface mining (note that separate underground and surface data were used in computing the policy weight). These data reveal large differences in productivity caused by geology, mix of mining techniques, and union- and government-legislated work rules in each state. The estimated elasticity of demand for each state, as calculated by our hypothetical bureaucrat, is shown in the third column. The variation is large, and knowledgeable persons may detect possible weakness and may dispute these estimates. The demand function actually employed to estimate demand in calculating the policy weights is in a slightly different form, a linear one. It represents a method which is consistent and which is derived from published data. The next six columns list the estimated effective tax rate per ton of coal in percentage terms and in absolute dollar terms based on simulated 1977 coal prices and tax policies. Details on how these tax rates were estimated from state severance, property, excise, and corporate income taxes are delayed until chapter 5. Note also that these rates are derived from only those taxes falling directly on coal extraction. Tax revenues are also indirectly produced from employee income taxes and from the secondary economic activity generated by the industry. However, these are undoubtedly directly related to output, and as such their effects are reflected in the estimated policy weights (for a detailed examination of these effects see Krutilla, Fisher, and Rice 1978).

When it is assumed that existing tax rates are equal to an optimal level and the complete information underlying table 1–2 is substituted into the formula for the optimal tax rate (given previously), then the policy weight is the only unknown and can be derived. These estimated policy weights are reported in the last three columns of table 1–2 for underground and surface technologies and weighted average of the two as determined by the relative proportions of underground and surface production in each state.

The estimated policy weights on the wage bill goal vis-à-vis the tax revenue goal reveal some interesting things. The first is that the weights fall within the expected range between 0 and 1, and they vary widely within this range. States revealing a high weight on the wage bill (Alabama, Illinois, Indiana, Ohio, Tennessee, Virginia, and West Virginia are all above 0.70) are the ones with long-standing or powerful labor and industry constituencies which are politically active in protecting their interests for continued and expanding development. The states with low weights on the wage bill (Arizona, Colorado, Kentucky, Montana, Pennsylvania, Utah, and Wyoming are all less than 0.40) are a mixed bag. Montana's low 0.22 is expected, and Wyoming (0.37) is a bit behind in its interest in plucking the golden goose of coal. Although North Dakota is a middle case (0.66), it has increased its taxes since 1976 substantially. In the West, feelings about coal development and associated power generation and transportation of coal are mixed. The proindustry constituencies are new and small and thus lack the political clout they possess in the Midwest and East. Pennsylvania and Kentucky are statistical anomalies caused by their unusually low estimated elasticities of demand. Were these estimates more in keeping with nearby competitor states, their revealed policy weights would have fallen into the middle category.

What may safely be concluded from this examination of tax policy? There appears to be strong evidence that preferences for the tradeoff between development (the wage bill weight $= \alpha$) and tax revenue (the tax revenue weight $= 1.0 - \alpha$) are not uniform among the states and coincide with preferences in the East versus West noted by many observers. The implications are important for fiscal federalism. States wish to regulate the development of their natural resources in different ways, and at some point these preferences come in conflict with national energy and resource policies. The battle for coal profits is taking place amont the coal-producing states, unions, and transportation companies. At some point, tax exploitation may infringe on interstate commerce, and the courts (witness the Montana case) or the U.S. Congress may impose limits.

The Canadian case may prove enlightening. The provinces enjoy greater independence in taxation and expenditure policy making than states in the United States, and in fact, the provincial governments owned the mineral rights on all lands, public and private. The 1973–1974 oil-price hikes produced potential "windfall" profits, which the Western producing provinces would not allow to escape in the form of higher profits and dividends to the owners of extracting companies or in low prices to consumers. They responded by increasing royalties and taxes on oil and gas producers. The federal government of Canada did not let these changes pass unnoticed and did its best to obtain its share of the pie, but about the only method it had at its disposal was to cancel the deductibility of provincial royalties and taxes against the federal corporate income tax and impose price controls on old oil. It should be noted that the U.S. Congress is not as constrained, and more avenues are open to it if state tax policies conflict with national energy and resource policy.

This excursion into tax policy and the political process is only a first step. It should also be mentioned that the approach just described is a new and unique one. The only similar work looks at the single tax goal of maximizing tax revenues or tax exporting. The technology of extraction presented in these models is so highly simplified as to be misleading, and the only tax considered is based on output. Virtually every stage and component of exploration, discovery, development, and extraction are potentially taxable. These taxes have not been analyzed. Furthermore, the model investigated how taxes affected only costs, price, and quantity demand. In actuality, the location and technology of investment can be affected in other ways not described here. Finally, the model assumes the tax increases prices to consumers as estimated from the elasticity of demand. A closer examination of the extracting firm is necessary before all the potential tax effects are understood. Only then can the actual burden of resource taxes be measured as to how much falls on consumers, owners of resources, the firms which extract resources, and labor. In order to understand the complexity of tax-induced effects and how these, in turn, determine who bears the final tax burden, the economics of the industry needs to be understood.

2 Basic Economics of Nonrenewable Resources

The model of resource markets developed in chapter 1 to reveal preferences for conflicting tax policy objectives in coal-producing states is elementary. The only economic adjustment mechanism which is allowed to take place is consumers responding indirectly to a tax on the production of a resource by purchasing less when the tax increases its cost. However, many taxes are levied on financial flows other than sales revenues, and the firm and resource owners respond to taxes in a variety of ways. What this implies is that the supply of nonrenewable natural resources should be analyzed in more detail, for technology is not fixed in the long run, as assumed previously. Only when the behaviors of both producer and consumer are analyzed simultaneously is it possible to determine who bears the final burden of a tax and how a tax may affect resource allocation and economic efficiency.

One factor which must be realized is that a nonrenewable resource is qualitatively and quantitatively different from other goods and resources produced and consumed by a society. In considering reproducible resources, it is possible to imagine a continuous production and consumption flow which takes place without end. This is only partially true in the case of reproducible natural resources, and it is false for nonrenewable resources. When exploiting fisheries and forests, people are constrained by the laws of nature (reproduction, growth, and death rates) to act within given bounds. If the stock of renewable natural resources is harvested at a rate in excess of the reproduction rate, exhaustion of a species will eventually result, as in the case of the American buffalo. However, in the case of a nonrenewable natural resource (for a thorough examination of the economic literature, see Peterson and Fisher 1977), depletion means that a portion of a fixed stock is forever used up and eventual exhaustion is that much closer. Thus Ciriacy-Wantrup (1952, pp. 35–38) has defined a nonrenewable resource as one which displays no economically significant rate of regeneration to distinguish it from reproducible natural resources. This definition clears up any misunderstanding about certain minerals whose reserves may be growing in terms of geologic time but whose growth is imperceptible in terms of the human clock.

The distinction between fuels and nonfuel minerals is also meaningful and should be made before we proceed further. The second law of thermodynamics dictates that when a fuel is converted to energy, entropy occurs. The released energy, although used in production is dissipated so that it is impossible for it to return to the previous state. The implication is that energy is nonrenewable and nonrecyclable (for a clear and literate explanation, see Commoner 1976). When nonfuels are extracted, fabricated, and incorporated into a commodity, the refined state of the mineral is achieved through the input of energy. In other words, the bauxite refined into aluminum ingots and recycled aluminum beverage cans converted to other commodities absorb energy. In recycling, nonrenewable energy is applied to collecting and refabricating cans into another aluminum product. Recycling does not alleviate the specter of exhaustion, but is governed by the laws of nature (recycling requires energy so the law of thermodynamics applies and some material is nonrecoverable) and economics. The virgin and recycled materials must compete with each other. Recycling is not possible for fuels (oil, natural gas, coal, and uranium) for once they are consumed, the second law of thermodynamics dictates that it would take more energy to reconstitute the original product than the fuel itself possesses.

Exhaustion implies that when a nonrenewable resource is consumed, there is both an economic and a physical loss. Thus efficiency becomes a critical issue for people. The concept of engineering efficiency reflects the laws of nature—more useful work must come out of a process than is given up to run it. Economic efficiency has a somewhat different orientation. To the purely physical process of inputs and outputs it adds an evaluation of how individuals and society value produced goods and services and scarce human and nonhuman resources. Economic efficiency entails getting the greatest "value" out of scarce and ultimately exhaustible resources.

Pareto Optimality Revisited

The formal definition of economic efficiency was first clearly stated by the Italian economist Pareto in the nineteenth century and was discussed in chapter 1. We expand this concept because of its central importance in policy evaluation. Until Pareto's codification, a large number of vague and inconsistent ideas involving engineering and economic efficiency were utilized by various writers in the eighteenth and nineteenth centuries. The classical liberal economists had sought to formally demonstrate that individuals behaving in their own self-interests would end up allocating resources in an efficient manner. It is this correspondence which Smith attempted to demonstrate in his 1776 *Wealth of Nations*. The formal proof could not be shown until Pareto's definition of efficiency was refined and understood and until various principles of economics could be stated in a succinct mathematical language. The latter did not occur until the advent of the twentieth century, and a formal demonstration of the correspondence did not occur until well into the twentieth century.

The lack of formal proof did not prevent economists starting with Smith from proclaiming the superiority of the free-market system. The inefficiencies associated with government regulation, whether in the form of seventeenth- or eighteenth-century mercantilism, the twentieth-century planned economies of the Soviet Union and Eastern-bloc countries, or the ponderous regulations promulgated by the U.S. Department of Energy are decried.

The formal proof of correspondence brought with it the ironic conclusion that a perfectly run, planned economy might also be economically efficient. Quite naturally, the chamber of commerce takes Smith's position, but has never acknowledged this potential efficiency of socialism. However, there are sufficient theoretic, pragmatic, and empirical arguments why a centrally planned economy can never live up to the theoretic equations which describe its potential efficiency.

Rather than becoming overly pedantic, we illustrate the conclusion that a market economy achieves Pareto optimality in a direct and mathematically free way. If this is true for most of the economy, then government intervention is necessary only when deviations or "market failures" occur. We first demonstrate that correspondence occurs for reproducible commodities and then go on to tackle the more involved case of nonrenewable resources by surveying the professional literature.

Recall that the Pareto definition simply holds that resources are being utilized in an inefficient manner whenever it is possible to reallocate them so at least one person is made better off without making any person worse off. This situation is called a *Pareto-suboptimal* one. A movement toward *Pareto-superior* resource allocation takes place when some agent alters the state of the world so that, given a fixed stock of human and nonhuman resources, the lot of one or more persons is improved without diminishing that of any other person. Pareto optimality is achieved whenever there is no Pareto-superior move which can feasibly be made. The definition appears on face value to be somewhat lacking in practicality, particularly for tax policymakers, and it is. As stated in the first chapter, virtually no tax policies can be implemented without hurting someone. However, the Pareto definition of efficiency does lend insight into which tax structure is preferable under various conditions, but that part of the story will have to wait for a bit.

Correspondence between Pareto Optimality and Competitive Markets

The twentieth-century formal demonstrations of Pareto-optimal resource allocations do not need to be investigated here, for a more intuitive and heuristic approach is permissible. It is clear that the bricks and mortar of the definition are people's preferences, the state of technology, and the stock of resources. Only the individual can tell us if she or he is worse or better off, and only state-of-the-art technology is able to reveal what is possible. For simplicity's sake, we resort to employing a dollar dimension to show preferences and the resource allocation which is Pareto-optimal. However, in the formal proof, prices or money dimensions are not required. The Pareto-optimal condition can be found by maximizing the welfare of one individual subject to the constraints that the welfare of all other members of society is held constant and all resources necessary to produce commodities are in fixed supply. In our approach, individual preferences are revealed by means of the willingness to pay. We take a simplified approach by analyzing a one-commodity world. There is good reason to believe that as more of any single good is made available, the satisfaction engendered by additional quantities diminishes. So a high price is paid for few goods, and lower prices are paid as larger quantities become available. This ocurs because individuals are able to substitute other commodities when something is expensive and are less inclined to do this when things are cheaper. It is the fundamental economic law of demand. This pattern of behavior can be summarized in the willingness-to-pay line shown in figure 2–1.

The willingness-to-pay function can represent either one individual or a group of individuals. We assume that it represents all residents of a state or the entire United States and indicates how many units in total they will buy at various prices. There is also a willingness to produce, or equivalently the cost of producing, additional units. This function rises as more units are produced because resources must be bid away from other activities. It is widely

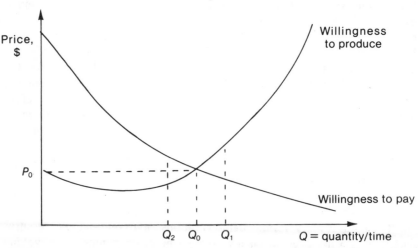

Figure 2–1. Willingness to Pay (Demand) and Willingness to Produce (Supply)

believed that at some level of production, technology becomes a limiting constraint, begins to lose its physical efficiency, and thus forces production costs upward. This is the fundamental law of supply.

What does this say about Pareto optimality? If output at level Q_1 in figure 2–1 is produced, it is Pareto-suboptimal. This must be since from Q_0 to Q_1 people are willing to pay less for each additional unit than the costs of production. This implies that the resources employed for production between Q_0 and Q_1 are better spent elsewhere. Conversely, if Q_2 is produced, people are willing to pay more than incremental production costs, and this implies waste. If more is manufactured and consumed, then preferences are satisfied by resources diverted from some other activity where people are just willing to pay for the cost of the last units made. Producing more would be a Pareto-superior change. Production and consumption at level Q_0 are Pareto-optimal. This may be confirmed by a mathematical demonstration which reveals that maximizing the area between the willingness-to-pay and the willingness-to-produce functions (as long as the former is greater than the latter) or maximizing the number of voluntary exchanges yields Pareto optimality. This occurs at Q_0. An analogous situation holds for every good and service produced in an economy.

Now suppose that the ability of people to pay for goods and services or income is altered. One factor lying behind the willingnes-to-pay function is an implied distribution of wealth and income. Should people's incomes be changed by shifting wealth among individuals in a society, then the total willingness-to-pay function is likely to change because tastes and preferences vary. If more purchasing power is awarded to people who prefer this commodity, the function will shift outward and to the right, and the opposite case holds as well. In either case, Pareto optimality will dictate a level of output different from the original Q_0. The point being emphasized here is that Pareto optimality or economic efficiency can be achieved with any distribution of income and wealth. However, once income and resources are distributed, then a unique level of output and resource allocation satisfies Pareto optimality.

Do Pareto-efficient outcomes have a relationship to resource allocation determined by the marketplace where all individuals act in their own self-interests? Yes—the relationship holds, for the willingness to pay corresponds directly to market demand, and the willingness to produce corresponds to the market supply curves when all industries and firms act competitively. Competition basically means that buyers and sellers act as if their own decisions and activities had no effect on market prices; in other words, they are pricetakers. The price P_0 corresponding to the intersection of the supply and demand functions (willingness-to-produce and -pay curves) represents market equilibrium. It is a price which clears the market where the quantity consumed is equal to the quantity produced. The price itself acts as an efficient allocation mechanism. Any other price is not stable, for either a shortage or a surplus will occur. In these situations, profit-maximizing and personal-preference-maximizing behavior will result in price changes. When supply equals demand, the stable or equilibrium price corresponds to the Pareto-efficient allocation of resources. Whenever an institutional barrier or other problem interferes with private markets and prevents the correspondence to take place, a *market failure* is said to exist.

Does the correspondence between Pareto optimality and competitive market hold for exhaustible as well as reproducible resources? In order to investigate this question, the behavior of resource firms and optimal public use must be investigated. This is necessary because the specter of exhaustion makes owners of nonrenewable resources act differently from owners of a firm producing a reproducible product. The mineowner must take the future into consideration, for exhaustion terminates that business. The willingness-to-produce function or the supply curve has a finite life. Thus investigating the correspondence between Pareto-efficient resource use and the market equilibrium is more complex and must take time into account. This question is approached by reviewing the literature.

Analyzing the Mining Firm

The owners and managers of a mining firm must make a large number of decisions which determine profitability and the very existence of a firm. The primary concern for the tax policymaker is how various tax regimes affect these decisions, what the consequences are of these tax-induced changes, and what the "proper" or "optimal" role for tax policy is. We must first understand how these decisions are made in a tax-free regime before the effects of taxes can be understood. This will also tell us if correspondence occurs.

The economists' approach has been to construct a highly simplified model of the mine so that one variable (most frequently the rate of extraction) or two decision variables characterize behavior of mineowners. However, geologists, engineers, and businesspeople have to make decisions involving a multitude of variables, where simplified but elegant theory is not directly applicable. They must decide whether to explore for a nonrenewable resource and, if discovery is made, whether to develop the resource and the best-suited technology and best-sized facility to build. Finally they must decide how much to extract from the resource in terms of raw-resource input, cutoff grades, and final output. The pragmatic decisionmakers have thus developed their own analytic techniques. Some of these correspond to the economists' models, and others are unique. The two groups worked independently, and only in the last few years has the knowledge generated by each group started to become synthesized.

What makes the economic model of exhaustible resources unique with respect to other economic models is that the capital stock (depletable natural resource) is fixed and nonrenewable. This means that a barrel of oil consumed today or a pound of copper extracted and used up now is forever lost to future generations. Decisions about the use of these resources, then, ultimately involve both present and future generations. The owner of property rights to these resources realizes that exhaustion is occurring. In the case of reproducible capital, a slice of time may be continuously reproduced, and a static model is appropriate. Depletion of a nonrenewable resource alters future time slices because the stock of resources is diminished. This means that a dynamic model where time is explicitly accounted for is required to describe the use of exhaustible resources.

The first investigators into the economics of the mining firm, such as Gray (1914) and Paish (1938), managed to depict this inherently dynamic decision process as a static one by means of a few simplifying assumptions and mathematical tricks. The best-known static approach was defined by Anthony Scott (1953, 1955) in the 1950s. He accounted for eventual exhaustion of a fixed stock of resources by the concept of *user cost*. This cost represents the future resources or resource-generated profits which are forgone when the resource is extracted today rather than at some future time. He specified that capital comes in two forms: reproducible (human-made or naturally growing) and nonrenewable (that which humans cannot augment). Reproducible capital loses value as it becomes obsolete, and frequently this is most sensitive to the passage of time. This is the familiar form of depreciation over which, aside from choosing schemes which meet the approval of the Internal Revenue Service, the business manager has no control. However, she or he does decide how intensively to use capital at each point, and this causes depreciation in addition to that dictated by time. The marketplace for used computers, trucks, and mining equipment dictates their value with respect to both age and use. However, once used up, they can be replaced. The businessperson is able to regulate the depreciation of exhaustible capital in its entirety by controlling the rate at which it is extracted from the earth's crust. Time alone has no effect on a resource in the ground (in situ). The business manager will extract it now or in the future depending on when profits are greatest. User cost represents the cost of extracting now rather than waiting. It equals the present value of potential future profits which present extraction forever eliminates. Another way of defining *user cost* is the amount that a fully informed, knowlegeable buyer would pay for the ownership rights to a resource in situ.

The OPEC oil cartel understands user cost all too well. In fact, the Saudi

Arabian oil minister, Sheikh Yamani, has alluded to the fact that their billions of barrels of oil in the ground are similar to a bank with fixed capital. Economists with a literary bent have drawn the analogy to Robinson Crusoe and his stock of hardtack recovered from the shipwreck. What one uses or eats today leaves less for the future. When the stock is fixed and known, then the owner will wish to extract it so that survivability or the value of total profits over time will be a maximum. If too much is extracted (eaten) today, then the value or price of the resource will fall and total profits (or survivability) may diminish. As long as the value or market price now is greater than user cost, more should be extracted. If price drops below user cost, then extraction should be slowed in order to maximize profits. Although computing the precise number for user cost may be impossible, its theoretical function is straightforward. User cost allows the mine owner to allocate the resource over time so that its total value is maximized.

Just imagine you are an oil minister in a Middle Eastern country having a few billion barrels of oil. Assume that the demand function for your oil is inversely sensitive to price, although it is not necessary to know this for you need only to receive current information on oil prices via Telex or some other means. You and your fellow OPEC ministers realize that if pumping proceeds too rapidly then an oil "glut" may take place and reduce the world oil prices. If pumping is cut back, then oil prices will firm up, and everyone is happy. Once it is pumped, you spend some of the revenue on dates from the Imperial Valley in California, Rolls Royces from England, and other necessities such as F-14 fighters and other armaments. But lacking imagination and worrying about your harem's retirement program, you squirreled away most of your money in bank accounts and other investments throughout the world. Your basic choice is whether to pump now and invest in financial securities or leave the oil in the ground for another day or year. Think a minute, and abstract from political and other motives for altering your production plan. Think only of wishing to maximize total profit, so that you live well and your harem and children and grandchildren may be given the largest legacy possible, whether it be in the form of oil reserves or monetary wealth.

You are also told that the cost of extracting each barrel of oil is $0.50 (not far from Saudi Arabia's costs), and this will remain constant until your oil reserves are exhausted. What to do? In this simplified case, you simply experiment a bit and observe world oil prices. If your bank accounts, financial assets, or gold holdings earn a greater rate of return than the percentage price appreciation of your oil in the ground (after the price of extracting each barrel is deducted), then it would be logical to pump faster. You could earn more with assets parked in bank vaults than oil lying underfoot. If all your friends did the same, it would reduce the oil price a bit, but the strategy would still pay up to a certain point. Should your nonoil asets earn less than oil profits per barrel (the percentage appreciation in oil prices), then your four undergraduate and three graduate years of training at Harvard Business School would tell you to reduce extraction. After you advised all your friends to do the same, the effect would be to increase prices because total extraction had slowed.

This is precisely the course undertaken by OPEC countries after the 1973–1974 oil embargo. The difference that the embargo made was that they realized a joint decision to curtail production had a significant impact on world price. This point marked the beginning of OPEC acting as an effective cartel and utilizing its monopoly power. However, the decision on how rapidly to produce was still determined by whether profits are greater with assets aboveground in the vaults of the Bank of England or below ground as oil reserves. Persistent inflation since 1974 reduced the effective rate of return from bank deposits and short-term securities, and the world's demand curve had shifted continuously outward. These facts encouraged the OPEC oil ministers during their meetings in 1978–1980 to adjust oil prices substantially upward. The net effect has been to increase their profits while maintaining roughly the same rate of extraction.

Nonmonopolistic firms are unable to control prices, but these firms can decide their production targets in much the same manner. Individual resource owners and competitive firms are aware that there are many other potential suppliers. Consequently, they take resource prices as given and extract as long as user cost is less than or equal to current profit per unit of output (price minus the cost of extraction).

Since user cost is a present-value calculation, this implies that user cost and profits per unit must increase at the rate of interest. If the rate of interest is less than the percentage increase in resource profits expected by resource owners, then user cost exceeds market price and extraction is slowed, because more profits are to be made in the future by holding the exhaustible resource as reserves. Since all producers act in a similar manner because of profit-maximization motives, supply is reduced. A higher extraction rate will occur only at higher prices. A reduced supply tends to increase price, which is exactly what the resource owners have been waiting for—a tilt upward in current prices. Wells will be uncapped and extraction speeded up to capitalize on current prices only when profits are greater than user cost. Overconservation may soften prices in the future because past curtailment has increased future depletable-resource supplies.

On the other hand, if the rate of interest on financial assets exceeds anticipated percentage oil price increases and future profits, again by assuming a fixed stock, profit is increased by pumping now. Again, this assumes future prices and costs are precisely known. The financial asset above the ground from resource profits is growing in value faster than having it underfoot. Oil companies will pump and invest in real estate, Hollywood

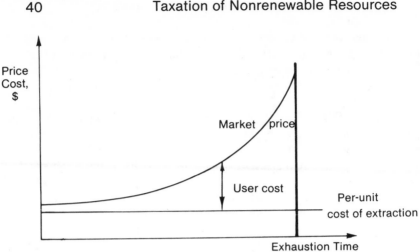

Figure 2–2. Resource Price and User Cost

film production, other commodities, or whatever else earns the highest rate of return. But when all follow suit, extraction is tilted forward, and the present increased supply will end up diminishing current price. At some point, resource owners will stop increasing current production because the profit from faster current extraction has disappeared.

The implication is clear. Individuals responding to prices and interest-rate differentials will allocate their scarce resource over time so that the value of the resource in the ground (in situ) grows at the rate of interest. Otherwise, there is an incentive to alter the rate of extraction, which in turn affects prices and profits. Equilibrium occurs when the same rate of return can be achieved through either financial means or maintaining resource stocks as investments. A difference between individual firms or resource owners and OPEC oil ministers is that the firms or owners respond to market-determined prices and interest rates, but OPEC production decisions affect prices directly.

The price behavior just described is shown graphically in figure 2–2. The unit cost of extraction, assumed to be constant through time, is depicted by a horizontal line in figure 2–2. Market price is the sum of the unit recovery cost plus the user cost or scarcity profit. As just argued, this component is rising over time at the rate of interest so that total market price, the sum of the two is rising less rapidly than the rate of interest. The price and cost behavior implies that the level of extraction falls over the life of the resource (for a more detailed explanation, see Herfindahl 1967) if demand is assumed to be constant. As user cost rises, market price rises (market price equals user cost plus unit cost of extraction), and at higher prices less is demanded.

How do individual owners respond to government price regulations? When the regulatory environment is such that current prices are less than future prices are expected to be, then it is rational for resource owners to curtail production. This is precisely the situation engendered by talk of natural-gas and oil decontrol in the late 1970s. Producers, who were convinced that government regulation was suppressing prices below market levels and that decontrol would be enacted within a reasonable time, capped wells and slowed exploration and development. Once deregulation was passed, prices increased in line with expected future interest rates, and a record number of rigs drilling exploratory wells were in the field.

What keeps the OPEC ministers who enjoy monopoly powers from continuously increasing their prices to even more exorbitant levels? The primary reason is obvious: They want to maximize profits and survivability—not prices. They know that an invasion and expropriation of their oil fields and facilities is a possibility. Also, when oil prices reach certain levels, substitute fuels become cost-competitive. At sufficiently high prices, solar energy and production of oil by synthetic processes from oil shale, tar sands, and coal become cost-competitive, and OPEC might be left with unsold oil in the ground. As one might suspect at this point, monopolists extract a resource more slowly compared to a competitive industry structure, where managers of firms act as if they are unable to affect prices. The monopoly rate of extraction will be lower, and the price will be higher. This will be confirmed shortly, but no matter how many firms make up the entire industry, they will extract from their fixed resources so that the asset in the ground yields the same return as an asset in any other investment, as in the example of bank deposits used here. This behavior will result in a resource price which increases over time to reflect the fact that in a riskless and unregulated world, every income-producing asset is yielding an equal rate of return after adjustments for risk and taxes.

The seminal work on the theory of the mine, incorporating a dynamic model utilizing the calculus of variations, was done by Hotelling in 1931. His work has been expanded on by Gordon (1966, 1967); Cummings (1968); Cummings and Burt (1969); Cummings et al. (1975); Kneese and Herfindahl (1974); Schulze (1974); Scott (1953, 1955, 1967); Smith (1968); Solow (1974); Solow and Wan (1976); and a large number of others, all of whom confirm that the theoretical time path of prices conforms to that shown in figure 2–1, namely that user cost grows at the rate of interest.

This assertion and the accompanying condition for a declining rate of extraction break down when either demand or cost changes over time. These conditions are added to the theoretical model of the firm in order to make it more realistic. Cummings and Burt (1969, pp. 207–209), Weinstein and Zeckhauser (1975, p. 380), and Heal (1976) allow for extraction costs to escalate as the resource is depleted. This is realistic because the most profitable reserves are extracted first, and costs are dependent on the rate of extraction. For example, pressures drop as oil is pumped, and eventually more expensive secondary and tertiary recovery methods must be employed. Mine shafts may have to be longer and more expensive to reach deeper reserves, and as quality declines, more raw material must be removed and processed. In this case, user cost grows more slowly than the rate of interest, in part because future costs will be higher and profits lower if present output is expanded.

Schulze (1974, pp. 71–73) analyzes the possibility that technical advances may lower future extraction costs. People are constantly developing new ways of doing things and the machines to carry out these ideas. In mining, technological advancement allows lower-grade ores to be processed more efficiently. Although user costs grow at the rate of interest in early years, falling extraction costs due to improved technology may mean that the market price falls. Eventually, rising user costs swamp cost savings, and market prices will begin to rise. This behavior may depict the historical behavior of prices more accurately than the simpler constant-cost, constant-technology model. Scott (1967) believes that costs rise as the rate of extraction is increased, and this alters the most profitable time path of prices as well.

Barnett and Morse (1963) conducted one of the first definitive studies of the behavior of resource prices and extraction costs over time. They concluded that costs fell over time for all the natural resources investigated with the exception of forest products. Peterson and Fisher (1977) summarize more recent studies which imply that the real costs of extraction and market prices have now started to rise for some nonrenewable resources. A glance at oil and precious-metal price behavior should confirm this finding. Smith (1979) edited a series of papers which update the Barnett and Morse work. The fundamental question of measuring scarcity is reconsidered; and although no consensus is achieved, evidence of rising costs and prices is presented.

The mining thus far has taken place from a natural-resource reserve of uniform quality and of known size. Geologists tell us that the grades of hard-rock minerals and other nonrenewable resources vary and that the cost of extracting them is dependent on a host of geologic and economic variables. Schulze (1974), Weinstein and Zeckhauser (1974, 1975), Herfindahl (1967), Solow and Wan (1976, pp. 365–367), Sweeney (1977), and Conrad (1978a) investigate the effects of multiple grades and costs on extraction and prices. All find that the richest (cheapest) ores are extracted first because present profits are preferred to future profits. Conrad (1978a) hypothesizes that market prices are cyclical as well. He finds that *high grading*—extracting the richer ores and bypassing lower grades—occurs when prices are highest; that user cost is sensitive to the design capacity of the extracting plant and equipment and shifts over time; that positive storage costs do not alter this behavior; and that the expected grades, resources, and prices all go into determining an optimal investment strategy.

Pareto Optimality and Resource Markets

The sophisticated, theoretical optimal control models which have demonstrated the behavorial characteristics of resource firms may also be used to derive extraction time paths that are Pareto-optimal. These socially optimal resource allocations over time are then compared to the conditions found for profit-maximizing firms, sheikhs, and others. Market and optimal depletable-resource uses can then be compared in order to ascertain whether competitive markets allocate resources optimally, as is the case for reproducible capital. These questions are investigated by Hotelling (1931), Anderson (1972), Vousden (1973), Schulze (1974), Solow and Wan (1976), Beckmann (1974), Dasgupta and Heal (1974), Stiglitz (1974), and Ingham and Simmons (1975). The consensus is that if firms behave competitively and all resources are privately owned, the correspondence exists between self-interest profit maximization by firms and resource owners and Pareto optimality.

Richard L. Gordon (1967) expresses reservations because producers may not produce at the point where marginal extraction costs are equal in all mines. Solow (1974, pp. 6–10) doubts that markets can predict future prices accurately and believes that distorted resource use among generations will result. He is joined by Nordhaus (1973a, pp. 536–537), Stiglitz (1975a), and Peterson (1978) on the same general grounds. The general problem here is that if myopia or some other aberration in the marketplace results in an interest rate which is higher than the true social rate of time preference (the social interest rate), then the resource will be extracted too quickly. Intergenerational inefficiency may result. Further analysis here is beyond the scope implied by resource tax policy, since this general myopia would distort all investment decisions. Bear in mind that, until this point, the future is assumed to be known with certainty and competitive spot and futures markets for the resource exist.

Monopoly

The OPEC cartel is not the only monopoly engaged in finding and extracting nonrenewable natural resources. Monopolies display many potential advantages, including their ability to curtail entry by controlling ownership of limited reserves, being able to assemble large amounts of capital for exploration and development more easily than smaller firms, and spreading risks over more projects, thus reducing overall risk. Hotelling (1931, p. 153), Kay and Mirrless (1975, p. 167), Weinstein and Zeckhauser (1975), Stiglitz (1975b), Sweeney (1977), and others have looked at the behavior of a monopolist who realizes that his or her behavior affects prices (a pricemaker as opposed to the competitor pricetaker) and who establishes an extraction plan in order to maximize profits. Hotelling (1931) and many others have demonstrated that in order to maximize profits, the monopolist would extract the resource more slowly than if the industry were organized competitively. The monopolist is a "conservationist," but this conservation is not in the interest of society. It is a Pareto-suboptimal use of the nonrenewable resource.

However, other investigators cited above have constructed examples in which the monopolist extracts at the competitive rate or at even a more rapid pace. However, these examples are special rather than general cases. It is well known that the monopolist will never restrict output in any single period below the point at which the change in total sales revenues falls (demand elasticity is less than 1), because no matter what his extraction and user-cost schedules, such an action would reduce profits. Under plausible circumstances, it can be concluded that cartels and monopolies extract the exhaustible resource more slowly than competitive industries.

Exploration and Discovery

As Exxon and Mobil advertisements may have taught us, the model of the extractive firm and industry developed thus far is unrealistic in assuming that the resource stock is known and fixed. The exploration for and development of natural resources require investment. These processes make the size of ultimate resources unknown because of yet undiscovered resources and because of changing technology that determines which portions of the ultimate resource stock are economically recoverable. The economics of this process were first discussed by Paisch (1938) and Scott (1955) and has since been considered by others. Paisch (1938), Scott (1955), Fisher and Krutilla (1975), Ise (1925), Peterson (1972, 1975a, 1978), Peterson and Fisher (1977), and Myers (1977) conclude that a rise in interest rates will reduce the amount invested in exploration and discovery. Because the decision to invest in exploration is closely akin to investment decision for reproducible capital (Koopmans 1973), reduced exploration will lead to reduced resource reserves. The short-run effect discussed in the simple model of increasing the rate of extraction as interest rates rise still holds, but in the long run both reserves and extraction will be reduced. Adelman (1970, 1972), Peterson (1975a), Peterson and Fisher (1977), and Myers (1977) look into the effect of the size of reserves on the cost of recovery. Peterson concludes that a rise in interest rates may cause "overconservation" of the resource, but Myers concludes that competitive firms maintain reserves at a Pareto-optimal level. However, in all cases, if a threat of expropriation or taxation which will increase mining costs is impending, the resource owner will extract the resource at the maximum feasible rate, because future user costs are reduced and time/price profiles are distorted. It is no wonder that extractive firms are so upset by the uncertainty surrounding regulatory delays and future energy legislation, for the decision on how rapidly to extract now and how much to invest in exploration and development depends on future prices, costs, technology, and government policy.

Exploration gives rise to another situation in which private markets may underallocate investment in discovery. This stems from the fact that once a discovery is made in an area, this information is of value to other resource owners. A discovery provides information about the value of their mineral rights. (Hotelling (1931, p. 144), Gaffney (1967, pp. 391–394), Kneese and Herfindahl (1974, pp. 134–136), Stiglitz (1975a), and Peterson (1975b) acknowledge this potential market failure and agree for the most part that too little investment may result. This occurs because the prospector realizes only a portion of the social value of the investment because of the generally free transfer of information to nearby landowners. Information is a property which is difficult to regulate and sell (witness the recent impacts of xerography) and which, from a socially optimal standpoint, should be available to all seekers. However, free access to information means that less investment occurs than if all the benefits could be captured by the original investor.

A number of graphic illustrations of this behavior exist. The search for uranium in the Southwest is initially carried out by electronic surveys. But uranium mineralization is localized in small pockets which are subsequently discovered and delineated by drilling and inspecting cores. It follows that the location and movement of drilling equipment provide information as to where discoveries are being made. An entrepreneur in New Mexico, where 50 percent of the known U.S. uranium reserves are located, capitalizes on the fact that this information is available to anyone who captures it. The method is to fly over areas being prospected every few days and take photographs. Tracing the location and movement of drilling rigs over a period provides information about areas where intense, and by implication successful, activity is taking place. These data are analyzed and sold to all interested parties without having to compensate those making exploration investments for the information they are generating. Owners of adjacent properties are better able to estimate the probability of discovery on their sites, and extracting firms use the information to demarcate areas which may be attractive to lease and explore further. Since those footing the bill for initial drilling surveys do not benefit, an externality occurs, and exploration may be less than if all benefits were captured by the original investor.

Similar information may be gathered in other ways. These uranium deposits in Valencia County, New Mexico, are found in tight formations in sandstone. Underground mining to extract resources was prevalent in the 1950s. A headframe is a piece of machinery located at the mine surface. Prospectors used to make both day and night sightings from one headframe to the next and thereby hoped to locate intervening uranium. Many leases and exploratory activity resulted from areas identified by this practice. The technique has also been used recently in the North Sea to spot additional drilling platforms.

The geologic characteristics of certain minerals also create common-property problems. Oil and natural gas are frequently found in reservoirs which underlie lands owned by many separate individuals. This gives each individual surface owner an incentive to drill and extract as rapidly as possible in the hope of securing the largest share of this resource before the neighbors extract from the common reservoir. Extraction takes place at a faster rate than is Pareto-optimal. Also, it is known that too rapid extraction of gas and oil reduces the total amount recovered, because reservoir pressures are reduced too rapidly. This is the avowed reason for Texas granting to the TRC the responsibility for conserving oil through pumping regulation. The pumping rates it established were reputedly designed to maximize total oil yield, but undoubtedly accomplished the important role of reducing the rate of extraction below the competitive market level and earning resource owners monopoly profits.

The same phenomenon occurs in the discovery process when property rights are created through the right of capture. This situation arises when, if you do not search for and claim the resource right now, someone else may. This "Easter egg hunt" gives investors an incentive to search for and claim resources at a faster rate than is socially efficient. This occurs because the resource owner is undefined or is not compensated for the mineral on his or her lands. Examples of the California gold rushes, Cripple Creek in Colorado, and the Oklahoma land rush engender visions of violent competition and indiscriminate resource use in the discovery process. Hotelling (1931) noted the undesirable and antisocial behavior of such rushes for windfall profits. Peterson (1978) concluded that the effects of the information spillovers discussed above and the "Easter egg hunt" were likely to cancel each other. "Right of capture" inefficiencies have led to a number of proposals for modifying the 1872 General Mining Law, which sanctions this behavior (Anderson 1976).

Risk

The very fact that nonrenewable resources are hidden away in the earth's crust means that the search, development, and extraction process is a risky one. Grayson (1960) simulated the behavior of oil and gas exploration by a few small producers in this context and concluded that they behave in a risk-averse manner. To be *risk-averse* means that the investor requires compensation to engage in activities which have uncertain outcomes, even though the average outcome from many attempts may be known with absolute accuracy. Gilbert (1975*b*) investigates private and public objectives in the risky discovery process. He concludes that private speculators generate more information concerning future prices than is optimal and less information on ultimate reserves. Myers (1977) finds the opposite is true in that private firms maintain resource stocks at the Pareto-optimal level in a regime of risk. It is virtually impossible to hold expected reserves at a zero level (just barely as much as you need to extract in the near future), because of uncertainty in discovery. It follows that firms must hold positive resources, and Myers determines that the levels are Pareto-optimal. Hoel (1978*b*) also examines the consequences of uncertainty in exploration on reserve size and extraction costs. Hoel concludes that as the uncertainty of size of the resource increases, extraction will be slowed and an increase in the uncertainty accompanying cost will speed up extraction rates.

Future prices and demand are uncertain variables. Weinsein and Zeckhauser (1975) conclude that uncertain future demand, but with a known and fixed resource stock, induces risk-averse firms to extract too rapidly (above the Pareto-optimal rate) and risk-neutral ones to behave efficiently. Another possibility is that technology in the future may produce cheaper substitutes. In the case of energy development, some people expect the breeder reactor and nuclear fusion to produce economic and virtually unlimited electrical power or for renewable-energy sources such as solar and geothermal to become economically viable. Dasgupta and Heal (1974, pp. 18–25; 1975) conclude that this uncertain event is not fully reflected by adjustments in the rate of interest, although, under certain conditions, risk adjustments are adequately reflected in the rate of return. They also find that a monopolist owning reserves may purposely delay the introduction of the superior substitute. Nordhaus (1974) has also considered the uncertain introduction of a "backstop" technology and concludes that changes in the interest rate do not produce a social optimum.

Recycling and Pollution

Although entropy precludes the recycling of energy, nonfuel minerals may be recycled, although the laws of nature dictate that some losses will occur. The economic factors that delineate the extent of recycling are examined by Schulze (1974), Hoel (1978*a*), and Weinstein and Zeckhauser (1974). They conclude that private competitive markets allocate virgin and recycled resources optimally. Extracting and processing of nonrenewable resources generate large quantities of air and water pollution which, in the absence of government regulation or taxation, the polluter freely disposes into the environment. Kneese and Herfindahl (1974) have investigated this market failure thoroughly and advocate a tax on pollution (an emissions-discharge tax) which is equal to the damages imposed on others. This matter is discussed by Schulze (1974) in the context of nonrenewable resources where he derives the necessary emissions tax which will make extracting firms behave in a socially optimal manner.

Market Failures

This review of the existing literature concerning the exploitation of exhaustible natural resources says little about the role which tax policy might or should play. In fact, if one conclusion is evident, it is that in an abstract, theoretical context of perfectly competitive markets, individual firms pursuing profit maximization ensure that nonrenewable resources will be utilized efficiently over time. The implication for tax policy is that any tax which affects decision making with respect to exploration, development, investment, rate of extraction, and ore cutoff grade will distort resource use away from this efficient time path. In this context, taxes should be levied only when publicly supplied goods and services utilize resources more efficiently than privately suplied goods and services. When the taxes necessary to support such activities are levied, they should be designed to have a minimum effect on competitive markets. This means that taxes should be levied on commodities whose demand is totally unresponsive to prices. In the case of price-sensitive commodities, taxes should be levied in inverse proportion to the sensitivity of commodities to price. This implies that extractive firms should be treated no differently from other sectors of the economy.

We know, of course, that this is not the case. Extractive firms are treated differently from other industries. We have only to think of depletion allowances to realize this. The effects of taxes on key decisions made by extractive firms and how efficiency is affected are discussed in detail in chapter 3. The relevant question asked here is whether there are occasions in which the theoretical literature reveals that decisions made by firms result in inefficient allocation of resources which taxes might correct. Virtually all students of the exhaustible-resource problem acknowledge that there are a number of such circumstances. However, there is a good deal of controversy over how significant these are and if tax policy is the proper tool to correct so-called market failures.

When a firm constitutes enough of a share of a market or an industry that its managers realize their decisions affect prices, then the basic characteristic of pure competition—price taking—falls apart. In the case of renewable resources, a firm which is a pricemaker will produce at a level that is less and charge a price that is higher than at the socially efficient levels. This conclusion also holds for extractive firms which have market power either because they have organized cartels or because government regulation has, in effect, created market power to set prices or production levels for the industry. Certain of the nonrenewable-resource industries are dominated by a few large firms. Of course, the best-known case is the seven sisters of international oil and their relationship with the OPEC countries. Before 1973, profit rates of oil producers were just a bit higher than for manufacturing as a whole in the United States. The 1973 embargo and dramatic oil-price increases of the 1970s resulted in dramatic increases in earnings for the international oil companies. It can be inferred from these instances that if the international oil companies individually possessed significant market power or operated in concert as a cartel, they had not been fully exploiting these powers prior to the 1973 or 1979 price events. If they had been maximizing profits, then the rounds of OPEC price increases would not have resulted in higher earnings. However, this ad hoc and circumstantial evidence does not refute the possibility that a certain degree of monopoly power exists and is not fully utilized because of political constraints or that the OPEC moves rationalized the power of these companies.

Hotelling (1931), Schulze (1974), and numerous others have confirmed that monopolists curtail output in order to increase profits. However, Weinstein and Zeckhauser (1975) and others have all derived counterexamples in which monopolists extract at the same rate as or faster rates than competitive firms and are not "conservationists," as others have labeled them. In the most likely cases, pricemakers do extract more slowly than is socially efficient. As Adelman (1964, 1972) has noted, they protect their market power by controlling reserves and engaging in preemptive exploration in order to deny potential reserves to competitors. Martin (in Gaffney 1967) has employed the ratio of reserves to output as a measure of monopoly power held by extractive firms. But higher investment in exploration and reserve inventories may well result in countervailing increases in production, so that the overall impact of the monopolist is unclear. It is rather hard to diagnose the precise ills that monopoly powers produce and to prescribe the corrective tax medicine. Although there is a good deal of circumstantial evidence that giant natural-resource firms control mineral leases in part to restrict entry into the industry by new firms, such as the long delays in producing coal from leases held on public lands, evidence of slower extraction rates has not been confirmed through rigorous analysis of the data.

Attanasi (1978) concludes that theoretical and empirical evidence indicates that large firms acquire large tracts and farm out exploration drilling to small independents in low-payoff areas. In frontier areas, large firms control and explore large blocks in order to restrict access to smaller firms. However, Attanasi is unable to conclude what the effect of restructuring the industry would be. This is caused in part by the reluctance of these firms to release information which they perceive might support uncomplimentary allegations against them.

If controlling potential and proved reserves in order to preclude entry by competitors is one way to maintain monopoly power, this behavior leads to purposefully delaying the introduction of a new superior technology or substitute commodity, which would reduce the value of existing reserves. For example, it is often alleged that producers of conventional forms of energy have delayed the introduction of solar and other forms of renewable energy. Nordhaus (1973a) and Dasgupta and Stiglitz (1975) have noted this possibility along with a few other observers. They concur that it is likely that firms and cartels possessing monopoly power will restrict output, control resources, and delay the introduction of new technologies, all of which result in an inefficient allocation of resources. An appropriate tax policy to cancel these effects is one which stimulates the rate of extraction and exploration for such firms. And if this is the case, the existing structure of the federal income tax may accomplish this goal. Percentge depletion allowance and immediate deduction of "intangible" drilling expenses stimulate production.

Another potential market failure identified by Hotelling (1931), Nordhaus (1973a, pp. 536–537), Stiglitz (1975a), Solow (1974b), Heal (1976), and others is that firms may be myopic, because they either discount future profits at a higher rate than the true social interest rate or formulate exploration and extraction programs based on faulty future-price information. The correspondence between competitive markets and socially efficient resource use over time assumes that no myopia exists. However, it is likely that managers of resource firms operate under a shorter time horizon and a higher interest rate than are socially optimal. Furthermore, the theoretical models assume that contingent markets exist for all future prices and risks. While certain commodities are traded in futures markets, this is not universally true, particularly for periods of more than a year or two in the future. Furthermore, future contingency markets do not exist. It has also been alleged that these speculative markets have produced greater price instability than is optimal, rather than fulfilling their presumed socially efficient role of dampening price fluctuations and providing consensus estimates of future prices. One reason for this erratic performance is that expectations about future economic growth and resource availability have themselves been erratic in the 1970s. The net effect of these factors is to make firms operate myopically, extract at a higher rate, and explore less than may be socially desirable. In this case, an appropriate tax policy would be one of discouraging production and encouraging exploration.

Richard Gordon (1967) feared that competitive firms would misallocate resources because they would not extract resources at each site at the level where the incremental costs were everywhere equal. This assertion is incorrect, for socially efficient resource use dictates that the extraction rate is increased until the point at which the incremental extraction cost plus the incremental user cost equal the price of commodity. Because costs vary with the location and quality of the deposit, the profit-maximizing firm operates at levels where incremental costs and user costs are not uniform. As Goldsmith (1974) points out, this is also the point at which this resource is being extracted at its socially efficient rate. Gordon (1978) has recently recanted his 1967 position and acknowledges that user costs are different for deposits of different quality. However, he believes that if the firm's manager perceives that prices are increasing at a rate faster than the interest rate, misallocations may occur. His argument is essentially that resource markets may continually be in a condition of disequilibrium which may result in misallocation of resources. It is unlikely that a tax policy could be formulated which could alleviate this problem.

Day (1978) expands on this theme by applying his concept of adaptive economics. This approach is more general than the ones taken in conventional economics; it combines sociology, psychology, and systems dynamics. Day hypothesizes that disequilibrium prevails and markets may not resolve deviations in supply and demand in a socially beneficial manner. He classifies the major problems as being those of inappropriate levels of intergenerational transfer of resources (by leaving it in the ground to await future extraction), "overshooting," and "surprise." Given these broad problems, the appropriate role for resource tax policy is not at all clear. In fact, Day does not address how taxes might affect resource markets, much less whether tax policy might be used to mitigate or exacerbate difficulties of how the market fails to converge to an efficient new equilibrium when exogenous shocks ocur.

The problem of the "proper" intergenerational transfers of renewable and nonrenewable natural resources and capital is beyond the scope of this book. However, it should be noted that the problem is one of both efficient resource use and transfer of income. Both problems are primarily resolved if firms use interest rates in their discounting calculations which represent the socially "correct" rate. For the sake of simplicity, this is assumed to be the case for the remainder of this book.

The final set of reasons why the market may fail in allocating resources efficiently is classified under the term *externalities*. An externality occurs when the behavior of an individual or a firm directly affects (not through the "market") another individual or firm and this effect is not reflected in prices. Another way of looking at externalities is involuntary transfers. Voluntary transfers are taken care of in markets. However, there are no markets for involuntary transfers. Two externalities have already been noted in this brief review of the literature. One is the "free" information to property owners supplied by investments in natural-resource discovery made by adjacent property owners. Since this valuable information is difficult to restrict, the full return of an investment is not captured by the actual investor, and exploration may be reduced below the socially optimal level. This is discussed by Stiglitz (1974) and Peterson (1975b). Uhler and Bradley (1970) analyze investment in exploration and conclude that risk aversion and the nonexcludability of exploration information will lead to less investment than may have taken place otherwise. Erickson, Millsaps, and Spaun (1974) refer to the fact that Alaskan North Slope oil leases jumped from $39 to $2,182 per acre after the initial find by Atlantic Richfield. However, overexploration by monopolies and the "Easter egg hunt" for mineral claims on public lands (wherein the right of capture (discovery) prevails and little or no compensation is paid to the government as property owner) encourage excess exploration. The net effect is uncertain, and tax policy is not well suited to adjust for these market failures.

The problem of the Easter egg hunt is primarily due to the federal government's failure to charge appropriate (market) prices for minerals found on its lands. This policy is contained in the 1872 General Mining Law. Note, however, that certain minerals, primarily fuels (oil, coal, and natural gas), do not fall under this law and are leased on public lands under the 1920 Mineral Leasing Act. The 1872 law has been under review for some time, and it is likely to be revised at a future date (Anderson 1976).

Another externality is the common oil pool and the resulting incentive for each surface lessee to capture as much of the oil as quickly as possible, in effect, another Easter egg hunt. Again, this problem is one which is not amenable to tax policy, but is more amenable to direct regulation on leasing policy and property rights. An externality problem which perhaps should be addressed through tax policy is one which at present is mainly rectified by direct regulation and the complex legal processes that accompany direct intervention—pollution.

Pollution

Emitting harmful by-products into the water and atmosphere and onto the land surface constitutes an externality because the polluter is not necessarily required to incorporate these social costs into his or her decisions. The solutions pursued in the United States consist mainly of direct regulation on automobile, manufacturing, mining, and power-plant emissions as well as simultaneous regulation and subsidies for building water-pollution control and treatment facilities. Economists have long argued that the costs of cleanup are not uniform among industries and environmental damages are not the same in each location. This means that uniform regulations cause too few or too many resources to be devoted to pollution control, depending on the relative costs and benefits of cleanup. The externality could be more efficiently regulated by means of a direct charge or tax on actual emissions irrespective of their source, based on health and property damage. When cleanup is less costly than paying the tax, emission control will be voluntarily undertaken by industry. Whan abatement costs exceed the tax, it would not be efficient to apply resources to the cleanup job. This approach has been advocated by Kneese and Herfindahl (1974) and Kneese and Schultz (1975) and numerous other economists and has been shown to apply equally well to extractive industries by Schulze (1974). Church (1978a) has shown that direct regulation of sulfur dioxide emissions from copper smelting and electricity production is inefficient at present because cleanup costs escalate rapidly as regulations become more stringent, and these cost patterns are not identical for the two industries.

Although emissions taxes are not in use in this country, a wide variety of taxes are levied by the federal and state governments. Natural resources have always been an attractive tax base. It is readily apparent that the resource cannot be moved in order to avoid tax liability, and resources are easily identified for special tax treatment. The pattern of differential taxes and tax rates applied to nonrenewable resources among countries and their various jurisdictions is explained most easily by observing which resources are the most prevalent and profitable. The apparent "ability to pay" has almost always seduced tax policymakers. At one time Minnesota singled out the rich iron ores found in the Mesabi Range for special tax attention; oil and natural gas produce generous revenues in Texas, Louisiana, and the other petroleum-rich states. Copper is singled out in Arizona; coal, in Montana and Wyoming; and uranium, in New Mexico. Pragmatic tax policy making is more closely attuned to the goals of economic development and raising tax revenues, discussed in chapter 1, than to a goal of correcting market failures, discussed above. However, this brief review of the economics of exhaustible resources is necessary in order to understand how taxes affect the industry. In chapter 3, theoretical models are extended so that the effects of taxes levied in a plethora of ways can be analyzed and evaluated with respect to a number of tax policy goals, including efficient use of resources, economic development, and a number of other criteria.

3 Taxes and the Extractive Firm

In chapter 1, a general analysis of tax policy was considered. A simple model emphasizing the tradeoff of encouraging development of the resource versus extracting the maximum tax revenue was explored for domestic coal production. Private markets and the downward-sloping demand curve constrained the ability of state tax policymakers to achieve goals of maximum resource development and high tax yields simultaneously. The theoretical literature pertaining to the extractive firm surveyed in chapter 2 reveals that it is difficult or impossible to identify optimal tax policies which ensure that conservation, pollution-control, and efficient-resource-use goals will be achieved. An alternative approach is to look at what taxes do to the economic system. In this chapter, we investigate how various tax structures affect the firm and how these behavioral reactions can be used to measure the effects of tax policy in a theoretical context. In other words, here we concentrate on the supply side of the market.

The analysis and results reviewed in chapter 2 are now extended to indicate the probable effects of various tax structures. It is then possible from this analysis to indicate how closely taxes achieve a wide variety of policy goals. The only conclusion about correct or optimal tax policy which could be drawn from chapter 2 is that taxes should be employed for resource reallocation only if special circumstances exist. It appears that private markets allocate natural resources optimally or close to this without government interference and coercion. Therefore tax policy should strive for neutrality—no dislocation from the market solution. Where tax policy can aid in ensuring that private decisionmakers utilize renewable and nonrenewable resources in the most efficient manner from the point of view of society, special taxes such as pollution-discharge fees should be designed and implemented. Other exceptions include monopoly, externalities, and myopia. However, when markets are competitive and buyers and sellers are price-takers rather than pricemakers, a Pareto-optimal or socially efficient allocation of resources is achieved by the private marketplace wherever these special conditions fail to exist.

Tax-Induced Changes in Market Equilibrium

The possible nonneutral consequence of a tax may be seen by examining figure 2–2. It may be recalled that efficient or Pareto-optimal allocation of resources is achieved when the maximum number of exchanges takes place, the point at which the willingness-to-pay and the willingness-to-produce schedules intersect. If output or the number of exchanges is smaller than Q_0, then resources are wasted because people are willing to pay more for additional output than producers require to produce. If output is larger than Q_0, it is inefficient because it costs more to produce than people are willing to pay. In the absence of coercion, neither nonoptimal level of output will occur for long in the private marketplace. In fact, the willingness schedules are the same as demand and supply schedules when competition prevails, and they are labeled as such in figure 2–2.

A *neutral tax* is imposed on firms, consumers, or workers and does not distort the market allocation of resources. In figure 2–2, this means that output remains at Q_0 and price remains at P_0. In a general equilibrium context with all goods and services considered simultaneously, tax neutrality occurs when the ratios of all prices remain constant.

Economists have spent a good deal of time and energy devising tax schemes which possess the characteristic of neutrality. While these grandiose schemes are not discussed here in all their magnificence, it is interesting to note that proposals for sales taxes on virtually all items produced and consumed [the value-added tax (VAT)], the universal proportional income tax, and certain other tax schemes are derived from the intellectual activity of attempting to construct a neutral tax. Our goal is more modest: It is simply to show that when a nonneutral tax is introduced, the person remitting the tax responds. These responses result in reallocating resources in an inefficient way and shifting part of or all the tax burden onto other people than the actual tax remitter. Let us look at this process in more detail by analyzing a tax levied on the natural-resource extracting firm.

When a tax is introduced, the firm is basically limited to three possible tax-avoidance actions. First, the firm can alter the rate of extraction and the level of output. This includes the decision to totally shut down or alter the cutoff grade that establishes which resources are left in the ground and which resources (those above the chosen grade) are recovered and processed. Second, the firm can alter its technology. In virtually every stage of the extraction and beneficiation process, alternative technologies are available. The firm's choice is determined by the productivity of each technology and its cost. Third, in the long run, as renewable capital depreciates and has to be replaced and the mine is exhausted, the firm can choose to change location to any other place which possesses the required resource. This decision may occur because of the discovery of a more profitable reserve or because of existing or anticipated state or local effective taxes or other costs which have affected profitability at the current location. When the firm makes a decision to change the level of output, technology, or location based on the effect of taxes, it is reallocating resources and the goal of tax neutrality has been violated. Changes in firms' decisions engendered by a tax results in *tax shifting*. This means that the party actually remitting tax payments may bear a portion of the tax, and other portions of the burden are passed onto those who are affected by the firms' reactions.

Tax shifting can be illustrated by means of elementary exposition of supply and demand. In figure 3–1, output per unit of time Q is depicted on the horizontal axis, and price or cost (in dollars) is on the vertical axis. Demand is labeled DD, is negatively sloped (more is demanded at lower prices), and is also the willingness-to-pay schedule. Industry supply is labeled SS and is positively sloped. Its positive slope is caused by two factors. First, as economies of scale are exhausted, the incremental cost of each additional unit produced then increases. Second, as the rate of extraction increases, lower-grade ores must be exploited. The industry supply curve is so depicted in figure 3–1 and is derived by adding the supply curves or willingness-to-produce curves of all firms in the industry. Market equilibrium and efficient resource use occur where demand and supply intersect at price $= P_0$, output $= Q_0$.

Voluntary exchange of all production takes place as long as the amount demanded is greater than or equal to the amount supplied at any given price. When the converse is true, supply is greater than demand at some given price. There are not sufficient transactions to clear the market because of the

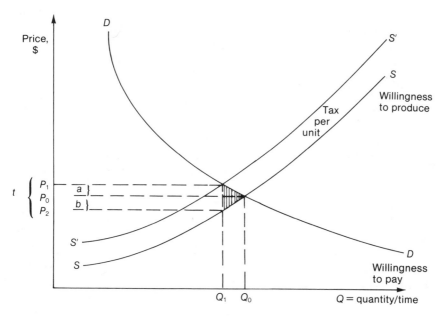

Figure 3–1. Supply and Demand with Tax Shifting

difference between producer and consumer behavior. The clear implication is that the market equilibrium (P_0, Q_0) corresponds to maximizing voluntary exchanges; and it equals the socially optimal use of resources, as shown previously. Also, the literature review indicates that the market supply curve corresponds to the supply curve for most beneficial rate of extraction from society's viewpoint when competitive conditions prevail.

Now introduce the government into the picture by placing a tax on the firm. This may be a tax on physical output, as is assumed here, and similar effects are achieved for any tax which affects either supply or demand. The net price that the firm receives in each transaction equals the market price minus the per-unit tax t. The effect of the tax—reducing the proceeds to the firm—is shown in figure 3–1 by shifting the supply curve vertically upward by the amount of the severance, an output-based tax t. The supply curve including the tax is shown as $S'S'$. It will result in a new market equilibrium at (P_1, Q_1), where price is higher and the amount extracted is lower. After the imposition of the tax, not only is less output produced at a higher gross price for consumers (P_1) than before the tax, but also the firm receives a lower net of tax price $(P_1 - t)$ than previously. The government collects tax revenues of $t \times Q_1$, although in their naîveté tax policymakers may have anticipated tax revenues of $t \times Q_0$, based on the pretax level of output.

It is now possible to see three effects of the tax. First, the former efficient allocation of resources is distorted because Q_1 is now produced instead of the socially optimal level Q_0. Second, the difference in price between what consumers pay P_1 and what firms receive $P_1 - t$ compared to the original price P_0 measures how income to consumers and firms is affected. Third, these price changes indicate the distribution of the tax burden. The tax is said to be driving a wedge between consumers' and producers' prices, and the effect is to reallocate resources because fewer inputs are required to productt Q_1.

This tax-induced reallocation of resources and resulting net loss to society may be measured by computing the difference between the total amount consumers are willing to pay for the lost production (the difference between Q_1 and Q_0) and the total cost of producing this output. The amount that consumers are willing to pay for this is the area under DD between Q_1 and Q_0, so that the net loss or difference is the small triangle filled with vertical lines in figure 3–1. This is called the *social loss*, or *deadweight loss*, due to the tax.

The only way that this net loss may be reduced is if the government program funded by the tax revenue $t \times Q_1$ is valued more highly than the amount which would have been spent on lost output $Q_1 - Q_0$. Thus, the social-loss triangle is likely to be the minimum measure of the distorting effect of a tax which drives a wedge between consumers' and producers' prices.

Any time that a tax drives a wedge between the prices paid and received in a transaction under perfect competition, a similar social loss is engendered. Thus, taxes on capital, land, labor, and all commodities produce social losses (analogous to the reduction in output in figure 3-1) unless no reallocation of resources takes place. If no reallocation of resources occurs, then a tax is said to be neutral.

The effect of a tax on income and how the ultimate economic burden of the tax is passed from the actual taxpayer to others can also be illustrated in figure 3-1. Even though extracting firms must remit tax payments to the tax collector, the economic effects caused by the shift in equilibrium from (Q_0, P_0) to (Q_1, P_1) reach beyond the firm. The price that resource users pay increases from P_0 to P_1. The ultimate burden of the tax is appropriately described as being borne by resource owners and owners of the firm in proportion to the changes in price: $(P_1 - P_0)/P_0$ for resource users is called *forward shifting*; $[P_0 - (P_1 - t)]/P_0$ is borne by owners of the firm and is not shifted. Tax shifting is depicted in figure 3-1 as distances a and b, which must total 100 percent of the tax t.

Although the effect cannot be shown in this diagram, further shifting is possible. When output falls, the demand for input also is reduced, which disturbs markets for labor, capital, and resources. The percentage of the tax borne by the firm in certain situations may be shifted onto inputs—this is called *backward shifting*. The explicit consideration of these subtleties is delayed until chapter 4 where a more complex model of the shifting of tax burdens is developed. While relative changes in net prices reveal the distribution of tax burdens, the absolute changes in prices times the quantities purchased and sold reveal changes in incomes. In the case of nonrenewable resources, tax shifting clearly goes further than the simple situation shown in figure 3-1. In virtually all cases, the extracted resources are transported, processed, and incorporated into goods or services which are eventually consumed, such as gasoline for transportation or electricity to heat water and to run the dishwasher. At every point in this long production path, the tax burden may be borne in part, and the rest may be passed along to others. The implication of this process is clear. The remitter of a tax may bear only a portion of its burden, and individuals who bear the ultimate tax burdens and effects are not easily ascertained.

The model illustrated in figure 3-1 represents the tip of an iceberg, for there is another story to tell about taxes. They affect decision making and the marketplace in many ways. For this reason, we look next at the decisions which the extracting firm makes and the effects of taxes on these decisions.

To do this, first we develop an abstract representation of the firm, and then we examine each major tax base and its effect on the firm.

Behavior of the Firm

Most observers and industry analysts agree that the sole or at least primary goal of the firm is to maximize profits. In order to accomplish this, managers must make a multitude of complex decisions which basically involve investment, financing, technology, production, and marketing. The choice of technology is a key decision, for once it is made and investment projects to carry it out are completed, managers are greatly restricted in what they can do. Once the mine or well is developed and the associated processing, transportation, and other on-site plant and equipment are installed, the managers' choices are limited to settilng the rate of production or, in extremely difficult times, to shutting down. Since technology and, by implication location are perhaps the most important long-run decisions, this is denoted in the central decision box in figure 3–2.

Perhaps the term *technology* as it is employed here could do with a bit of exploration. Technology simply means ways of doing things. The industrial revolution and subsequent acquisitions of theoretical and practical knowl-

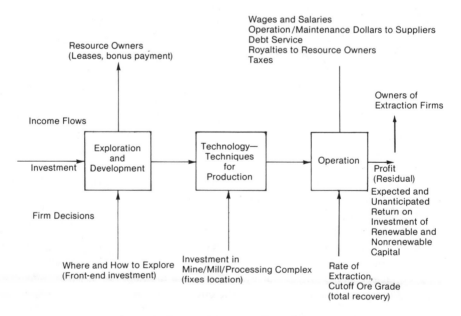

Figure 3–2. Decisions by Firm Managers

edge confirm that there are more ways than one to skin a cat. For example, copper may be mined underground or from the surface with basic hand tools and various explosives or in mines of a wide variety of configurations and sizes up to the immense open pits found today in many parts of the world. The ore can be transported by animals, rubber-tired vehicles of a great variety, or trains. The crushing, extraction, and concentration process also is variable both in technique, such as flotation versus leaching, and in scale. Finally, smelter operations of wet-charge reverberatory furnace, flash furnace, and Noranda reactor are all competitive today depending on the location, environmental regulations, and cost of electricity, fuel oil, and other inputs. Economically efficient copper extraction and processing involve a best-suited technique for each specific site and geologic configuration and for different envisioned capacities.

The managers of mining firms must first decide how best to search for and control property rights in areas which may yield economically exploitable resources. In all but a few cases, nonrenewable natural resources are widely distributed in the earth's crust and are defined in a number of ways. A *geologic reserve* is one which is economically exploitable. Given existing prices and technology, it is the most profitable of those which the firm hopes to identify and legally control. The development process is one of gaining further knowledge about the characteristics of the reserve by drilling cores or development wells and determining the most profitable technology, engineering design, and capacity to bring the reserve into production. Sinking shafts, building milling and beneficiation plants and equipment, providing transportation facilities, gaining access to water and energy utilities, and, in some cases, building housing and other off-site facilities for workers may take five to ten years for a large mine site and ten or more years for Alaska and North Sea oil and may involve hundreds of millions of dollars.

Costs which have grown faster than all others are the direct and indirect ones associated with complying with government regulation—environmental impact statements; public hearings; health, safety, and environmental regulations; and a host of others. The delays and abandoned projects caused by these regulations are now a major component of the costs of start-up or expansion. All these costs are part of the central technology box shown in figure 3–2.

Labor requirements are sensitive to technology. In the case of underground coal mining, they are a major cost component, whereas they are of less importance in surface strip mining, particularly in the Far West where large earth-moving machinery is used extensively. There are comparable technology-induced variations in operation and maintenance costs. Finally, taxes are undoubtedly a consideration in deciding in which state or county to locate a new facility.

Tax payments are shown in figure 3–2 as an outgoing arrow. However, it is clear that no mine could be developed or could operate without government-supplied highways and the complete infrastructure of public safety, national defense, a judicial system, education, and a host of other services. The simple arrow does not reveal the ingenuity or, from industry's viewpoint, the perversity of tax policymakers. Taxes are levied on nearly every operation of the extractive firm, and the temptation to tax natural-resource extraction is rarely resisted.

Profits are the amount of money left over after the firm meets its payroll, tax liabilities, and all other costs, including principal and interest payments on short- and long-term debts. To accountants and taxmen, *net profit* is defined as sales revenue minus all legal obligations, such as wages, salaries, taxes, debt, service, and other incurred costs. Although "creative" account-ants and the taxmen often conflict over what some of these items are and how long-term capital investments should be depreciated, this concept of profits is, by and large, well defined. Economists, however, have developed their own concept of profits: unanticipated income or increases in wealth. On the surface, this concept comes as almost a slap in the face to outsiders who observed the fantastic "increases of wealth" which the large oil companies suffered so embarrassingly after the quantum oil-price hikes of 1973 and 1979. To the economist, these reported accounting profits are partially pure profits and partially anticipated returns. The economist takes the position that stockholders of the firm deserve recognition and have bought shares in anticipation of earning a positive rate of return and preserving their capital. The economist reasons that equity ownership in a firm takes place only in anticipation of a return. Thus, anticipated return on equity capital is a cost rather than a profit, for the return is indeed expected. However, unanticipated profits such as those engendered by a "lucky" resource find, the capricious-ness of the OPEC oil ministers, or changes in government legislation are *economic* or *pure profits*.

Economic profit is sometimes related to cash flow. This includes both anticipated and unanticipated profits. The primary advantage of this defini-tion of economic profits is that it can be measured. Economic profit equals accounting profits net of taxes plus depreciation and depletion expenses (because these are "fictional" and not actual payments) minus expenditures made currently on investment projects. Cash flow does not correspond to the strict definition of economic profit unless a number of restrictive assumptions are made about how capital markets function. For simplicity's sake, we define economic profit to be cash flow minus the return necessary to raise and maintain equity capital. The problem with breaking this definition or any other into anticipated and unanticipated returns is that there is no objective way to perform this type of analysis. The reason is that anticipated return on investment is not uniform or, by definition, predictable.

At a minimum, investors expect a project to yield a rate of return somewhat above the anticipated rate of inflation. Returns from risky investment should be commensurately higher. In chapter 4, the effect of risk and taxes on rates of return in extractive industries is examined empirically. The problem is that it is difficult to anticipate future risks and to know precisely how much higher the rate of return needs to be in order to induce investors to purchase more risky assets. In any case, we attempt to always distinguish when accounting profits are discussed, for those are the ones which are taxed as corporate income, and when economic profits are being discussed.

Let us return to the decisions that the firm must make. Clearly, when investment is being contemplated in a new site (*greenfield investment*) or when modernization and expansion investment is being contemplated in an existing site (*incremental investment*), the anticipated revenue resulting from the investment must be sufficient to cover all costs, including a sufficient return to investors, so that equity capital can be raised. This situation is denoted as the *long run*. Naturally, the projects with the highest anticipated rate of return are pursued first.

When plant and equipment are fixed, the situation is different and is denoted as the *short run*. The major decision is the rate at which to run the mine or well, including whether to shut it down entirely. As long as the site produces sufficient revenue to cover all costs which vary directly with the quantity of the resource extracted, such as labor, energy, operation and maintenance expenses, then management will maintain production even though the site is losing money from an accountant's standpoint. The only qualification which needs to be made is that it also pays to keep the mine operating as long as negative income is less than the costs directly associated with plant shutdown, which can be substantial in the case of mining operations. The accountant is still computing fixed costs and overhead, which are charged to operations, but these must be paid regardless of whether the plant is operating. Thus, they have no bearing on the short-run decisions of managers. In the short run, shutdown is chosen only when operations cannot cover variable costs (and, therefore, less money is lost by ceasing production than by continuing). Although this depiction is highly simplified, these rules of operation are borne out in the real world—Chrysler or American Motors does not shut down completely when it loses money in an accounting sense.

However, if in the long run the enterprise cannot show an acceptable rate of return, then the rational decision is to dismantle the operation and liquidate or to invest in another location and technology which are anticipated to be profitable. This behavior shows why taxation of natural-resource firms may be a veritable cornucopia of tax revenue for a while. The firms will sustain accounting losses after taxes and still operate over the short haul. But over the long haul, as the developed resource is depleted, greenfield investments may very well be made in other locations with a more attractive tax environment. This brief discussion of the concept of economic profits is closely related to another economic concept—economic rent.

Economic rent is any payment which is above the amount necessary to induce a person to voluntarily sell her or his services or possessions. Thus, the salary that Steve McQueen receives which is above the minimum amount that would be necessary to get him to act in your planned extravaganza picture is economic rent. The reason he can secure the rent is that you feel that this motion-picture production will make you the next Darryl Zanuck, and therefore you are willing to pay Steve's asking price, up to some limit. If you are an effective negotiator, it will be possible to bid his price closer to the minimum acceptable offer, and you secure more of the economic rent produced by the motion-picture production. If McQueen's lawyers are insightful and tough, they will end up negotiating terms which are close to the maximum you are willing to pay. Thus, an important characteristic of economic rent is that it can be transferred in whole or part; yet McQueen's services will still be offered, and you will still produce your picture.

Economic rents are potentially generated by any unique resource—Steve McQueen, a rich uranium deposit, a Picasso painting, a unique location for a shopping center. The unanticipated portion of economic profits (since they are unique in that they are not expected) is frequently classified as economic rent along with some of the other examples of economic rent just cited which can be anticipated. Strictly speaking, this portion of economic profit is not rent, in the sense that it is not known beforehand and therefore is less easily negotiated for. This explains why negotiations over activities with potential but risky future profits are complex and usually made on a contingency basis—where profits are divided according to different formulations with a key breaking point frequently being made after investors have recovered their capital plus a minimal rate of return.

The primary problem with economic rents is that they are difficult to actually quantify even when they can be anticipated—you do not know McQueen's minimum price for your picture, even if you sign him at that rate, nor is it possible to discover which portions of the accounting profits earned by General Motors are economic rent and which are returns anticipated by stockholders. Thus, economic rent may be both anticipated and not anticipated, and it is difficult to know which portion of a salary or an accounting profit is economic rent and which is the necessary compensation for someone to sell services or to invest capital. While it is possible to talk about economic rent theoretically, it is virtually impossible to measure it either before or after negotiations have transpired.

The concept of economic rent is easily related to our description of the theoretical extractive firm. Davidson, Falk, and Lee (1975), Carlisle (1954), and Brooks (1964) estimate that rents in extractive firms are positive in the long run and short run. Theory yields the finding that in the absence of technological change and with fixed demand, user cost is positive and increases at the prevailing rate of interest. This result was shown graphically in chapter 2 and is repeated in figure 3–3. However, two sites are depicted, one with a low-cost, high-profit reserve (panel A) and the other with a high-cost, low-profit reserve (panel B). The time path of price must be identical, for both sites are assumed to be owned by different firms and to be covering all variable costs once extraction begins. Since their output must be competitive, price will be equal for both firms. Economic rent is the difference between costs of extraction and the initial (first-time-period) price. Recall that user cost must be increasing at the rate of interest, so that the time path of price is increasing at an increasing rate. When the high-cost mine begins extraction, this price path begins increasing at a faster rate. The gap between the initial price and future prices is not economic rent because this return is necessary to induce the firm to allocate the resource over time. If they were absent, then the firm would have no incentive to allocate the resource optimally. If the high-cost mine (panel B) is making no economic profit, no other mines with higher operating costs will be developed. If costs increased or a tax were imposed on this high-cost mine, it would have a zero worth and a shutdown would be indicated. It earns economic rent equal to the distance between initial price when extraction starts and the costs of production, and in this case it is zero. On the other hand, the low-cost, high-profit mine, shown in panel B, yields an economic rent equal to the cross-hatched area.

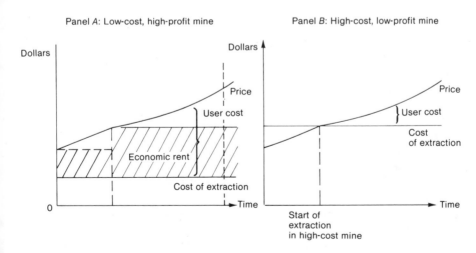

Figure 3–3. Low-Cost/High-Cost Mine: Prices

A Tax on Economic Rent Is Neutral

The concept of economic rent is so intriguing because government may try to secure it along with you, Steve McQueen, or anyone else, and no matter who ends up with the economic rent, the resource owner will continue to operate as before. To the extent that a tax falls on economic rent, it is neutral—a goal that economists hold in high esteem. Tax revenues of part of or the entire economic rent may be collected and the original owner's income reduced by a commensurate amount, but no resource reallocation results from the appropriation of economic rent. Although tax policymakers usually refer to the term in passing, if at all, their noses are attuned to signs of rent and pursue it with vigor. State legislators with low-cost reserves know that resource taxes can be increased and the rent extracted, and as long as some of the rent remains with the producers, the firm will never leave its location. Governor Thomas Judge of Montana obviously believes that the state's coal is producing economic rent to which the state is entitled when he quotes a Detroit Edison fuel purchaser as stating that over the twenty-five-year life of that utility's contract with Decker Coal of Montana, his utility will save $1 billion over the purchase of Eastern coal—even with the 30 percent severance tax in place.

However, the state is not the only one who may extract rents. Another is the owner of the land in which the resource is found, and others are those who have bargaining power over the extracting firm. The resource owner will want to renegotiate the mineral lease, if it is possible, when there is rent to be extracted. For this reason oil companies and other extractive industries usually negotiate long-term leases, so that they may retain windfall returns if a lucrative deposit is found. In the case of coal extraction, discussed in detail in chapter 6, railroads have near-monopoly power in many areas where economical alternatives such as barge transportation are not available. Western railroads have requested and usually received large increases in unit transportation rates for coal from the Interstate Commerce Commission (ICC) in apparently successful efforts to capture economic rents from low-cost, low-sulfur Northern Plains coal (*Business Week*, July 23, 1979, and Zimmerman 1979*b*). The Western railroads know that the best way to maintain monopoly power is to utilize government "regulation" as a device to prohibit or at least control potential competitors, and thus they fought the Carter administration and congressional proposals for decontrol.

Entry by potential competitors is precluded not only by manipulating regulatory bodies. A number of enterprises have proposed coal-slurry pipelines for transportation of Western coal to proposed and existing power plants adjacent to large Midwestern population centers. A necessary prerequisite is that rights-of-way be secured, and the railroads have been steadfast in refusing to allow pipelines to use railway rights-of-way. They have carried their fight to Congress by lobbying against proposed legislation which would grant the power of eminent domain to pipeline companies.

Competition over wresting economic rents from the coal producers is not limited to the railroads. Montana enacted a severance tax on coal production which goes up to 30 percent, the highest of all such tax rates in the United States. Coal producers and consumers (for some of the tax is certainly forward-shifted) are not prone to give up easily, and they have attempted to have the courts rule against the tax as being an improper imposition on interstate commerce. A similar battle has been fought over the windfall profits (economic rents) created by decontrolled natural-gas and oil prices. The battle over the windfall-profits tax has been bloody. States have been able to exempt production on their lands from the tax, thereby saving hundreds of millions of dollars of state revenues.

Tax Effects on the Firm

Now that the concept of economic rent and some of the basic economics of extractive firms are out of the way, it is possible to look at what operations of the firm can be taxed, the economic effects of such taxes, and how this information may be used to compare taxes with respect to how closely they achieve various tax goals. Conceivably taxes can be levied on virtually every activity of the extractive firm and resource markets. Constitutional, practical, and other constraints restrict tax bases somewhat, although the variety and complexity of existing tax bases and rates are substantial enough at present to provide employment for thousands of attorneys, accountants, economists, government bureaucrats, and other tax professionals. We look at taxes levied on output of the extractive sector, various property taxes, and corporate income taxes. Those taxes levied on wages and certain operation expenses are not analyzed since they are identical to those levied on all business enterprise and do not, therefore, especially affect the extractive sector of the economy. Leasing is similar to a tax in certain respects, and this is reviewed as well.

A Numerical Example of Severance and Property Taxes

Before these various taxes are discussed in detail, we develop a numerical example which illustrates the major criticism of taxes levied on nonrenewable resources—they are not neutral. Virtually all analysts have concluded that the effect of a tax levied on output is to slow the firm's rate of extraction, and a property tax on the value of the resource in the ground (in situ) will encourage the firm to extract it faster than the firm would have without the tax. These nonneutral effects are advantageous only when there is strong reason to believe that private market allocation of resources is not efficient and privately determined rates of extraction are not socially "correct." The numerical example, although highly simplified, indicates why a tax on output and a tax on property alter the firm's decision regarding the rate of extraction. A version of a neutral property tax is also constructed.

Consider an oil-extracting firm which owns an underground reserve. We expect the value of minerals in place, which is the present value of future profits that are forgone when present extraction occurs, to increase at "the" rate of interest. Let us assume that this interest rate on alternative investments is 10 percent, so that the price of oil per barrel in place increases at 10 percent per year—$5 now, $5.50 at the end of year 1, and, since the interest is, in effect, compound interest, $6.05 at the end of year 2, and so forth. The $5 oil price is chosen for nostalgic reasons. Assume, for the purpose of simplification, that the cost of extraction is zero. This assumption is not totally absurd, since Saudi Arabian oil can be pumped to the surface at the negligible cost of $0.25 to $0.30 per barrel. The resource firm in a taxless world has determined that 1,000 barrels should be extracted per year until economic exhaustion occurs at the end of the third year. Annual net revenues, or total anticipated profits, would be $5,000, $5,500, and $6,050 at this pumping rate. Since future income can be discounted to give current dollars, at the assumed rate of interest (10 percent) this income stream has a present value of $15,000. That is, the property rights to this oil are worth $15,000 right now. This numerical example is shown in table 3–1 for the situation in each of the three years for the price per barrel, the yearly extraction, profits before taxes, and the present value of those early profits.

Suppose a property tax on the property rights to the oil is enacted. What actions might the owner take to avoid taxes? Under the resource owner's original plan to extract 1,000 barrels per year, the value of the remaining resources would diminish each year as the reserve shrank. Suppose the property tax is contrived in a manner which ensures that its effect on the

Table 3–1
An Oil-Extraction Example
(dollars)

Year	Price per Barrel in Place	Extraction without Tax (Barrels per year)	Profits before Taxes	Present Value before Taxes
0	5.00	1,000	5,000	5,000
1	5.50	1,000	5,500	5,000
2	6.05	1,000	6,050	5,000
3				
Total (present value)				15,000

property owner's decision concerning the rate of extraction is zero, so the tax is neutral. A tax on present value of all three years' profits ($15,000) of 10 percent yields total tax revenues of $1,500. The oilman is given the option of paying the tax as a lump sum ($1,500) or in installments spread out over the life of the oil well, so that the total of annual tax payment equals a present value of $1,500, the total tax bill to be collected. This may be calculated by dividing $1,500 by the number of years over which extraction takes place and then increasing each annual payment at the rate of interest. Since the present value of profits is maximized with an extraction rate of 1,000 barrels per year, the firm will not find it advantageous to alter its plans because it retains 90 percent of all profits, no matter what pumping plan is followed, as can be seen in table 3–2 for a three- and four-year plan.

A three-year extraction rate or a four-year rate (or, for that matter, a one-, two-, or n-year extraction rate) does not alter the present value of the property tax on the rights to the resource, as long as the tax payment must be made in accord with the extraction rate and the resource left in the ground. This tax scheme is then neutral with respect to the use of the resource, and its only effect will be to extract income from the resource owner and deliver it to the tax collector. However, this specially contrived tax on ownership of commodity reserves is not in actual use anywhere. The reasons are discussed in a bit, but let us now look at the property tax in the form that some taxing jurisdictions do try to impose it.

The property tax on nonrenewable resources is most frequently thought of, but not actually administered, as based on the value of the reserves remaining in the earth's crust. In this form, the tax is not neutral, as can be seen by adjusting the numerical example so that a three-year extraction plan yields $1,500 in tax revenues as measured in present-value dollars. A 5 percent property tax on the value of minerals remaining in place (in situ) under a three-year extraction plan, as shown previously, yields yearly current-tax revenues of $750, $550, and $302.50. These are reduced to

Table 3–2
Tax on the Property Right to the Resource
(dollars)

Year	Three-Year Extraction, 1,000 Barrels per Year			Four-Year Extraction, 750 Barrels per Year	
	Net Income on Extracted Oil	Annual Tax	Present Value of Tax	Annual Tax	Present Value of Tax
0	5,000	500	500	375.00	375
1	5,500	550	500	412.50	375
2	6,050	605	500	453.75	375
3				499.13	375
Total (present value)			1,500		1,500

present values in the numerical example shown in table 3–3. The tax is assumed to be levied on the value of oil remaining in the ground at the beginning of each year. Obviously this value depends critically on the firm's decision about the rate of extraction. Pumping the oil faster will reduce property-tax liability in present-value terms. For example, a two-year rate of extraction would produce tax revenues of $750 and $412.50, which sum to a present value of only $1,125, compared to a three-year plan producing $1,500. There is a $375 incentive to increase the rate of extraction from 1,000 to 1,500 barrels per year in order to mine out from under a property tax levied on the value of the resource remaining in place.

This example illustrates an oft-cited criticism of the property tax on nonrenewable resources: It encourages uneconomic and rapid depletion (Harberger 1961).

A tax policymaker can devise alternative methods for assessing the property tax in addition to the tax on the present value of profits and on the value of the resource remaining in place. If the property tax rate were made dependent on the rate of extraction, the property tax on the resource in place could be neutral. In our example, a tax rate of 6.66 percent under two-year extraction would make the present value of property taxes for three- and two-year extraction plans (a 5 percent tax rate) the same, $1,500. However, the administrative and enforcement costs of this complex system could be overwhelming. The property tax levied on the present value of profits from extraction escalating at the rate of interest, which is also neutral, would be difficult to accurately assess and administer, because the true net present worth of the resource is known with certainty only after exhaustion and can be only crudely estimated on an a priori basis.

Incentives to increase the rate of extraction also result if a firm's right to extract natural resources is taxed each year, such as a license tax on a corporation. For the conventional firm, lump-sum taxes have no effect on output as long as profits remain adequate for the company to remain in

Table 3–3
Property Tax on the Resource Remaining in Situ
(dollars)

	Three-Year Extraction			*Two-Year Extraction*		
Year	*5 Percent Property Tax on Resource in Situ*		*Present Value of Property Tax*	*5 Percent Property Tax on Resource in Situ*		*Present Value of Property Tax*
0	(15,000)	750.00	750	(15,000)	750.00	750
1	(11,000)	550.00	500	(8,250)	412.50	375
2	(6,050)	302.50	250			
Total			1,500			1,125

business. However, a firm extracting a nonrenewable resource can reduce its extraction-right tax, by increasing extraction rates and exhausting the resource at an earlier date than in the nontax situation (Burness 1976).

A tax on output, frequently called a severance tax, is used by some jurisdictions in lieu of a property tax. Its effects on the extractive firm can be shown by returning to our numerical example. Assume that a three-year extraction rate of 1,000 barrels per year is chosen as the optimal one in a tax-free regime. A physical-output severance tax of $0.5483 cents per barrel would yield tax revenue with present value of $1,500. This rate was chosen for the convenience of having the numerical examples equivalent. However, if the resource firm decided to extend extraction over four years rather than three, taxes would drop in the earlier periods, because the rate of extraction would be less and the present value of total taxes would fall to $1,433.88, giving the firm an incentive to slow the rate of extraction.

When the marginal cost of extracting oil increases with the rate of extraction, as most theoretical and empirical evidence supports, the incentive is larger than in the constant-cost case, as illustrated in an example of a severence tax on the value of output shown in table 3–4. Assume, now, that it costs the oil-field developer $2 per barrel to extract the oil, and the severance tax is levied on the value of extracted oil (including this $2 plus the value of the oil in place). Given a three-year extraction rate, a severance tax of 7.33 percent would yield $1,500 in present-value terms.

However, slowing the extraction to a four-year rate reduces the present value of this severance tax to $1,482.35, again encouraging a slowdown in the rate of extraction. However, if the incremental cost of extraction is not constant, as assumed in this example, but is an increasing function of output, the incentive to slow extraction is greater. When the incremental cost of

Table 3–4
Severance Tax on Value (Ad Valorem)
(dollars)

Year	User Value of Oil in Place	Extraction Charge	Total Market Value of Oil Extracted	7.3274 Percent Severance Tax at 3-Year Extraction Rate	7.3274 Percent Severance Tax at 4-Year Extraction Rate	With Increasing Marginal Cost of Extraction
0	5,000	2,000	7,000	512.92	384.69	370.95
1	5,500	2,000	7,500	549.55	412.17	398.43
2	6,050	2,000	8,050	589.86	442.39	428.65
3					475.64	461.90
Total (present value)				1,500.00	1,482.35	1,434.45

extraction increases, this means that the average per-barrel costs must also increase. Assume the average cost is $2 per barrel at the 1,000-barrel annual rate of extraction, but is only $1.75 per barrel at a four-year rate at 750 barrels per year. The effect of moving to a four-year rate of extraction now reduces the present value of the severance tax to $1,434.45. These examples indicate that a tax on physical output or the value of output is not neutral with respect to the rate of extraction. This effect of slowing extraction is widely recognized, and in fact, taxes on the output of natural resources are called conservation taxes by some jurisdictions.

Under some conditions, a severance tax may be neutral with respect to rates of extraction. If the user cost or value of the resource in situ is increasing over time at the rate of interest, as we have been assuming, and if the incremental cost of extraction also increases at this rate and is independent of the rate of extraction, then the severance tax is neutral. But if mineral prices and extraction costs remain roughly constant (presumably because of unanticipated increases in market supplies), then the severance tax rate must increase over time at the rate of interest in order for the tax to be neutral. If the tax rate increases at less than the interest rate, it slows the extraction rate, and vice versa (Burness 1976). The effects of property and severance taxes on the rate of extraction are well documented in theoretical literature. We next survey this literature in order to find out the likely effects of taxes on output, property, corporate income, and leasing of public lands. These findings are subsequently used to rank these taxes by a multiple set of criteria.

Taxes on Output

Taxes levied on the output of an extractive firm are attractive because they are relatively easy to administer and different rates can readily be applied to separate resources. Although the manner in which the taxes are structured and enforced is wide-ranging, it is possible to classify them into three categories: those falling on the physical level of output, on the quality of output, and on the value of output. The names given to these taxes tell little about them—severance, resource excise, conservation, proceeds, gross-receipts, and sales taxes are frequent titles. Royalties paid to government for resources extracted on public lands, or in the case of Commonwealth countries from the majority of land, are similar to taxes on output.

Taxes on output frequently have been instituted by states in lieu of property taxes because of their administrative simplicity. The historical and legal justification for them is based on the argument that the state retains certain property rights to resources extracted from the ground which may be exported and consumed elsewhere (Bingaman 1970). This ownership notion is adhered to far more closely in other countries where the crown or government owns all subsurface mineral rights (England, Canada, and so on). This rationale, coupled with public concern over environmental damage and loss of an irreplaceable resource, is the reason why extractive firms have been singled out for these taxes.

Beginning with Gray's analysis in 1914 (Gray 1914), economists have held that the severance tax induces the firm to restrict output and utilize higher-grade ores. The numerical example demonstrated how a tax on output reduced after-tax revenues. Another way of seeing this is that the tax, in effect, makes the incremental cost of extraction higher since it is levied directly on output. If incremental costs increase as extraction takes place more rapidly (largely because economies of scale are fully exploited), the firm's manager maximizes after-tax profits by slowing the rate of extraction. This tax effect has been cited by Hotelling (1931), Schulze (1974), Lockner (1962b), Burness (1976), Peterson (1976, 1978), Conrad and Hool (1978a), Vickrey (1967), and Steele (1967). Industry output falls not only because each firm slows extraction rates, but also because marginal firms become unprofitable and therefore cease production.

The severance tax has sometimes been called a conservation tax, and the conservation label is justly deserved, because the resource is extracted more slowly. However, this conservation may well prove to be inefficient because the economic growth rate is reduced. Furthermore, if the effect of the tax is to increase the ore cutoff grade, as students from Gray (1914) to Lockner (1962b) and Conrad and Hool (1978a) maintain, then conservation of the resource is itself forfeited because lower-grade ores are bypassed and, once isolated in low-grade pockets and trapped under collapsed ceilings of underground mines, may never be economically retrievable. Helliwell (1978) uses a different approach to analyze how taxes reduce the ore grade in a copper mine, but comes to the same conclusion as the authors cited above. Lower-grade resources may be irreversibly lost if they are not recovered simultaneously with richer ores (see Gillis and Bucovetsky 1978, chap. 7, for a discussion of tax effects on "high grading"). Scott (1955, 1967) concurs by pointing out that if low- and high-grade ores can be extracted selectively, the tax will reduce the total amount extracted. In cases of technologically required ore grade blending demanded by mill and concentrator requirements, the firm's ore-grade response is uncertain.

Burness (1976) and Conrad and Hool (1978a) reveal that alternative forms of taxes on output yield somewhat different effects. The tax frequently is placed on the physical unit of output, and adjustments in the tax rate are sometimes made for resource quality. The other method is to place the tax on the value of output, which in most cases results in a greater reduction in the rate of output and reduces the value of the unexploited resource reserve more than the tax on physical output. Burness (1976) investigates tax rates which vary over time and derives rates that have no effect on the extraction decision as well as ones that increase it. Conrad and Hool (1978*a*) and Attanasi (1978) derive results which indicate that a fixed severance-tax rate reduces investment in discovery and development; mine capacities will also be distorted. Attanasi also shows that as lower-grade ores are extracted in each mine, the tax becomes a larger burden to the firm because user costs become a smaller and smaller portion of sales revenues.

Since there is almost universal agreement that the severance tax distorts investment and extraction decisions, why is its use widespread? Taxes on the output of extractive firms came into existence in 1846 (in Michigan) and spread rapidly during the early part of the twentieth century (to Pennsylvania, Minnesota, Texas, West Virginia, Oklahoma, Louisiana, Kentucky, and Alabama). Today, twenty-nine states utilize them. They are major sources of revenue in mineral-rich states, are invariably applied to the most profitable minerals, and were instituted in large part because of their simplicity and ease of administration.

In 1924, the Supreme Court held that Louisiana's severance tax was not discriminatory (under the Fourteenth Amendment) with respect to nontaxed products imported from other states. In other cases, the courts have held that the severance tax does not violate the commerce clause of the Constitution and should be interpreted as an excise tax and not a property tax. However, after a relatively long period of legal calm, litigation over output-based taxes has again arisen. The most notable example is the suit brought by Montana coal consumers (Midwestern electric utilities). This group maintains that the tax structure enacted in 1975, with rates going up to 30 percent of the mine-mouth value, restrains interstate commerce and is thereby unconstitutional. It is important to realize that taxes on output are shifted in part to consumers, because each firm maximizes after-tax profits by cutting back on the rate of output. Altogether, this shifts the industry supply curve and results in higher final resource prices. However, some of the tax is absorbed by extractive firms since tax forces the cost of extraction upward.

Property Tax

The property tax is complex in its theoretical form and even more complex in its practiced form. It is possible to classify it into three subdivisions, each having somewhat different economic effects. The theoretical literature generally defines it as falling on the present value of the resource in place. When the property tax is levied on the land value of extractive sites, it is precisely identical to the tax on the resource in place. Recall the discussion of economic rent and figure 3-3, and then visualize the tax as being levied on the present value of user cost plus economic rent for each particular mine site. The numerical example illustrated that it pays the firm to speed up extraction so that it can mine out from under the tax. A second way to impose the property tax is on the economic rent accruing to each mine site, and at a 100 percent rate this would equal the cross-hatched area (economic rent) in panel *A* of figure 3–3. The numerical example illustrated how this tax would be neutral for the rate of extraction, as long as the payments always yield the same tax revenue in present-value terms. Finally, property taxes may also be levied on the current value of renewable capital—physical plant and equipment.

Property Tax on the Value of the Resource in Situ

Hotelling (1931) formally shows that a tax on the value of reserves, if anticipated, will induce the firm to extract more rapidly, and if it is unanticipated, it will have no effect because it is like a negative windfall profit. His conclusion has been upheld by Burness (1976), who emphasizes that a property tax will have no effect on a profitable firm producing a renewable commodity because it is a lump sum deducted from profits with no effect on production costs. The extractive firm is unable to live forever and plans for the ultimate death of each site as the reserve is exhausted. When time becomes relevant, the property tax is not a fixed lump sum but one which can be terminated earlier by premature exhaustion. Steel (1967) and Conrad and Hool (1978*a*) concur with Burness. The latter authors find that the tax may encourage the firm to extract high-grade ore earlier in the mine life and to mine low-grade, high-cost ore in later periods. When planning investment, the firm will make the mine capacity larger than in the absence of tax.

Bush (1972) emphasizes the negative impact of the tax on exploration and development, since carrying larger reserves increases tax liabilities. This will, in turn, lead to eventual decreases in reserves and lower total industry output. Peterson (1976, 1978) makes similar conclusions, along with DeYoung (1978) who surveyes empirical data on taxes and exploration activity. By increasing near-term output and reducing long-term output, the tax will have the effect of making present resources cheap and future ones more expensive. This form of the property tax, then, will benefit current consumers at the expense of future ones. But the tax will fall on economic rents to a significant degree.

Tax on Economic Rents

A property tax falling solely on economic rents will not alter the rate of extraction. Steele (1967) analyzes a tax similar to the one employed in the numerical example which could be paid in one lump sum or in a series of annual payments having the same present value. Gaffney (1967, in editor's notes) has argued for many years that property taxes can be structured so that they fall solely on economic rents. However, he acknowledges that numerous pitfalls and problems are associated with this approach. Tax rates must vary with the cost of extraction; therefore, they must be structured for each mine site and each period. If a mistake is made, sites which are barely able to compete (but earning a profit by the accountant's methodology) could be forced to shut down by a positive tax. A further problem is that resource firms anticipate that economic rents associated with future discoveries will be taxed away in part. Since this will reduce the rate of return on investment dollars, exploration will be reduced.

Property Tax on Plant and Equipment

A property tax falling on the current value of plant and equipment is similar to the portion of property taxes levied on any human-made improvement. When it applies to a developed mine site, there is nothing the firm can do to minimize its effect. However, as reserves at the site are depleted and the plant and equipment wear out, the firm has the choice of not reinvesting or investing in another taxing jurisdiction. The long-run effect, then, will be to discourage exploration and development of reserves in jurisdictions which levy higher taxes because the tax reduces expected return on capital and economic rents. The overall effect is to reduce discovery and reserves in low-tax jurisdictions. The vast literature on this conventional form of the property tax leads one to conclude that the tax will be partially shifted forward to consumers in the form of higher prices and partially backward to owners of firms and resources.

Administering the Property Tax

A brief survey of techniques to administer taxes levied on the value of reserves in situ is in order for two reasons. First, although vast amounts of intellectual and legal energy, accounting finesse, and geologic and mining knowledge have gone into these attempts, these efforts have proved the tax to be administratively awkward. Second, the history of tax administration illustrates how hard it is for the tax collector to identify and tax that quantity which is so easily described in theory but so elusive in practice—economic rent.

The most widely known method of valuing resources in place comes from the late nineteenth century and has been the subject of much litigation and misunderstanding. Mine value is based on the present value of its net profit stream, calculated from the 1877 formula attributable to Hoskold (1877). This has been utilized for evaluating potential mine-site investment as well as by tax revenue agencies. Rather than simply capitalizing a net income stream (profits), used in the case of replaceable capital, the Hoskold formula takes into account the unique characteristic that the asset (the reserves) is exhausted and each unique site is irreplaceable at the end of its life. Hoskold argued that a sinking fund should be accumulated during the life of the mine which would recoup the initial purchase price and development cost of the mineral property.

A contemporary rationale for this fund is that it amounts to internal depreciation or a capital-recovery factor. The depreciation rate for human-made capital is determined by external market forces which evaluate improvements based mainly on age and only partially on use. Renewable capital, such as buildings and machinery, is depreciated because of obsolescence (primarily a function of the passage of time), whether or not the asset is used. A nonrenewable resource "depreciates" when extraction takes place. Of course, improvements in technology and discoveries of more profitable reserves are external forces which will affect the value of the resource. At the end of its life, a replacement must be found which most likely is of a lower grade and more costly to extract. The user cost is simply an internal charge made by the firm owning the resource which indicates the present value of forgone future profits when extraction takes place today rather than in the future. It is user cost which ensures that the resource is allocated over time to maximize the present value of total profits. The Hoskold approach may be considered a rough approximation of user charges, since user cost per unit increases at the rate of interest, whereas the Hoskold sinking fund is computed as an annuity payment earning a "safe" rate of interest. Hoskold argued that the accumulating sinking fund would be invested in "safe" securities (at interest rate r), as opposed to the more risky mining-venture rate (r') that is used to estimate the present value of the mine reserves. It is argued that the net income flow from the mining activity, or user cost in a modern interpretation, should be capitalized at a higher interest rate r' because of its risky nature. Thus, the Hoskold formula for the present value V for a mine site may be calculated from

$$V = \frac{A}{r/[(1 + r)^n - 1.0] + r'}$$

where A is the total annual net return from the mine, n is the life of the mine in years, r is the safe rate of interest, and r' is the higher, risky interest rate. Mining engineers and financial analysts used this formula for years until the advent of discounted-cash-flow techniques.

A second widely applied capitalization technique is attributable to Morkill (Baxter and Parks 1949). This formula is similar to Hoskold's except that the sinking fund itself is assumed to earn no interest and the income flows to the investor are discounted at the risky rate of interest (r'). Furthermore, the payments to the sinking fund are proxies for the mine's rental value or annual user cost, and thus they increase at the risky rate of interest. If the sinking fund is calculated to recover the initial purchase price of the mineral property at exhaustion, the Morkill formula yields

$$V = \frac{A[(1 + r')^n - 1.0]}{(1 + r')^n(r')}$$

Inspection reveals that the Morkill formula is simply the present value of a fixed annuity payment for n years. Inspection also reveals that if one assumes that the safe and the risky interest rates in the Hoskold formula are equal ($r = r'$), then these two formulas are identical.

The major problem in discounting future income streams, whether by the Holskold or the Morkill formulation, is that the future must be accurately perceived. Future prices and costs of extraction (including taxation) must be known. Extraction costs depend, in turn, on the method of amortizing exploration and development expenses, which is currently not standardized in the industry, as well as the prices of future productive inputs, including labor, operation and maintenance materials, and capital. Additionally, the entire time stream of extraction must be known along with the date of exhaustion, which has been adjusted to reflect the effect of the property tax on the resource in place. Finally, appropriate interest rates (those which reflect risk premiums) must be known.

In a widely reported article, Grossman (1953) reviews the geologic, locational, and economic difficulties associated with mineral land valuation and cites the advantages of capitalizing royalty and lease payments, so that the difficulty of predicting prices and future extraction costs may be avoided. In this advocacy he assumes that owners of mineral rights have full knowledge of the value of their economic rent and bargain it away in full from extractive firms buying the resource in situ. Keith (1938) argues for the use of the Hoskold formula since sales of mineral properties, which are market indicators of value if full knowledge is assumed, are few and the large number of locational and geological variables makes it difficult to make comparisons from the few sales of mineral rights which do take place. He also states that the effect of taxes on net income must be considered and that prices of publicly traded stock are too sensitive to changes in expectations to use in valuations. He recommends an interest rate of 8 to 10 percent under normal circumstances for use in the Hoskold formula and 15 to 20 percent in particularly risky cases. Baxter and Parks (1949), in perhaps what has been the most influential reference among mining engineers, explain the Hoskold, O'Donahue, and Morkill formulas and rcommend a 14 to 25 percent risk rate for domestic investment and a 25 to 35 percent rate for foreign projects. They note that rates in use in 1949 varied from 6 to 25 percent, with 8 to 10 percent generally applied to valuing incremental investment of operational mines. Johnson (1953) recommends a 10 percent rate for petroleum valuation and discusses methods to forecast the inevitable decline in extraction rates. He also allows for a 25 percent hazard factor in most of his calculations. The high interest rates to be used in discounting, notwithstanding that inflation was generally low during this period, indicate the high risks associated with exhaustive natural-resource development and production.

Although the Hoskold and Morkill capitalization techniques have been widely used by industry in evaluating alternative investments and by states in determining assessed values for the property tax, industry has turned from these techniques to discounted-cash-flow analysis and various simulation techniques (Church 1977). Virtually all states have abandoned capitalization techniques for property tax assessment because of the problems with gathering the relevant data and defending them in the courts. In order to determine annual net income for the life of the mine, accurate forecasts of revenues and costs must be made. This entails knowing future prices and the costs of extraction, processing, and transporting the resource. Additionally, the rate of extraction and economically exploitable reserves determine mine life, and this must be estimated as well. Once the net annual profit is calculated from these data, it must be capitalized with one or two interest rates called for by the various formulas. The literature cited above indicates that there is no unanimity as to what the safe and risky rates of return should be.

McGeorge (1970), Roberts (1944), and numerous others emphasize that the information necessary to utilize the capitalization formulas is not easily accessible or verifiable, and most jurisdictions lack the engineering and economic expertise necessary to accomplish such data collection and analysis. This method of valuation for property-tax purposes appears to have broken down, so that in practice either simplified formulas based on previous levels of output are used by tax appraisers or, in many cases, the law has been changed to define property value explicitly in terms of such a simplified approach. Some states and jurisdictions have replaced the property tax with another tax base, usually an output-based tax. In many jurisdictions, mine value is self-assessed and not verified by taxing authorities, or the value for property-tax purposes is established after negotiations between mineowners and the tax collectors.

Extensive review of published materials and personal interviews indicates that Arizona is the sole state which still attempts to actually establish mine value for property-tax assessment by using the Hoskold capitalized-value approach. The targets are copper mines, which have battled the state's techniques and data over the years in the courts. Copper producers abhor this tax, which is collected regardless of the level of output and for three years after shutdown. However, it remains in large part because it produces stable revenues and has withstood numerous court challenges.

The alternatives to the capitalized-income approach have been the method based on previous years' output, to tax land at some arbitrary value per acre, and to assess mine equipment and real improvements at replacement cost less depreciation. These approaches are utilized in a wide variety of combinations. Another approach taken by some states has been to rescind the property tax or to leave it solely in the hands of local taxing jurisdictions. Clearly, the property tax is not a simple one in practice or in theory when the various assessment techniques are considered.

Corporate Income Tax

The corporate income tax is placed directly on the acounting profits of incorporated enterprises. Utilizing this tax base creates two problems, a practical one of administration and a theoretical one of being able to determine how the firm responds to the tax and who bears its final burden. The tax is difficult to administer and enforce because lawyers and accountants manipulate business and accounting practices so as to minimize tax payments. The Internal Revenue Service (IRS) and state tax-agency lawyers and accountants employ their energies in ferreting out tax evasion and being on guard to see that avoidance techniques are legal and permissible.

The tax code is especially complex when applied to the extractive industries. The complexity arises because economists define profits differently from accountants and the taxing authorities. Consequently, the corporate income tax falls partly on the expected return to equity capital, since this is not a deductible item, and partly on economic rents and unanticipated economic profits.

The consensus among economists who have analyzed the effect of the corporate income tax on enterprise as a whole is that in the short run as human-made capital is held constant, the tax has little or no effect on the behavior of the firm; capital bears the tax burden. However, as capital depreciates, management must decide how to invest. Since the tax falls initially on equity capital, owners will divert their capital to activities with tax preferences because the after-tax return in these will be higher. The key factor which prevents the straight transfer of these concepts to the nonrenewable-resource firm is that the depletion of reserves must be clearly defined so that it is analogous to depreciation of human-made capital. Conrad and Hool (1978*a*) maintain that the theoretically ideal depletion is the decrease in present value of the reserve as it is depleted. If this could be calculated and were the method used by the IRS, then they conclude that the corporate income tax would introduce no distortion in the allocation of resources. Burness (1976) adds that if the tax rate is expected to increase over time, then the firm will accelerate its rate of extraction in order to minimize tax liabilities. It should be pointed out that tax revenues will decline as extraction takes place, because it can be shown from the abstract model of the mine that incremental profits will decline as exhaustion takes place if extraction costs increase, as shown by Cummings and Burt (1969), Weinstein and Zeckhauser (1975), and Heal (1976).

The long-run situation for the extractive firm is somewhat different. If the corporate income tax reduces after-tax returns to invested capital, as it surely must to some degree, then investment in exploration and development for new reserves and investment in extracting and processing equipment will be discouraged. Ultimately, this will mean, as Kay and Mirrless (1975) and Peterson (1978) point out, fewer reserves and a distorted rate of extraction.

The depletion provision, based on the historical cost of discovery, was incorporated into the federal corporate income tax in 1916 and expanded in 1926 to include the option of depletion based on historical costs or percentage depletion computed by deducting a portion of revenues. Peterson (1976), Burness (1976), Conrad and Hool (1978*a*), Agria (1969), Harberger (1955), and Steiner (1959) all concur that the depletion allowance is akin to a negative severance tax (or subsidy on output) in the manner in which it reduces the income-tax burden. The economic effect is to encourage current extraction, decrease the present value of the mine, and thereby introduce yet another tax distortion. A number of observers have advocated a progressive corporate-income-tax rate structure. Lockner (1962*b*) maintains that such a rate would be nondistorting, but Conrad and Hool (1978*a*) show that it may encourage high grading and result in a loss of low-grade ores.

A related issue is that the depletion allowance and other tax preferences granted to firms extracting and processing virgin materials are denied to those recycling scrap and materials already mined, such as from abandoned slag heaps and tailings dumps. Anderson and Spiegelman (1977) investigated existing tax laws and industry cost structures for several materials, and they concluded that the distortion is preent but is most likely minimal in its actual impact. Hoel (1978*a*) analyzed the social cost of pollution engendered by extractive and processing activity and held that since recycling activities emit fewer residuals than virgin materials, there should be a differential in the tax structure to encourage recycling.

Corporate Tax Preferences

There are three primary sources of differential tax treatment for the extractive firm: (1) preferential depreciation of the nonrenewable-resource stock as it is exhausted (normally designated as *depletion allowances*); (2) immediate deduction (rather than amortization) of capital expenditures for exploration and development activities; and (3) differential tax rates applied to certain nonrenewable resources when property rights to them are sold.

The depletion provision began as a deduction for actual resource depletion and gained in importance as tax laws changed and the federal corporate income tax rate rose from its original low level to its current levels. A brief history of the depletion allowances will give some insight into how tax policy is made and modified over time. The Revenue Act of 1913 (the first act levying an income tax after the ratification of the Sixteenth amendment of the Constitution) provided for "a reasonable allowance" for capital consumption (McDonald 1963). Both depletable resources and human-made capital were treated equally for depreciation purposes. However, in the case of mines, a maximum of 5 percent of the gross value of mine output was allowed as a deduction, whereas this limitation did not apply to other industries. The Revenue Act of 1916 shifted depletion to a historical value based on the market value of the asset as of 1913. The 1913 specification was made to place corporations on an equal footing, but effectively it benefited those mineral producers having reserves discovered before 1913. Because these generally cost less to discover and develop but were worth more on the valuation date, the firms holding old reserves garnered a tax benefit for income from assets acquired before 1913. The Revenue Act of 1916 allowed market-based value (as of 1913) or the original cost as a basis for depletion and rescinded the 5 percent limitation. The depletion deduction was limited to the total value of the mineral at the point of extraction. The 1916 act was the first to use the term *depletion allowance*. The congressional intent was to encourage exploration by allowing for the higher deduction for depletion whether on a discovery-cost or market-value basis.

The inequity for pre- and post-1913 discoveries was rectified in 1919 when all properties were allowed to utilize discovery value (estimated within 30 days of the discovery), as opposed to the previously allowed investment cost, as the basis for depletion. The effect was to shelter income up to the level of its market value at discovery, as opposed to the cost basis applicable previously. The differential could be substantial, for once a discovery is made at a site, typically the market value far exceeds the costs of exploration. The Revenue Act of 1921 restricted total depletion to 100 percent of the property's net income. This change was due to a sharp fall in mineral prices in 1921, which resulted in depletion for earlier discoveries exceeding incomes generated from those properties. This limitation was further tightened to 50 percent in 1924 and remains in effect to this day.

Difficulties in assessing market value and other administrative difficulties led the 1926 Congress to institute a rule-of-thumb measure of depletion. The adopted figure for gas and oil was the infamous 27.5 percent of gross revenue from sales of output, and the limitation of 50 percent of net income was retained. More importantly, all properties now qualified for this percentage-based depletion, and an allowable option was to employ depletion based on actual costs prorated over the life of the reserves.

Since the statutory federal corporate income tax has been increased over the years from a relatively low level to the near 50 percent marginal rate (46 percent) found today, the importance of the depletion allowance has increased commensurately. Criticism of this provision made by economists (among the earliest critics is Harberger 1955), public interest groups, and others led the Congress in 1969 to rescind percentage depletion for the largest oil corporations. The small extracting firm still has the option to choose either percentage depletion, which is a deduction from gross income, or cost depletion, which is based on the cost of the discovered resource.

The tax advantage or sheltering effect of percentage depletion is computed relatively easily. If the tax rate is t and the percentage-depletion rate is d, then the tax payment for a gross revenue of y and total costs of production C is

$$t \left[(1 - d)y - C \right]$$

because gross revenues are reduced by d for tax purposes. It is relatively easy to show that the percentage-depletion allowance d benefits income of the firm by a percentage equal to $td/(1 - t)$. If the corporate tax rate were 50 percent, then the depletion allowance would be equivalent to a subsidy on the sales price of output equal to the depletion rate. However, the maximum deduction remains limited to 50 percent of net income, and so its maximum effect is to halve the effective income tax rate to 25 percent, if the nominal tax rate is 50 percent.

Mining firms may also elect to take cost depletion, in which the costs of acquiring, discovery, and developing a mine are depreciated over the life of the asset. The annual rate of cost depletion is determined by the first year's output divided by the total resource expected to be extracted. The capital investment is reduced by this proportion, and in each succeeding year the cost depletion is recalculated in a similar manner. At exhaustion, total cost depletion will equal the initial-cost basis. However, it should be noted that with percentage depletion no limitation other than 50 percent of taxable income in each year on total deductions exists, which makes cost depletion unattractive. Percentage depletion linked with the second major tax advantage for extractive firms, the current expensing of exploration and development costs, is highly advantageous because more income can be sheltered from taxes.

The law governing depletion allowances was modified beginning with a 1917 U.S. Treasury decision that allowed the "incidental expenses of drilling wells" and expenses that "do not necessarily enter into and form a part of the capital invested on property account" to be immediately expensed and deducted from gross income. This was the beginning of the current expensing of exploration, discovery, and development costs, which constitutes the second major tax-saving device available to extractive firms. In 1933, the IRS by regulation allowed deductions for unproductive wells. By the 1950s, greater latitude was given to discovery expense of "intangible" drilling expenses for petroleum production. These intangible expenses are now interpreted to be costs of labor, fuel, power, materials, supplies, tool rental and repairs of drilling equipment, and those items not salvageable when the well is abandoned. These ordinarily account for two-thirds or more of total drilling expenses for oil and gas wells. Lease acquisition and geological survey costs can be written off at abandonment.

The effect on tax burdens of the current expensing of discovery and development costs is to shift expenses considered to be amortizable capital costs for other industries into a current expense. This alters the timing of tax payments by reducing current taxes (as a result of a capital item which is expensed rather than amortized) and deferring tax payments to the future. These released funds can be invested and used to increase the profits of the firm and are analogous to an interest-free loan from the U.S. Treasury. This advantage is significant when the firm is an ongoing and growing one which has other income to shelter. The effect of the current expensing and percentage depletion is a subsidy on sales, since the capital costs associated with discovery and development can simultaneously be declared as expenses and not as a depreciable capital cost; and with percentage depletion, up to 50 percent of its net profit is sheltered as the resource is extracted. Furthermore, large-scale underground and strip-mining operations take several years to develop, and the expensing provision allows for deductions for the firm before any income is flowing from the project. The timing provision is then particularly helpful to large corporations with positive net income flowing from other projects which can be sheltered from the tax. Recent tax revisions have restricted the use of immediate expensing of exploration and development costs for oil and gas operations. The current rules are too complex to be detailed here. Basically they are that dry-hole costs must be amortized over production from successful wells based on historic data for each firm.

Another tax advantage for extractive industries is the classifying of income for preferential capital-gains tax rates. Coal-lease royalties and a limited number of other lease incomes can be treated as capital gains and taxed at a lower rate than the ordinary income tax rates for individuals (40 percent of the marginal tax rate) or corporations (the 30 percent corporate capital-gains tax rate versus the 46 percent corporate income tax rate). This differential tax status permits significant tax savings when ordinary income is converted to capital-gains income. The overall savings is substantial. The gain from a successful discovery which is subsequently sold is taxed at the more favorable capital-gains tax rate. Since the corporation had engaged in exploration activities to make a discovery, the cost of unsuccessful ventures is expensed as a deduction. Consequently, every expensed dollar shelters $0.46 of ordinary income. For the successful activities, the net gain or discovery value (market value less costs) is taxed at the lower capital-gains rate if the resource is sold. The net effect is that the government absorbs 46 percent of all losses, but shares only 30 percent of the gain for successful adventures. The response by industry is to maximize this tax advantage by engaging in risky ventures which have a high probability of losses but a large gain for the successful projects. The government, in effect, is encouraging risky exploration activities by means of this tax advantage; thus, yet another tax distortion is introduced into the economic system. Similar tax savings are possible for private individuals because of the capital-gains versus ordinary income tax rate differentials, although recent tax changes (1976) have limited the tax advantages of drilling programs for individuals. It is also notable that royalties from coal leases are treated as capital-gains income.

The effect of income tax advantages for the extractive firm is to lower its effective tax rate vis-à-vis other industries. Nonuniform effective tax rates mean that in the long run, investors will reallocate capital away from more heavily taxed enterprises to the lower-taxed firms until *after-tax* rates of return are equal. The degree of capital transfer depends primarily on the extent to which profitable projects exist in the industries being taxed at the lower rate and the size of the tax differential. This transfer is limited, for one would expect rates of return to decline as more projects are taken on in the tax-favored industry. This process ceases when posttax rates of return are in equilibrium. The resulting inefficient allocation of resources has led Vickrey (1967) to accuse these preferential federal income tax provisions as being "notorious."

Taxes and Leasing

Although leases held by government agencies are not strictly taxes, they are a means of raising government revenues (Crommelin and Thompson 1977). If competition prevails in resource markets and if the extent and quality of reserves along with future prices and costs are known, then sale or lease of government mineral rights would capture the entire economic rent. However, as discussed before, it is naive to assume that these prerequisites prevail. Bidding on leases takes place under conditions of uncertainty, and this fact has been emphasized in the literature on the subject. Institutional arrangements established under leases are wide-ranging. But the most widely used form is for the lessee to pay the lessor a percentage royalty on the material extracted, and in some cases, an annual minimum amount is established or a fixed yearly fee per acre is agreed upon along with the royalty. The federal government makes extensive use of the bonus bid, wherein the lessee pays a fixed fee at the start of the lease and assumes the risk whether economically exploitable reserves are discovered or not. However, most North Sea leases are not auctioned, but are negotiated with a few competing firms. This process is tailored to the needs of the countries owning these resources, not only for revenues but also as to the manner in which and the speed with which the resource is developed.

Auctioning and bidding methods are varied, ranging from closed-door negotiating sessions to public oral or sealed bidding. The economics of leasing and bidding process are primarily concerned with describing the uncertainty of bidding for an unproved potential reserve and the cost and methods of acquiring information which affects those risks. Gilbert (1975a), Hughart (1975), and Gaskins and Teisberg (1976) have all looked into information-gathering strategies of potential lease purchasers. The consensus holds that risk aversion leads to more seismic measurement and exploratory drilling than is desirable from the standpoint of economic efficiency.

Tax Effects and Tax Policy

The economics literature describing the extractive firm and how taxes affect its decision making is a bit afield from the central theme of tax policy. However, it is necessary to understand how taxes affect the firm and resource markets in order to be able to judge the desirability of various taxes. The primary conclusion to be drawn from the literature is that a tax policy goal should be neutrality. This position has been taken by virtually all economists specializing in the field, including Harberger (1955), Agria (1969), Steiner (1963), Peterson (1976), and Stiglitz (1975a). Only in cases where there are reasons to believe that the allocation of resources determined by private markets has been inefficient or Pareto-suboptimal should a tax policy be purposely designed to create distortions. In those cases, policies other than the taxes reviewed here may be better designed to achieve those ends. In virtually all cases, tax policy should be designed to be neutral, which implies that pure economic profits are appropriate targets for tax policy.

The problem with economic rent is that when it is fully anticipated, it becomes the economic prize that is pursued by all segments of society which have leverage in the resource battle. Government is one of the participants in this battle, but so are other groups which gain bargaining power as a result of economic change or changes in the legal and institutional environment. Such groups include labor, which exerts monopoly power by means of union organization and the strike threat; the owners of mineral rights, who discover the value of the resources and are able to negotiate successfully with extractive firms; individual transportation firms, which in certain cases are the sole economic way to move the commodity; and the extractive firms themselves, who individually or in concert have some degree of monopoly power. Predicting the distribution of rents among these actors is nearly an infeasible task; little is known about the reservation prices, such as anticipated profit, necessary to secure the use of equity capital or minimum wages necessary to lure workers in remote sites. Negotiating power and cleverness are difficult to predict. In chapter 4, we analyze profit rates in the extractive industries and find that they are variable according to the industry and firm. Cummings and Mehr (1977) have shown that there is a tradeoff among higher wages, site isolation, and the availability of publicly supplied goods and services which are necessary to induce labor to move to remote Western boomtowns. Thus, economic rent varies by site and even by individual, making it a complicated and subjective quantum which has never been measured successfully. Therefore, taxes designed to ferret out and expropriate economic rents for the state are difficult to design and harder to implement. The saga of attempts at possessory interest and property taxes on the present value of the mine site should be enlightening in this regard. The more complex and selective a tax becomes (and this is precisely what must be done in order to capture economic rents), the more difficult it is to administer and enforce.

Ranking Taxes

Now we evaluate the tax bases discussed so far with respect to their suitability to the needs of government and their effect on private-sector markets. The results are summarized in table 3–5. Tax revenues are necessary to sustain programs and projects carried out by state governments. State and local governments utilize tax revenues to pay for various goods and services produced or contracted by the government. This takes planning and advanced budgeting. Thus, a tax whose revenues are sensitive to the business cycle is generally less desirable to these units of government than one which produces predictable and stable revenue from year to year. However,

Table 3–5
Evaluation of Natural-Resource Taxes

Tax	Effect on Government			Effect on Markets								
	Stability	Income Elasticity	Administrative Base	Rate of Extraction in Short Run	Current Output (Total) in Short Run	Employment in Short Run	Efficiency Loss—Distortion in Long Run	Level of Recovery, High Grading, in Long Run	Exploration and Discovery in Long Run	Capital (Outmigration) in Long Run	Labor (Outmigration) in Long Run	Cost of Industry Compliance
Severance Tax												
On physical output	3	7	1	−5	6	5	3	−3	−3	−4	7	2
On quality of output	3	7	3	−4	5	4	4	−3	−3	4	6	2
On value of output including royalties (ad valorem)	4	3	2	−4	5	4	2	−2	−3	5	6	2
Property Tax (economic rent of the resource on the capitalized value of the resource)	2	2	9	10	1	2	1	10	−4	1	1	5
On value of resource in place	1	5	8	+6	1	1	5	20	−4	4	4	4
On value of plant and equipment	1	6	7	+5	2	2	6	−5	−5	8	4	4
Corporate-Profits Tax	6	2	5	−2	4	3	4	−4	−5	7	4	2
Effect of *decreased* depletion allowance	5	3	5	−3	3	2	2	−6	−3	6	5	2
Effect of *decreased* exploration and development expense deduction	5	3	5	−3	4	3	3	−1	−6	6	3	2

Note: Rankings are in order of desirability.

demands for public goods and services have grown at a rate equal to or exceeding the rate of inflation plus the real growth rate of the economy. Thus, a tax which produces revenues that increase commensurately with real and inflationary growth is generally desirable. If the tax has an income elasticity equal to or greater than 1, then its growth rate equals or exceeds that for the overall economy.

Tax administration and enforcement should place a minimum burden on the economic system. From the viewpoint of government, this means that the tax should be easily assessed and collected. A vague tax law places the administering agency in the position of having to formulate tax policy goals through detailed regulations. A common result is continuous, expensive haggling and litigation over the meaning of various statutes as interpreted by the regulations.

A similar situation arises from the taxpayers' side in the form of compliance costs. If the tax law and regulations are complex, then the firm may have to maintain accounting records dictated by the tax code. Furthermore, a tax code containing legal uncertainties or one which is administered with wide discretion is vulnerable to situations where the firm can benefit by going to the courts. If the firm wins its case, its effective tax rate is reduced; if it loses in court, the delay in complying with the disputed tax code may also produce benefits. Whenever a tax payment is deferred to the future, the firm is, in effect, granted a loan by the government. Consequently, low costs of compliance are desirable characteristics.

Government should foster economic growth and at the same time ensure that economic efficiency is achieved. As we know, a tax itself may foster a distortion away from the efficient use of resources. Certain tax bases have greater effects on the level of output and production and alter the most efficient technology more than others; the resulting distortions are greater. A tax which is less distorting is ranked lower, in table 3–5, than a more distorting tax. Distortions are measured by efficient or deadweight loss, which equals the cross-hatched area shown in figure 3–1. The larger the deadweight loss, the more undesirable a tax. If firms change the rate of extraction in response to a tax and no firms enter the industry and some shut down, then total industry output falls, and in the long run employment must fall as a result of an upward shift in the industry supply curve. Changes in output and employment are measures of how economic development is affected by a tax, and these changes are evaluated in table 3–5 along with the expected net social or deadweight loss. Another effect indicating a distortion is whether the firm adjusts to a tax by changing its ore cutoff grade and reduces the total level of recovery. This may result in an irretrievable loss of the natural resource; thus, a tax which does not reduce the level of recovery is preferred and is given a low ranking. Tax distortions may also affect the exploration and discovery process, and so taxes that are neutral in this respect earn the lowest ranking. Finally, tax avoidance by means of an industry changing location indicates a nonneutrality, and this is measured by capital and labor migration. A low ranking indicates a tax to be preferred in that it does not encourage migration.

Results of Tax Ranking

Stability refers to the sensitivity of tax revenues to fluctuations induced by the business cycle. The standard property tax is considered stable because tax rates are determined by state and local needs, and the tax base is established by appraisals which are adjusted infrequently. The property tax on economic rent is ranked second, but is similar. The severance tax on physical output is ranked third because this base fluctuates less than business profits, which are very volatile. The most erratic tax base is accounting profits, so the corporate income tax is ranked at the bottom (a high numerical number in table 3–5).

The *income elasticity* of tax revenues refers to the response of a tax base to long-run increases in economic activity. Because lease contracts and business profits are a function of business managers' investments and expectations, which in turn are based on income trends, these tax bases are quite closely linked to changes in national income. A larger depletion allowance would reduce this somewhat by reducing taxable income. The ad valorem severance tax behaves similarly. However, the severance tax on physical output is less responsive because it does not reflect the effect of increases in resource prices over time. The property tax on resources in place does reflect the price of minerals, but long reappraising intervals make it relatively unresponsive. Taxation of extraction plant and equipment is the tax base least responsive to income changes because it is based on depreciation of value.

As to *ease of administration*, the easiest tax base is physical output, to wit: severance taxation. Estimates of free-on-board (f.o.b.), mine-mouth prices usually are made by taking the first valid point of sale and deducting transportation and processing costs and taxes to arrive at an estimated value. Corporate-profits taxes follow, because a federally defined tax base must be adjusted for profits generated by out-of-state operations. Appraisal costs and errors render the property-tax administration expensive and often inequitable in practice. The property tax on rent is a most difficult tax to administer.

In the short run, capital is immobile. Firms are relatively hamstrung in what adjustments they can make to tax changes, since their recovery and processing equipment is already in place and depreciating. However, the firm can alter the extraction rate and hire or lay off workers. Since a higher property tax on the mineral remaining in situ stimulates the *extraction rate*, it is ranked first in table 3–5, because both current output and employment would increase. The direction of change is also denoted in table 3–5, with a plus for a positive change, a minus for a negative change, and no designation for "no change." A property tax on equipment would have the same effect but to a lesser degree than the depletion-allowance provision of the corporate income tax. This occurs because the tax expense, if based on market appraisal, and the depletion allowance make natural-resource investment more attractive; the firm can shelter profits by increasing output. The tax treatment of exploration and development expenses under the corporate income tax has a similar but smaller effect, particularly in the short run. The property tax on the present value of profits or economic rent would have virtually no effect on short-run firm employment and output behavior. In fact, this tax is neutral for time and extraction rate, while the severance tax slows the rate of extraction and thus reduces current output and employment. This explains the severance tax's relative ranking in table 3–5.

In the long run, the ranking of taxes is considerably different because increasing the present rate of extraction eventually decreases future output and most likely reduces the total amount of the mineral recovered. In the long run, capital is mobile because nearly fully depreciated plant and equipment can be abandoned and new investment made elsewhere. Furthermore, the firm can alter not only the amount and time path of output, but also technology and the capital/labor ratio. Thus, a tax bearing more heavily on capital will encourage both a substitution of labor for capital and, more importantly, an *emigration* of investment to lower-taxing states.

The method for ranking *deadweight* or *efficiency loss* corresponds to our initial explanation. Since the deadweight loss increases as substitution in production or consumption is more easily undertaken, the property tax on economic rent engenders no loss and is ranked first in table 3–5. The property tax on minerals in situ is nearly similar except it alters the time path of extraction, and this earns it second place. The severance tax on physical output is ranked third, because it is expected to be less distorting over time than the other two forms of this tax. A property tax on equipment and the corporate income tax are ranked fourth, because these taxes bear most heavily on capital rather than on both labor and capital (and consequently produce greater losses).

The rankings of effects on *rate of extraction* in table 3–5 are almost the reverse order of those for short-run output. This occurs because any tax altering this rate from the tax-free condition is undesirable, and a tax increasing the current rate is the most undesirable for the long run. This factor earns the property tax last place and the neutral tax on economic rent first place, with the severance tax and value-added taxes second because both would tend to slow the rate of extraction. Virtually the same arguments can be made for the effect on the total amount *recovered*. It is well known that the best ores are extracted first and that too rapid a rate of extraction often diminishes total lifetime output, as in the case of oil-field production. A severance tax on physical output, rather than on value, discriminates against low-grade resources and total recovery.

The tax effect on *capital migration* was alluded to previously. A tax falling more heavily on capital induces investors to search for higher returns elsewhere. Thus, the property tax on equipment and the corporate income tax induce the greatest capital emigration, and a tax on immobile land (the tax on economic profit or property tax on resources in place) or on labor (the personal income tax) minimizes capital migration.

Table 3–5 next shows *labor-mobility* effects. This is a result of the substitutability of labor for capital. The property tax on economic rents has no effect, while a tax on output most likely reduces output, hence employment, and is ranked last. Taxes bearing on capital induce labor substitution.

While utilizing the literature and reasonable extensions of theoretic knowledge to rank various tax bases as to their effects on certain economic variables is instructive, it does not produce usable guidelines for making tax policy. It is generally agreed that a neutral tax is to be preferred when markets are competitive. Neutrality does not come without associated costs, such as complex administration and high costs of compliance by industry. The taxes ranked as being neutral and preferable are infeasible. Furthermore, many of the rankings are tenuous. First, only acurate empirical tests and estimates would show whether theoretically derived effects are significant. Then this requirement is magnified by the fact that these theoretical models are based on many assumptions which restrict the models' conclusions. For example, market imperfections undoubtedly exist to some degree, and accurate assessment of future prices, costs, and other factors is impossible. Further, these models are based on partial-equilibrium analysis, which limits their applicability. For the sake of simplicity, many economic variables are assumed constant. In the real world, there are few constants, particularly in the economic realm. A general equilibrium analysis in which all important quantities are treated as variable would improve these conclusions. Unfortunately, few general equilibrium models exist which are usable. However, when the second major policy question posed in this tax analysis is asked— Who ultimately pays for taxes and bears their burden?—a general equilibrium model becomes a virtual necessity. A simplified general equilibrium model is developed in chapter 4, where the burden of taxes on nonrenewable resources is discussed.

4

Who Ultimately Pays: The Incidence of Taxes on Nonrenewable Resources

As Supreme Court Justice Holmes said, taxes are part of the price of civilization. However, people prefer to have others pay taxes while nevertheless benefiting from government expenditures. Determining just who actually bears the final burden of tax is a complex matter because taxes induce individuals to change their behavior. The effect of taxes on behavior and on the distribution of income is discussed in this chapter. As argued in chapter 1, politicians are sensitive to how they believe tax policies (and how they think their constituents believe tax laws) affect incomes. Analysis and information on tax shifting and its effect on income could help develop a fairer, more rational tax policy.

Tax Shifting

Because individuals alter their behavior in an effort to minimize tax payments, resources are reallocated and these reallocations are reflected in prices. This process is called *tax shifting*. Taxes which do not affect resource allocation cannot be shifted and fall on economic rent and the price of the service or commodity which is taxed. All other taxes are shifted, at least to some degree. Price changes of the taxed commodity and all other commodities affected by resource reallocation measure the final burden of a tax. The tax burden is borne by those who are adversely affected by changes in relative prices. Prices include the effects on income sources, compensation for labor or property, and income uses, consumption, and savings. All price changes are traced to the people who are affected, for only people can bear taxes. To the extent that wages fall, labor bears the tax burden; to the extent that the price of mineral rights falls, landowners bear the tax burden; to the extent that profits fall, owners of capital bear the burden; and to the extent that prices of output increase, consumers bear the burden.

First, the way in which tax shifting ultimately affects the final burden of taxes is analyzed in a theoretical model of an extractive industry. Next, empirical tests for shifting the tax burden of corporate income tax are made for a number of resource industries and individual firms. Finally, data on Western coal-mining leases on federal lands are analyzed for evidence of whether user cost bears the burden of state taxes.

Before tax shifting of nonrenewable resources is analyzed in detail, it is necessary to clarify two factors. Only individuals can bear a tax burden, and tax revenues do not necessarily equal the burden of a tax. Although a corporation or any other legal entity may remit tax payments to the government, the ultimate burden of paying for taxes must rest on individuals—consumers of the product or owners, workers, or suppliers to the affected industry. The tax burden originates from two sides: the individual qua consumer and the individual qua income recipient. When a tax is shifted forward to consumers, individuals who purchase the particular commodities whose price is affected bear all or a portion of the tax. Here the incidence falls on the so-called uses of income. When a tax is shifted backward to a source of income, such as wages, royalties, or corporate dividends, the income recipient bears the burden of the tax. In this case, the incidence falls on the so-called sources of income. In the case of one specific tax, the correct way to determine tax burden is to measure tax-induced changes in commodity prices and changes in the price of inputs to the production process (sources of income) net of tax payments. Relative changes in these prices from the tax-free situation reveal how the tax burden is shared.

Although the implications of tax shifting may be understood by tax policymakers, they seem to conveniently ignore them when involved in the give and take of negotiation. The individual entity identified as a taxpayer in tax legislation bears the initial impact of a tax. It is not unusual for legislators caught up in their own rhetoric to proclaim that a tax will be painless because only corporations or the oil companies will be affected. In fact, tax shifting occurs, and the new equilibrium prices and resource allocation tell the tale concerning who bears the final burden of the tax. As shifting occurs, resource allocation is affected and distortions and inefficient use of resources may result. In their calmer moments, policymakers acknowledge that shifting occurs, but often they describe it in more convenient terms which fit into their own ideologies and preconceptions. Ergo, tax legislation is frequently passed because special interest groups and coalitions believe that the measure benefits them, even though rigorous analysis might prove otherwise.

The second important fact about tax burdens is that tax revenue does not necessarily equal the total burden. When a tax results in the inefficient use of resources, an "excess tax" burden or distortion occurs. This excess burden, shown in figure 3–1, is above and beyond the total tax revenue collected (equal to $t \times Q_1$ in figure 3–1) because it represents a net loss to society. *Excess burden* is a measurement of the loss in economic efficiency which the standard measure of tax burden neglects. Economists measure the burden of a tax by evaluating relative changes in prices occurring before and after short- or long-run adjustments by individuals and firms. Absolute tax burdens are either precisely equal to or approximately equal to tax revenues, depending on the method of analysis employed. The discrepancy is a result of taxes inducing firms to produce less (or more) and inducing individuals to work more (or less). When such tax-induced distortions in resource use occur, tax revenues are less than the total burden of a tax.

The study of tax burdens has another side. Taxes are spent by the government. The complexities of dealing with the expenditure side of the problem are normally dealt with in three ways. First, in *absolute tax incidence*, a tax base and rate are changed without any commensurate changes in expenditure policy. In the case of a tax increase, since the additional revenue is assumed to be retained in the public treasury, the deficit is reduced or a surplus is created. This engenders reduced aggregate demand, increased unemployment, and a reduced price level. The absolute-incidence approach is generally discarded as being unsatisfactory for the obvious reason of not keeping immaterial variables constant.

Second, in *differential tax incidence*, it is assumed that total expenditures and total tax revenues are held constant. The analysis is done by lowering one tax and increasing another so that the overall economy is neither inflated nor deflated and the issue of redistribution of income engendered by tax policy can be focused on. A nearly infinite number of financially equivalent, offsetting tax changes can be envisioned. Thus, measures of tax burdens made with this methodology are not unique.

Third, in *balanced-budget tax incidence*, offsetting changes in taxes and expenditures are made so that full employment and macroeconomic stability remain constant. For example, if a tax is increased, the additional revenue is assumed to be spent by government. The problem of uniqueness remains because of the change in expenditures, but it can be minimized by assuming that the government spends tax revenue in the same pattern as individuals spent private income before the tax change took some of that income away. In effect, this assumption neutralizes the expenditure side of the relationship and reduces the dimension of the analytic problem. The balanced-budget approach is followed in the model of tax incidence developed in this chapter.

The analyst generally is unable to chose an ideal solution to the tax-incidence methodology problem, and the approach most often taken is dictated by simplicity. In order to be able to say something about the relative incidence of various taxes on nonrenewable resources, we construct and analyze a simple model which permits various forms of tax shifting. Although the model is abstract, it does supply a certain amount of insight into the shifting process, when provided with plausible empirical guesstimates.

A Model of Tax Incidence

The model developed here allows all relevant prices to the consumer and to the firm to change and is so-called general equilibrium model. If the model were simplified and some prices were fixed, then it would not be able to measure the burden of taxes accurately. Historically, models in which only one commodity or input to production and its price are allowed to vary were utilized to measure the burden of taxes. Although something could be said about the relative portion of the tax borne by purchasers vis-à-vis suppliers of that particular commodity, the conclusions drawn from such models were questionable and in some cases patently incorrect, since all other prices were implicitly held constant. An example of simple market tax shifting is shown in figure 3–1. It was not until Harberger's (1962) analysis of the corporate income tax in the context of a general equilibrium model, borrowed from the theory of international trade, that the burden of most taxes could be accurately analyzed. We draw on a simplified Harberger model developed by Mieszkowski (1972) to analyze taxes on the production of nonrenewable resources.

Mieszkowski's major simplification is that changes in economic variables at the location being taxed are analyzed in detail and the incidence of taxes on the "rest of the world" is largely ignored. When used for analyzing extractive resources, the model implies that there are alternative locations which possess unexploited resource stocks that may be extracted and processed at higher costs than in the taxing jurisdiction. The model relies on the balanced-budget incidence methodology, which involves the simplifying assumption that government spends its tax revenues on the nonrenewable resource and on all other commodities in the same proportions as consumers and capital and resource owners. The tax then transfers resources from the private sector to the public sector, and the public sector spends this money in the same way that private individuals had. Thus all prices changes are due solely to the tax.

The taxes to be considered in this model are a severance tax on output; a tax on the value of land; a tax on human-made or renewable capital (in our case it is a tax on mining plant and equipment and mine development)—this tax might be levied as a property tax on plant and equipment or as a corporate income tax where the firm is assumed to be earning no economic rents; and a tax on the value of the labor input (which might be levied as a payroll tax or personal income tax on wage earners). When one or more of these taxes are enacted in one region, producing firms respond by immediately changing their level of output and, in the long run, via the location of new investment and modifying the production technology used in all locations. Only the long-run distributional effects of each tax are analyzed here. The short-run effect is not analyzed, for the only adjustment which would occur is the rate of extraction. In the long run, both capital and labor are mobile, so that the technology and location of extraction as well as the rate of extraction are variable.

The model of tax incidence concentrates on the last marginal investment dollars made by an extracting firm. The firm will invest in existing sites (sometimes called *incremental investment*) and undeveloped sites (sometimes called *greenfield investment*) as long as the rate of return from the investment exceeds the cost of capital. Investment ceases when the profit rate to the firm just equals the rate at which the firm can borrow funds, and this rate is the (risk-adjusted) market rate of interest. The model predicts when this process ceases and equilibrium is achieved. However, only two photographs are taken to show the change: one before the tax is instituted and a second after long-run equilibrium is achieved. Thus, the last marginal investment dollar earns no economic rent, because it costs just as much as it yields in net revenue. Only land earns economic rent in this model because it is assumed that land (mineral property rights) is in fixed supply in each region, that land has no alternative use, and that landowners bargain away rent from other economic agents. Thus, any price greater than zero is a net return to land. In reality, there are alternative uses for land (for example, agricultural use), but these uses only place a floor on land prices and including this floor would not alter the results drawn from the model.

The structure of the extractive industry is such that landowners usually possess limited knowledge about the value of their resources and thus are hampered in their ability to extract economic rent. What is likely in the real world is that firms own both renewable-capital and nonrenewable-resource rights and that through informational and other advantages are able to capture a share of the economic rent in negotiations with landowners. However, it is impossible to include these subtleties in this quantitative model.

Although severance taxes are expressed as a tax per unit of output (T_X, where X denotes output), the results of the model would not change for an ad valorem severance tax on output. The tax per unit on land R is denoted by T_R, the tax on labor L by T_L, and the tax on human-made capital K by T_K. The portion of the corporate income tax falling on returns to capital, a property tax on plant and equipment, and an excise tax on capital goods are all examples of taxes on human-made capital. The model is summarized in six equations:

$$X = f_1(L, K, R) \tag{4.1}$$

$$X = f_2(P_X) \tag{4.2}$$

$$K = f_3(W + T_L, P_R + T_R, P_K + T_K, X) \tag{4.3}$$

$$L = f_4(W + T_L, P_R + T_R, P_K + T_K, X) \tag{4.4}$$

$$L = f_5(W) \tag{4.5}$$

$$(P_X - T_X)(X) = (W + P_L)(L) + (P_R + T_R)(R) + (P_K + T_K)(K) \tag{4.6}$$

where X = amount of resource extracted per period

   L = labor services

   K = capital services

   R = Land services (mineral rights)

   P_X = price of nonrenewable resource

   W = wage rate

   P_K = price of human-made (renewable) capital

   P_R = price of land (mineral property rights)

   T_X = tax on output

   T_L = tax on labor

   T_R = tax on land (mineral rights)

   T_K = tax on renewable capital

Equation 4.1 summarizes all economically feasible production technologies. It states that output of the resource X is a function of services produced by the inputs: labor L, capital K, and land or mineral rights R. The relationship is assumed to be linear homogeneous, which means that if the number of units of land, labor, and capital were doubled, output would double. Furthermore, the flow of services from land R into production is assumed to be long-lived. The implication is that the stock of the resources is large and user cost is zero. In other words, exhaustion is not accounted for in this model. If depletion were formally included, the model would become unmanageable. Not a great deal is lost if, as Kay and Mirrless (1975) and others contend, user costs are small for many nonrenewable resources. Equation 4.2 summarizes the demand schedule for output, where the amount demanded X is a function of its gross price, which includes the tax P_X. The demand curve is assumed to slope downward, so that as price increases, the amount demanded decreases. Since the government buys the resource with tax revenues in exactly the same manner as taxpayers had, income does not have to be included as an argument in the demand function (because there will be no change in demand due to income shifts). This also gives balanced-budget incidence. Since the firm in this model does not experience economies of scale, if input prices were constant, marginal cost and average cost would be equal and constant in the long run. Other taxing jurisdictions in which resource reserves are located are potential producers. Since taxes and the cost of extracting resources in other areas are assumed to be constant, they are included implicitly but not explicitly in the demand relationship. The elasticity (price responsiveness) of the demand curve for the taxing region depends on these static conditions in other competing jurisdictions. If the resource base in the taxing jurisdiction being considered explicitly in this model is unique because of its low cost of extraction, the demand curve is inelastic. If there are numerous competitive locations, then the demand curve is elastic and sensitive to price changes.

The output decision of the firm is determined by maximizing profits, which implies that inputs are utilized until the cost of the last unit of each input purchased (including taxes on each) just equals net revenue received from increased production and sales. This process also implies that, for any level of output, the technology and combination of inputs which minimize costs are also selected. When we combine this with the assumption that no economies of scale are present, the firm makes no unanticipated economic profit. Accounting profits are just sufficient to induce equity and borrowed capital for this enterprise. If all firms are identical and competition prevails, then equations 4.1 through 4.6 depict aggregate supply and demand conditions for the region, since this is merely the result of adding identical firms.

Equations 4.3 and 4.4 summarize the region's demand for labor and capital. Each is a derived demand, and so the demand for capital shown in equation 4.3 is a function of the price of each input with its respective tax added, which gives a gross price, and the level of output. The determinants of each input demand function come from the profit-maximizing behavior of the firm. The demand for labor shown in equation 4.4 is analogous and states that as output X goes up, more of the input is needed. When the gross wage rate $W + T_L$, the price per unit of labor and its tax to the firm, the price of land $P_R + T_R$ to the firm, or the price per unit of capital $P_K + T_K$ goes up, other inputs will be substituted for the more expensive input, and these effects are summarized in the capital and labor demand functions.

Supply considerations for each input are described next. First, we assume that pure competition prevails, so that total revenues to the firm just equal total expenditures on inputs where land absorbs all economic rent. This assumption is shown in equation 4.6, where net of tax revenue from output, figured as the net revenue per unit to the firm $P_X - T_X$, just equals land, labor, and capital costs. This zero-profit condition is long-run equilibrium where no firm has an incentive to expand and no new firm has an incentive to enter or leave the industry. The supply functions for inputs are simplified by assuming that capital is perfectly mobile and that land is in fixed supply. There is no explicit land-supply function because its supply is perfectly inelastic, which means all land will be offered at any positive price. Equation

4.5 expresses that the supply of labor L is an increasing function of the wage rate W. Beacuse of its mobility, the supply of capital is perfectly elastic, or unlimited, at the going rate of return P_K, and consequently this price is unaffected by the extracting industry. This rate is determined in the economy as a whole and is assumed to be independent of investment made by extracting firms in this region. Thus the model assumes long-run equilibrium with perfect capital mobility, wherein investment occurs and production is expanded until the last dollar spent yields zero net return. Whenever conditions change, investment is reallocated to different technologies and locations until this condition is met.

The firm pays all taxes on output, capital, labor, and land, but final tax incidence must be determined from resulting changes in prices. First, depending on the demand conditions for output, the firm and resource industry may pass a portion of any tax forward to consumers. If demand is perfectly inelastic, where the same amount is demanded at any price (which implies that the resource can be recovered in other regions only at exorbitant costs), a major portion of any tax can be passed on to consumers through commodity price increases. However, the ability to substitute one material for another, technological change, and competing supply from other regions mean that no demand curve is perfectly inelastic. Even OPEC realizes that the less oil purchased, the higher its price, everything else being constant, because alternative sources of oil and other energy sources are available at least in the long run. Since the model assumes a purely competitive firm with respect to factor payments, all revenue is distributed to the factor inputs of land, labor, and capital. This distribution is based on derived demand for inputs equaling supply at an equilibrating price. Capital supply is assumed to be unlimited at the prevailing rate of return P_K, land is in fixed supply and always fully allocated to mineral production (measured in constant quality units), and labor supply increases as wages increase (equation 4.5). To the extent that a tax reduces the amount produced and consumed, taxes are shifted backward to factor inputs. Reduced production lowers the demand for the input and thus lowers its price, except for capital whose price is determined in the national economy.

A tax on any productive input alters its effective (gross) price to the firm and initiates a series of adjustments leading to tax shifting. In the short run, adjustments are achieved primarily through altering capacity utilization (output or the rate of extraction) and labor input. Thus, to the extent that a tax on input or a tax on output is shifted backward to input utilization, capital and land bear the brunt of the tax since their availability is fixed in the short run. This occurs because labor is the input which is most easily changed. It follows that labor bears some of the burden in reduced wages and through workers displaced to other sectors of the economy. Consumers also bear part of the burden if reduced output results in higher resource prices.

In the long run, which this abstract model depicts, capital is assumed to be perfectly mobile. Since capital markets are national and even international, resource taxes have no effect on interest rates. Thus, changing demands alter land prices and wage rates, but leave the price of capital unaltered. However, when taxed, returns to capital in the taxing jurisdication initially fall by the amount of the tax. Investors and firm managers will shift capital out of the taxed location until the projects remaining there yield an *after-tax* return equal to or above the rate of interest established in capital markets. Managers will also revise technological relationships of resources and labor until these inputs are combined with capital to yield the going rate of return. This occurs over the long haul as plant and equipment wear out and new investment is made. Thus, the outward flow of capital induced by a tax ensures that capital will not bear any tax burden in the long run. This result comes from simplifications inherent in the model concerning capital mobility and national capital markets and corresponds to the real world over long (ten- to thirty-year) periods.

Recall that economic rents in this model accrue only to land. Capital invested at the margin earns no economic profits which could bear a tax burden. More productive capital investment (mines) earns only a competitive return because economic rents are assumed to be bid away in negotiations with resource owners. Because resources found under the land surface earn economic rents (in fact, the entire price is economic rent in this model), they will bear a large share of virtually any tax. However, should labor demand fall as the result of a tax, workers will share some of the burden. If alternative employment is readily available, which means that labor demand in other economic sectors is elastic or sensitive to wage rates, then wages will fall little and labor will not suffer because of a tax. However, if the supply of labor to the extractive sector is inelastic or not price-sensitive because few employment alternatives are available, a tax-induced drop in the demand for labor could engender a significant fall in wages.

Adjustment by the profit-maximizing firm to taxes engenders two effects. Since a tax on an input increases its cost or a tax on output reduces net revenue to the firm, managers must reduce output, which according to economic theory and empirically verifiable data reduces unit costs (ore grades are increased, which in turn alters milling and processing procedures). And as output is reduced, price will increase according to the elasticity of the demand curve.

The firm also mitigates tax costs by changing technology. As output is reduced, the reduction in the use of the resource, capital, and labor is not uniformly proportionate because input markets and prices respond differently. Since we are talking about the long run, the cost of capital and its after-tax rate of return remains constant. The firm will reduce capital investment and invest only in profitable projects until the after-tax rate of return equals the national rate of return P_K. However, the technology of the firm allows for substitution among the natural resource, labor, and capital. There are a number of ways in which this may be accomplished, such as using fewer but more efficient machines to mine, load, and transport ore; reducing the speed of elevators and conveyers; and sinking fewer ventilation and access shafts in underground mines. Proportionately more labor and the natural resource can make up for reduced capital, even though technical efficiency may have to be reduced in the process, such as having workers waiting for more scarce equipment or mining higher-grade reserves and abandoning low-grade ores. Furthermore, as demand for labor and the resource falls when total output is reduced, wage rates will fall somewhat and the price of acquiring future mineral property rights may drop precipitously. At this point, the theoretical model tells us little except that changes in technology by substitution among inputs or changing their intensity of utilization and changes in the level of output will result from the imposition of a tax. At best, pure theory can tell us only the probable direction of tax-induced changes in resource allocation and the resulting new equilibrium market prices.

Using the Model to Estimate Tax Burdens

Once a tax is imposed, extractive firms will make the adjustments just described until after-tax profit is again maximized. This change can be solved algebraically from equations 4.1 through 4.6 and represented in a set of simultaneous differential equations. However, this exercise itself is only of technical interest and is relegated to Appendix 4A. In order to give some insight into the specific results of tax shifting and burdens generated in this general equilibrium model, guesstimates are made for key parameters which are necessary to know in order to make the solution of the model spit out quantitative answers.

This process of guestimating is called *sensitivity analysis*. Basically, we assign plausible quantitative relationships for the theoretical relationships and then assume that each tax is increased by 10 percent. Then the symbolic solutions shown in Appendix 4A can be used to estimate the tax-engendered percentage change in wage rates P_L, the price of land or mineral rights P_R, and the price of the output of the resource sold to users P_X. Relative changes in these prices reveal how the burden of various taxes is borne by labor, owners of mineral rights, and consumers of the resource. Recall that the net-of-tax price of capital (the interest rate) is assumed constant in this model. The taxes which are simulated to increase by 10 percent fall on output, mineral rights, human-made capital, and labor. The results of this exercise are shown in table 4–2, but before we review these, the guesstimates about key relationships should be described.

As stated previously, one major effect of a tax on output or inputs is for each firm to reduce output in order to increase after-tax profits to the national average. The elasticity of the demand for output (the sensitivity of demand to price changes) indicates how production curtailment affects the price of output and, in turn, determines the portion of the tax burden passed on to consumers and secondary users through higher prices. A zero elasticity of demand means that consumers must have the product at any cost and are willing to pay any price for it; thus, a tax is totally passed on to consumers because there is no resistance to taxes being added to the price of output. A unitary elasticity of demand (where a 1 percent price increase is accompanied by a 1 percent fall in quantity demand) is a standard benchmark wherein total expenditures for the resource remain constant regardless of price. However, when the producing region makes up a small portion of the industry, there are many regions which possess the resource and there are potential competitors, and so demand elasticity for any one taxing jurisdiction is large and approaches infinity. In this case, changes in the output from the region under consideration have little or no effect on national or world market prices, and the taxed regions' output can be sold only at the prevailing price. Except when a taxing jurisdiction possesses a resource which is unique and of low recovery cost, the demand elasticity is large. A 10 percent increase in taxes on output, capital, land, and labor is simulated when the elasticity of demand is 0 (perfectly inelastic), 1, and a number chosen to be very elastic, 500. Another elasticity measurement, in this case for the labor supply, is required. It is assumed here that a 1 percent increase in wages increases labor supply by 1 percent (labor-supply elasticity is 1).

Other quantitative requirements for the model are the ease of substitution among inputs, which is governed by production technology, and the initial proportions of gross sales revenue going to capital, land, and labor (see equation 4.6). Although available information on these data is sparse, some rough approximations are possible. Labor's average share of value added for U.S. industry is 70 to 75 percent. For purposes of the model, the 70 percent level is selected. While the share of costs going to land is low for the average manufacturing firm (most likely in the vicinity of 5 percent), it is higher in the case of extractive industries which must acquire leases and pay royalties for mineral rights. Table 4–1 indicates the total assets, depreciable assets, and percent depreciable assets for various extractive-industry groupings. Non-depreciable assets include land and mineral rights, accounts receivable, certain other financial instruments, and goodwill. Mineral rights and land make up an unknown portion of the percent nondepreciable assets shown in table 4–1. Land's factor share of the total revenues received by the firm is estimated to be 10.5 percent. Capital's share must then be the residual, 19.5 percent, since all revenues are paid to factor inputs of labor, land, and capital.

The solved form of the model requires nine specifications or guesstimates

Table 4–1
Total and Depreciable Assets in Selected Industries

Industry	Total Assets (Million Dollars)	Depreciable Assets (Million Dollars)	Depreciable Assets (Percent)
Mining (total)	26,042	17,139	65
Metal mining	7,062	4,001	56
Coal mining	3,819	2,817	74
Crude petroleum and natural gas	11,045	6,968	63
Nonmetallic mining (except fuels)	4,116	3,353	81

Source: U.S. Department of the Treasury, Internal Revenue Service *Statistics of Income—1971 Corporation Income Tax Return* (Washington: GPO, 1976).

for the partial elasticity of substitution between each input weighted by the factor's share in total cost. The elasticity of substitution is determined by feasible production technologies; it measures how sensitive the relative proportions of two factor inputs are to relative changes in their prices. Extractive production technology incorporates substitution between labor and capital by varying techniques from labor-intensive ones to capital-intensive procedures. However, it seems equally plausible that it is somewhat harder to substitute labor or capital for the natural resource, for the resource must be present and substitution involves altering technology and the cutoff ore grade.

For example, technological substitution in large part explains why labor costs in coal strip-mine costs are from one-half to one-fifth the cost of underground mines. In deep mines, labor must work in a cramped, dangerous environment, and machines must be designed to operate in thin seams. In strip mining, the ability to apply capital in the form of giant earth-moving machines to remove overburden and then extract and transport the natural resource is almost without limit. Thus changing location and technology from labor-intensive, underground mining to capital-intensive, surface mining techniques implies a substitutability of capital for labor. However, the natural resource must be present to make extraction possible, no matter what the technology. Although the average crustal and seabed abundance of nonfuel resources is sufficient to supply people almost in perpetuity, high concentrations are necessary for economically viable mining. Theory and practice both point out that the richest of known reserves are extracted first. Thus, it is seemingly harder to substitute land (the nonrenewable theory requires that the sum of the weighted partial elasticities of substitution equal zero). We assume that elasticities of substitution for labor and capital equal -1, which means that a 1 percent change in relative input prices will change the labor/capital ration by 1 percent. In the case of land or mineral rights, the elasticity of substitution is assumed to be much lower at −0.1, for it seems plausible that it is far more difficult and expensive to substitute the exhaustible resource for either capital or labor than to substitute capital for labor.

The results of simulations utilizing the guesstimates just described are presented in table 4–2. Recall that the model assumes perfect competition in input and output markets in the sense that the firm acts as if it can do nothing to affect prices. Also these measures of the tax burden do not account for resource inefficiencies (the excess burden of a tax). Furthermore, the model assumes full capitalization in the sense that the prices of inputs are determined by supply and demand and that long-run equilibrium prevails when a marginal investment earns the same (risk-adjusted) net-of-tax rate of return throughout the economy.

The severance tax or output-based tax is presented in the first row of table 4–2. When demand is perfectly inelastic (elasticity equals zero), a tax on output is fully shifted to consumers so that the price of output increases by the same 10 percent as the tax increase. However, as was shown in chapter 2, the mining firm also has an incentive to slow the rate of extraction. The effect of depletion is not incorporated in the model for reasons of tractability; but if it had been included, the degree of shifting and percentage price increase would be greater than the 10 percent shown because the supply function would shift and thus market price would increase by an additional amount. With unitary

Table 4–2
Simulation of Tax Incidence of Resource Taxes

10 Percent Change in Tax on:	Demand Price Elasticity	Percentage Change in Wages	Percentage Change in Price of Mineral Rights	Percentage Change in Output Price (to Consumers)
Output				
	0	0.0	0.00	10.00
	−1	−4.71	−4.94	1.17
	−500	−5.54	−5.82	−0.40
Mineral rights				
	0	0	−6.16	0.40
	−1	0	−6.96	0.32
	−500	0	−9.99	0.00
Capital				
	0	3.04	−13.10	2.93
	−1	2.16	−22.30	1.02
	−500	1.54	−28.80	0.12
Labor				
	0	−9.52	16.88	17.56
	−1	−9.52	8.46	1.22
	−500	−9.52	−0.46	0.53

demand elasticity, only about a 1 percent rise in the price of the resource results from a 10 percent increase in the tax on output. In the case of a highly elastic demand curve, no forward shifting to consumers should occur. The −0.4 percent decrease in the price of output shown in table 4–2 results from rounding errors and can be ignored. When a state or region is not in a dominant position as the supplier of a natural resource or human-made commodity, the major portion of tax burden is borne by reductions in wage rates, price of land, and other input prices. In the severance-tax case, wages and the price of mineral rights fall by approximately the same percentage, so that backward shifting is equally borne by workers and owners of mineral rights.

As explained previously, the model assumes that the national capital market is unaffected by long-run shifts in the location of investment. Thus, the price of capital P_K, which is its marginal rate of return, is constant and exogenous. In the short run, any tax which reduces the net-of-tax return to capital will place a tax burden on capital owners. When capital is mobile, as in the case studied here, in the long run capitalists will mitigate the effect of an accounting-profits tax by relocating investment as old investments become fully depreciated. This process continues until the rate of return, net of taxes, is everywhere equal for the final dollar invested in each project. However, existing capital in the taxing jurisdiction may earn a lower net rate of return (which will be capitalized into the market value of those capital goods). In the interim, either the owner suffers a lower rate of return until the capital is fully depreciated, or, if the assets are sold, the market value is, by capitalizing future tax payments, less than it would otherwise be. In the long run, however, the return on capital is equalized by means of capital mobility. Thus, changes in the location of capital and changes in the price of output are the engines which bring this model back into equilibrium after the imposition of a tax. If demand is highly elastic (500 is the assumed elasticity guesstimate), then capital flows are the sole equilibrating force. Capital leaves the taxing jurisdiction, no matter what type of tax is instituted, until the output and the demand for labor (supplied at unitary elasticity) and mineral rights (supplied no matter what the going price) fall sufficiently that their respective prices fall. In this way, tax burdens are shifted part to the price of output and part to the prices of inputs.

In the case of a tax on an input, shifting patterns are complicated by the assumption that the industry can alter its technology by substituting nontaxed inputs for the taxed input. If the ability to substitute is less than perfect, as is assumed here, then the cost of production must increase; when the demand for the commodity is inelastic, a portion of increased production costs can be passed forward to consumers. But when the elasticity of demand is high, these costs cannot be passed on to purchasers, and the returns to labor and land must fall by greater amounts until a profit-maximizing equilibrium is reached. The effects of different demand elasticities show up clearly when a 10 percent tax on land (mineral rights) is imposed, as reported in table 4–2. Since mineral rights are a relatively small portion of production costs (10.5 percent), it is difficult to substitute labor or capital for the nonrenewable resource; and since the supply of that resource is totally inelastic, the price of mineral rights bears the bulk of a tax imposed on it. Thus, a tax on the present value of economic rents is not shifted to any significant degree. However, the extractive firm does have an incentive to increase output in order to mine out from under the property tax imposed on the resource remaining in situ. The effect would be to reduce the tax burden shifted forward to consumers or backward to other inputs, notably labor and capital. Given the quantitative parameters assumed here, in this case labor bears no tax burden, and only a small amount of tax is shifted forward to consumers.

When a tax on capital is levied, outflow of capital and substitution of labor and land for capital take place. The net result is that the demand for labor shifts upward and wages increase. This conclusion depends critically on the guesstimates. The process occurs at the expense of resource property rights, which fall between 13 and 29 percent in value depending on the elasticity of demand for output (as seen in table 4–2). The large changes occur because the entire price of land is economic rent. The long-run structure of the model precludes measuring the tax burden borne by existing capital in the taxing jurisdiction; but in the short run with capital fixed, it would equal the entire amount of the tax.

A tax on labor induces industry to substitute land and capital for labor. Since an infinite amount of capital can be substituted at the going interest rate but it is more difficult to substitute mineral rights for labor. Land prices increase significantly as long as a portion of the tax can be passed forward to consumers. A totally inelastic demand curve for the extracted product means that prices to consumers increase by a greater percentage than the tax and that landowners are granted windfall profits.

Little is known about the actual elasticities of supply, demand, factor shares, and substitution, which had to be guesstimated for the model. Another limitation is created by assuming perfectly mobile capital and an exogenous price of capital. Harberger (1962) concluded that the corporate income tax, a tax on capital, is not shifted because the economywide net rate of return to all capital falls whereas other prices remain relatively constant. The specification of the modified Harberger model presented here precludes this result. Since extractive activities represent a modest share of all business enterprise, it is unlikely that a complete Harberger-type model would indicate that a tax on capital in an extractive industry would reduce the overall rate of return.

It should also be remembered that capital shifts slowly in the real world. For example, an open-pit coal or metal mine may take from five to ten years to develop and then may stay in production for twenty or more years. Mine-mouth electric power plants, dependent on nearby coal, will last from thirty to forty years. A tax which dissuades the location of capital will have immediate effects on greenfield investment. But investment in new mine sites represents a small portion of total investment in the mining industry, and a number of years may have to pass before the tax effects on capital become obvious. However, investment in exploration activity is a sensitive barometer of the tax and political climate of a taxing jurisdiction. DeYoung (1978) attributes the 20 percent annual decrease in exploratory expenditures in Canada from 1974 through 1979 to a succession of federal and provincial tax increases. British Columbia was the most aggressive province, and 50 percent of all the decreases occurred there. Empirical tests for shifting the tax burden of the corporate income tax are made later in this chapter, along with the effect of state taxes on federal coal leases.

This model does permit several important insights. The first is that the elasticity of demand for the resource is a critical determinant of the ability to shift taxes forward. If the taxing state or country represents a small portion of the market and if numerous alternative sources of supply are available or could be developed at comparable or slightly higher costs than in the taxing region, then the elasticity of demand is near infinity and forward shifting is near zero. The model also indicates that simultaneous forward shifting and backward shifting occur with both taxes on output and taxes on inputs. When an input is supplied at any price, as is the case for mineral rights, its elasticity of supply is zero and demand determines price. In this case, the entire price is an economic rent which can itself be taxed without incurring efficiency losses or excess burden, and the burden of the tax remains on the owners. Taxes levied on other inputs and outputs evoke large changes in economic rents. All the taxes simulated here, except for the payroll tax, produce large decreases in economic rents. It is interesting to speculate on who owns these mineral rights in the real world and, thus, who in all likelihood bears much of the burden of state taxation. When the residency of owners of capital and mineral rights and the location of consumers are known, then it is possible to determine the geographical incidence of taxes—how much is borne by out-of-jurisdiction residents compared to local residents.

Tax Exporting

When taxes are borne by persons not living in the taxing jurisdiction, the tax is said to be exported. The model of tax incidence just presented indicates the ability of a state or country to export its tax burden. One way that this occurs is if the tax is forward-shifted and the nonrenewable resource is exported from the region for consumption elsewhere. When backward shifting occurs, exporting is possible either when owners of mineral rights bear the tax burden and are nonresidents or when capital bears the burden and capitalists are nonresidents. The belief in tax exporting explains recent tax-policy behavior by states and countries exporting their natural resources (Church 1978*b*).

Gillis (1979), Gillis and McLure (1975*a*), and McLure (1975) point out that as the demand elasticities in our model indicate, tax exporting is unlikely unless the taxing jurisdiction dominates the market. If production costs are constant and one jurisdiction is dominant, both Gillis and McLure conclude that consumers and capitalists bear the tax burden in roughly equal proportions; but if capital is immobile, resource owners will bear the full burden of the tax. In the case of competitive markets for output, demand curves are elastic and backward shifting falls on capitalists (unless capital is perfectly mobile), resource owners, and labor. (See Harberger 1955, 1957, 1962; McLure 1967, 1969, 1974, 1975, 1978; and Mieszkowski 1966, 1969, 1972.)

One example of how the tax-exporting ability deteriorates over time as depletion occurs and renewable capital is depreciated comes from Minnesota. The rich iron ores created a monopoly for Minnesota, which taxed them heavily. After World War II, the ores were nearly depleted and industry had decided on the new taconite resource for iron production. Since other regions possessed reserves, industry insisted on a constitutional amendment limiting taxes for twenty-five years before taconite investments were made. Upon ratification, $500 million of taconite investment was committed.

Brown (1978) discusses Canadian federal and provincial mining-tax policy of the 1970s. He concludes that increasing income tax rates and interjurisdictional tax competition for perceived economic rents have discouraged exploration and development. He notes that lucrative tax advantages for processing facilities are insufficient to overcome the tax disincentives. The British Columbia Mineral Royalties Act of 1974 was structured so that large, highly profitable mines could be taxed at a marginal rate exceeding 100 percent. Brown notes that the extractive industries had strongly criticized the 1967 Royal Commission on Canadian Taxation Report (The Carter Report) for its advocacy of a tax neutrality goal. They preferred to be accorded preferential tax treatment. However, OPEC-induced price changes in oil and natural gas along with discontent over environmental damage and industry tax advantages led to a withering of public support for tax preferences. In the Canadian case, industry complaints and threats of migration had little effect in slowing provincial and federal governments in their quest to socialize economic rents. The long-run effect on the extractive industries appears to be, as DeYoung has reported (1978), a decrease in exploration activity. The ability of the provinces to export taxes on oil and gas is strong because of inelastic demand by consuming provinces and by the United States.

Empirical Estimates of Tax Shifting

Real-world conclusions about tax shifting and tax exporting can never be made without hard empirical evidence. The problem is that complete and reliable data are difficult to come by. Industry representatives consider much of their data to be proprietary and are fearful of having them fall into the hands of competitors or government. Although the data which are made public by industry and government agencies are wanting, the data should be exploited for any information they might provide as to tax distortions, shifting, and final tax incidence. We look at two sources of data: the performance of extractive firms on publicly traded stock markets for evidence of corporate income tax shifting and lease bidding for coal on Western public lands for evidence of shifting of state taxes.

Estimates of Corporate Income Tax Shifting

In the theoretical model of tax shifting just analyzed, a tax on human-made capital represented the corporate income tax. In the real world, corporations extracting natural resources own both human-made capital and property rights on natural resources. Thus, their reported earnings include return on equity capital, which is made up of expected and unexpected profits, and user cost and economic rents attributable to the resource. When this fact is coupled with the complexity of the corporate income tax code and its special tax provisions for extractive firms, economic theory becomes a Swiss cheese. One is unable to use it to forecast precisely how much of the tax is shifted. We analyze actual data from individual firms from 1960 through 1974 for evidence to determine whether the federal corporate income tax is borne by owners of the firm as a result of reduced earnings or whether this tax is passed on to others who are consumers of the product, labor, or those who have sold mineral rights to the firm.

The first individual to attempt to measure shifting of the federal corporate income tax in a general equilibrium model was Harberger (1962). He utilized a model similar to the one just described, although it explicitly included both the taxed sector of the economy (corporations) and the untaxed sector (partnerships and proprietorships). He also guesstimated important quantitative relationships for the corporate and noncorporate sectors of the economy. In part because he was analyzing a firm making zero profits, as we have just done, Harberger concluded that the overall rate of return to capital in *both* corporate and noncorporate sectors of the economy decreased. He concluded that capital bore the full burden of the tax. His findings were not universally accepted, and others developed alternative models and empirical means to test for corporate tax shifting.

Beginning in the 1950s, a number of empirical tests for evidence of shifting of the corporate income tax were reported in the professional literature. The methodology involved testing to see if pretax rates of return on capital respond to changes in the corporate income tax rate. If gross rates of return decrease as tax rates increase, the implication is that capital bears the burden of the tax. If the evidence shows that corporate accounting profits increase by the same amount as the tax, it implies that the tax is completely shifted to consumers and other productive inputs. The best-known empirical test for incidence of the corporate income tax was reported by Krzyzaniak and Musgrave (1963). They analyzed the economywide average pretax rate of corporate return, tax rates, and other variables from 1935 to 1959. Since the data were yearly observations, Krzyzaniak and Musgrave tested for the effect of short-run shifting by statistically isolating the effect of corporate tax rates from other influences on the pretax profit rates. They concluded that shifting exceeded 100 percent, but acknowledged that their estimates might have overestimated shifting owing to inflation and government-expenditure effects. Their study indicated that for every $1 increase in the effective corporate tax rate, average pretax corporate profits increased by $1.34. Because of their surprising conclusions, a number of criticisms were made by Harberger and others which center on the fact that critical cyclical measures were left out of the econometric model.

Cragg, Harberger, and Mieszkowski (1967) have been the most persistent critics of the Krzyzaniak-Musgrave study, and the professional journals were peppered with various criticisms and replies for the next decade. Cragg, Harberger, and Mieszkowski added several variables which measured wartime mobilization and the economywide rate of employment to the econometric model. They concluded that capital bore approximately 100 percent of the tax, for no strong statistical evidence of shifting could be found. The debate continued for several years with the most recent exchanges among the authors taking place in 1970. As with many arguments held in public, this debate has not been settled, and there is no unanimity even with respect to the direction of shifting. Even though the controversy still goes on about the degree of shifting [for example, Oakland (1972) and Turek (1970) find evidence of no shifting], the method of econometrically testing for the direction and extent of shifting by relating measures of effective tax rates to measures of pretax rates of return has been accepted by all those involved.

Musgrave and Krzyzaniak also looked at shifting, industry by industry, and found consistent evidence of 100 percent shifting. Although they analyzed a number of industries individually and found from 111 to 157 percent shifting, they did not analyze the extractive industries. This investigation of the extractive industries is original. We replicate the Musgrave-Krzyzaniak methodology for individual firms in seven extractive industries, whereas they looked at aggregate data.

Musgrave and Krzyzaniak and their followers and critics analyzed data on rates of return and tax rates from economywide data from the 1930s through the mid-1960s. The data analyzed here come from individual firms for the more recent period of 1960 through 1974 as opposed to their aggregate numbers for the entire economy. We use a pooled, cross-sectional (encompassing eighty-two individual firms within seven extractive industries), time-series (fifteen years) data base which could well produce different results from those found by people using economywide and industrywide time-series data.

Two major differences immediately distinguish our data. The first is that nominal tax rates did not fluctuate widely over the fifteen-year period (1960–1974) compared to the twenty-year to thirty-five-year periods analyzed by others. However, effective tax rates did fluctuate because of changes in the tax code, primarily with respect to deductions and credits. For example, during this period the investment tax credit was initiated, and its rate changed on several occasions; the percentage depletion allowance was reduced or modified for oil and natural-gas producers; the depreciation guidelines were rewritten; and loss-carry-forward provisions were modified. Another factor which increases the variation in effective tax rates among extractive firms is that accounting practices are not standardized. Alternative practices are permitted under IRS guidelines, which means that firms possess some discretion in reporting the level and timing of their profits, tax deductions, and tax credits. Consequently, effective tax rates vary significantly among firms and over time.

The second major difference distinguishing our data is that problems associated with aggregation and time-series data with an intervening world-war period are avoided because the period excludes World War II and builds from observations of individual firms. Specifically, we chose corporations listed on the American and Canadian stock exchanges with complete reporting for 1960 to 1974 as classified into seven industries: miscellaneous-metals mining (sixteen firms), copper mining (fifteen firms), lead and zinc mining (eleven firms), gold mining (nine firms), bituminous-coal mining (ten firms), gas and oil producing and exploring (eleven firms), and miscellaneous nonmetal mining (eight firms). In addition to analyzing the corporations grouped by industry, we analyze eighteen randomly chosen, individual firms which reported for all fifteen years and were not acquired by others during the period.

The models utilized by Musgrave and Krzyzaniak to test for shifting of the corporate income tax are relatively straightforward and far simpler than a complex general equilibrium model such as the one presented. If no shifting is possible and the owners of capital bear the full burden of the tax, then the change in the corporate tax rate t will not change the pretax rate of return on capital Y_g from a state of the world where taxes are nonexistent Y'. Thus, $Y_g - Y' = 0$ occurs when no shifting takes place. However, if the tax is shifted completely to others, total pretax income $Y_g K$ from a stock of capital K will increase from a tax-free state of the world by the amount of the tax payment T:

$$Y_g K - Y'K = T$$

where T = firm's tax liability

   K = capital stock of firm as measured by book value

   Y' = rate of return in absence of tax

   Y_g = pretax rate of return after imposition of tax

or, by rearranging,

$$Y_g - Y' = \frac{T}{K}$$

The effective tax rate t on the capital stock equals the tax bill T divided by capital stock K. This leads Musgrave and Krzyzaniak to define a variable indicating the degree of tax shifting S (lying between 0 and 1.0),

$$S = \frac{Y_g - Y'}{T/K}$$

If the pretax rate of return varies with the effective tax rate, then a statistically estimated regression coefficient for the effective tax rate equals the degree of shifting (0.0 equals no shifting and 1.0 equals 100 percent shifting). This simply means that if the firm's pretax rate of return increases by the same amount as the effective tax rate increases [$\Delta Y_g = \Delta(T/K)$], then the firm shifts the burden of the tax 100 percent. The statistical measure, the regression coefficient, for the effect of effective tax rate is equal to the shifting measure S:

$$\frac{\Delta Y_g}{\Delta(T/K)} = S = \text{regression coefficient for effective tax rate}$$

The major complicating factor is that a myriad of other factors affect the pretax rate of return in addition to a firm's trying to shift its corporate tax burdens onto others. These factors can be accounted for statistically by specifying the significant ones and including them in the estimated relationship. This technique is known as *multiple regression analysis*. Before these other specific factors are discussed, it is worthwhile to review how shifting might take place.

The traditional theory of the firm constructed by economists holds that the firm establishes a level of output which maximizes profits. If a tax falls on pure economic profits and economic rents, a corporate income tax will not affect output but will reduce the profits of the firm. This portion of the tax is absorbed by the owners. The portion of the tax that falls on the return necessary for capital to remain in the firm can be shifted in the short run only if the firm has some power over the price of output or the price of inputs going into the production process. If the firm had this power, it would have maximized profits originally and a profits tax would not affect its output/price decision. However, if the firm is following a markup pricing policy, consisting of cost per unit plus a profit margin, it would attempt to pass the corporate income tax on to others. Also if the industry is oligopolistic or markets are imperfect, a change in corporate tax rates may cause a firm to change prices, and all others would follow suit in a price-leadership fashion. Thus, short-run tax shifting would be possible only if a firm possessed some market power.

In the long run, firms and owners of fully depreciated capital will shift investments to new projects and corporations yielding the greatest posttax rates of return (risk-adjusted). As this occurs, pretax rates of return in industries which are taxed more heavily will rise, and returns in those industries being taxed less heavily and receiving shifted investment will fall as they expand and less profitable projects are funded. Thus, in the long run, posttax rates of return will adjust until these rates are everywhere equal. If rates of return before the imposition of an additional tax equal the posttax rate of return, then the tax shifting parameter $S = 1.0$. This indicates that full tax shifting has taken place. This is the same as saying that the pretax rate of return was increased by the firm by an amount equal to the higher tax. Partial shifting takes place when firms are unable to increase pretax rates of return by the full amount of the higher tax. Then the shifting parameter S will lie between 0.0 and 1.0. No shifting is indicated by $S = 0$.

The data base we utilize combines short-run shifting (the time-series portion of the data) and long-run shifting (data from different firms within each industry). Short-run shifting occurs when the firm is able to make output and price decisions only in response to a change in tax rates. Long-run shifting occurs as firms change their investment decisions over time in response to changes in tax rates. At least some firms in the sample in any particular year have achieved long-range profit maximization, net of taxes. If the firm anticipates future prices and technology correctly, it can maximize long-run profits by adjusting investment over a number of years. Since accounting conventions allow certain latitude in the timing of tax payments, the firm can anticipate future tax burdens and compensate for these by altering management decisions. Although not all firms can do this accurately, they most certainly are attempting to do so, and consequently some firms are closer to maximizing long-term tax shifting than others. The shifting parameter estimated from the series/cross-sectional data used here represents a mixture of the short-run and long-run tax shifting.

The detailed nature of the data on individual firms also permits us to include variables not considered by others. The Musgrave-Krzyzaniak study utilizes a rate of return and an effective tax rate based on economywide aggregate measures. The estimate of total capital stock is, by necessity, crude and is based not on security markets' estimate of the value of capital stock, but on accountants' estimates. We measure the rate of return on both the book value for each individual firm (which is determined by accountants) and two market-derived rates of return (the earnings on equity as established in the marketplace and the equity-plus-debt rate of return). The price of a share of stock (market-based equity per share) is determined by investors' expectations of the firm's future earnings and the investor's attitude toward risk. Past and present posttax rate of return, or net earnings per share, and future expectations for the firm and its industry in large part determine these attitudes. Total equity for the firm consists of the market value of all shares outstanding (both common and preferred stock). Pretax earnings include both distributed and undistributed accounting profits, and when divided by the market value of market equity, they equal the market rate of return on equity (ROE). The return on equity plus debt (RODE) is calculated by the ratio of interest payments plus earnings to the market value of stock plus interest-bearing debt. We calculate the rate of return on the book value of the firm's assets, which corresponds to the economywide measure used by Harberger and Musgrave and Krzyzaniak in their tax-shifting studies as a rate of return that equates earnings discounted for fifteen years to the firm's book value of capital (ROR).

Shifting of the effective corporate income tax rate is estimated for the rate of return on book value (accountants' definition of the value of the firm) and the equity and equity-plus-debt rates of return based on the value of the firm as determined in the marketplace. The market-based rates of return are related to two measures of the effective tax rate. The first is with respect to the firm's book value (actual tax payments divided by book value). The alternative measure is the effective rate per dollar of profits (actual tax payments divided by earnings).

A tax-shifting parameter estimated by statistically relating (regressing) the rate of return on the effective tax rate on earnings [defined as corporate income tax payments divided by reported earnings (profits)] is distinctly different from a shifting parameter measured by regressing the firm's rate of return on the tax rate on capital stock (defined by tax payments divided by the firm's total capital stock as used by Musgrave and Krzyzaniak). If the stock of capital K is assumed constant, which corresponds to the short run, and initially taxes are zero, the Musgrave-Krzyzaniak tax-shifting measure, denoted by S_1, reduces to the following (where Δ signifies a change in a variable and Π denotes the pretax profits):

$$S = \frac{Y_g - Y'}{T/K} \sim \frac{\Delta(\Pi/K)}{\Delta(T/K)}$$

When market-based rates of return are utilized to measure tax shifting, the market value of equity cannot be assumed to be fixed. It equals the capitalized value of expected future profits, and its value fluctuates from moment to moment as determined in investors' behavior in stock markets. If we assume that current total gross profits Π remain constant in the future, then the market equity value E of the firm equals the capitalized value net of tax on profits:

$$E = a(\Pi - T)$$

where E = market value of owners' equity

a = capitalization rate, or price/earnings ratio

Here $E = a(\Pi - T)$ is the relationship between market value and net profits. The variable a is determined by investors' expectations and behavior of stock markets and is commonly called the *capitalization rate*, or *price/earnings ratio*. When investors' expectations are unaltered by events other than a change in earnings Π or taxes T, this variable remains constant. Two alternative shifting measures (denoted by S_2 and S_3) are used to test for tax-induced changes in market rates of return. The first is the change in the market rate of return with respect to changes in the effective tax rate on fixed capital stock S_2:

$$S_2 = \frac{Y_g - Y'}{T/K} = \frac{\Pi/E - Y'}{T/K} \sim \frac{\Delta(\Pi/E)}{\Delta(T/K)}$$

The second market-based shifting parameter S_3 is similar to S_2 except that the effective tax rate is measured with respect to profits Π, instead of the accountants' estimate of the value for capital K:

$$S_3 = \frac{Y_g - Y'}{T/\Pi} = \frac{\Pi/E - Y'}{T/\Pi} \sim \frac{\Delta(\Pi/E)}{\Delta(T/\Pi)}$$

The three shifting parameters S_1, S_2, and S_3 can never be measured directly because no observations are possible of the rate of return in the absence of any corporate income tax Y'. What the data collected from 1960 through 1974 for the eighty-two firms separated into seven industries are able to show is how changes in effective tax rates are related to changes in rates of return, and this definition is denoted by the symbol for approximately defining shifting parameters S_1, S_2, and S_3. Changes in effective tax rates to each firm occur because of tax-induced changes in accounting practices and in investment and operating decisions. The degree to which the firm is able to reduce tax liabilities and to increase before-tax rates of return indicates how successfully it is able to pass the burden of the corporate income tax on to others. The shifting measures can be expressed in terms of these changes (note that Δ representes a change):

$$S_1 = \frac{\Delta(\Pi/K)}{\Delta(T/K)}$$

$$S_1 = \frac{\Delta(\Pi/E)}{\Delta(T/K)}$$

$$S_3 = \frac{\Delta(\Pi/E)}{\Delta(T/\Pi)}$$

Three extremes are possible:

1. All the tax is shifted to others.
2. None of the tax is shifted to others, and investors immediately recognize this, so that the market value of ownership equity falls until the original (posttax) rate of return is achieved.
3. None of the tax is shifted to others, and the market value of owners equity remains constant; but the posttax rate of return is reduced by the amount of an increase in the effective tax rate.

Full Shifting

The exercise of applying these three polar outcomes to the shifting parameters indicates what numerical values to expect in each case. If 100 percent shifting occurs, the pretax rate of return is increased by the amount of the effective tax rate. Should full shifting take place and if investors anticipate this, the market value of the firm will remain constant. If this is the case, it follows that the shifting parameters show (since $\Delta\Pi = -\Delta T$)

$$S_1 = 1.0$$

$$S_2 = K/E \sim 1.0$$

$$S_3 = \Pi/E = Y_g$$

With full tax shifting, the parameter $S_1 = 1.0$, S_2 will equal the ratio of book value to the market-determined value of ownership equity (approximately 1.0), and S_3 will equal the pretax rate of return on the market value of equity (averaging .05 to .20).

No Shifting and Full Capitalization

The second distinct outcome assumes that investors have complete knowledge about the operation of the firm and its profitability. If a change has occurred which increases tax liabilities, investors will immediately take this into account. The potential reduction of the posttax rate of return never occurs because investors sell shares and shift to other investments whose posttax rates of return are less likely to decrease. This tax-induced selling reduces the market value of the firm and therefore increases the posttax rate of return to its previous level. In this case, owners who retained the stock bear the full burden of the tax as a result of a loss in market value. This scenario implies that the pretax rate of return is unaltered ($\Delta\Pi = 0$), but that the market value of the firm falls. The outcome of no shifting and full capitalization of tax rates implies the following for the shifting parameters (where T' denotes a new and higher tax liability) [note that if full shifting had occurred and investors had anticipated it, there would be no need for selling and the resulting adjustment in market values (capitalization)]:

$$S_1 = 0$$

$$S_2 = \frac{\Pi K}{a^2(\Pi - T)(\Pi - T')} \sim .10 \text{ to } .40$$

$$S_3 = \frac{\Pi^2}{a^2(\Pi - T)(\Pi - T')} \sim .005 \text{ to } .015$$

The complexity of the expressions for S_2 and S_3 with no shifting and full capitalization of tax liabilities into market value precludes us from knowing a priori what the parameters will be numerically. However, reasonable values for the capitalization rate a, profits, and taxes imply that S_2 will likely fall in the range of .10 to .40. This contrasts with an expected value of 1.0 if the tax is fully shifted away from capital. Plausible numerical values for the components of S_3 imply a value of .005 to .015. This contrasts with a value equal to the market-based, pretax rate of return if the tax is fully shifted (approximately .05 to .20).

No Shifting and No Capitalization

Finally, if the tax is fully borne by capital but it is reflected in lower profits rather than in a decrease in the market value of owners' equity, then all three shifting parameters S_1, S_2, and S_3 will equal zero:

$$S_1 = 0.0 \qquad S_2 = 0.0 \qquad S_3 = 0.0$$

How can the shifting measures S_1, S_2, and S_3 be interpreted insofar as whether they measure the long run or the short run? When corporate income taxes increase, the reduced after-tax rate of return motivates investors to shift investment into more lightly taxed industries and firms until after-tax rates of return are equal. Once this has occurred, pretax profit rates in the taxed sector will have increased precisely by the size of the tax increase, and the shifting measure $S_1 = 1.0$. However, Harberger and others maintain that as this occurs, capital will be reallocated inefficiently. The reallocated capital will go into projects that, although more lightly taxed, are less productive and less profitable, which will result in reducing the rate of return throughout the economy. We do not accept Harberger's position in his corporate income tax study that overall return to capital declines when applied solely to the extractive industries. These industries in toto make up a small portion of the gross national product. Therefore, it is improbable that lighter or heavier taxation of these industries will distort the average return to capital for the entire economy. Thus, our data, which are mixed short-run and long-run, will support the contention of full shifting if $S_1 = 1.0$. Should capital bear the burden of corporate taxes in the mineral industries, two alternatives are possible. If market value of ownership equity reflects tax liabilities, then the widely accepted efficient-market hypothesis holds that this information is capitalized immediately. Thus, the long run and short run are one and the same. If capitalization does not occur and pretax profits are unaltered by changes in tax rates, then the distinction between long and short run is also irrelevant.

The Regression Estimates

We empirically estimate shifting measures S_1, S_2, and S_3 by statistically regressing measures of rates of return on measures of effective tax rates and other variables identified as being critical determinants of the rate of return. The addition of other explanatory variables to the statistical model is made in order to remove spurious effects which might otherwise prevent achieving a reliable estimate of the shifting parameters (for an example of other studies utilizing similar variables, see Petry 1975 and Martin and Scott 1974). These other determinants of rates of return are discussed next.

The factor affecting the market of return which is most significant, yet is the hardest to specify and quantify, is the *risk* associated with owning property rights in a particular firm. An extensive literature in economics and finance shows that investors choose investment portfolios which maximize their rate of return subject to individual risk preferences. When investors are risk-averse, they must be compensated with a higher expected rate of return in order to bear the risk. One way in which risk has been measured is by the variability of the rate of return over time. We measure risk as the statistical variance (VAR) of the market rate of return on equity for the fifteen years of data collected for each extractive firm. Theory indicates that a regression coefficient for this variable is expected to have a positive sign and to be constant. This indicates that as variance increases, the firm stockholders will buy or hold shares only when they receive a higher rate of return to compensate them for higher risk.

Another measure of risk often employed in financial studies of firm behavior is the *degree of leverage*. The greater the debt (both long- and short-term) undertaken by a firm relative to its total assets or its equity, the greater is its leverage. Greater leverage increases the firm's ability to earn profit when yield on assets owned by the firm exceeds the interest paid on debt. However, high leverage is a double-edged sword. When the profit rate on assets falls, rates of return on equity decrease by a multiple of the debt/equity ratio. Thus, the debt/equity ratio D/E measures leverage and potential profitability and indirectly measures risk. Investors balance risk and return, and consequently some may prefer to hold stocks in firms which have the potential to earn high profit. Consequently, the expected sign for the regression coefficient for this variable is uncertain, particularly when the variance measure of risk (var) is also included in the regression equation. Theory also indicates that the firm chooses a financing scheme (equity versus debt) which maximizes its present value of profits. If this ratio does not equal the firm's optimum level, then changes in the debt/equity ratio D/E undertaken by the firm over time should increase gross rates of return. If this is the case, the expected sign for the regression coefficient on the debt/equity ratio will be positive.

Another variable which is established at the firm's discretion is its depreciation, depletion, and amortization policy. Although federal tax laws make rapid depletion, depreciation, and amortization advantageous because of their sheltering effect, reported profits of the firm will decrease commensurately when these accounting expenses are greater. However, avoiding taxes increases net rates of return, and this should be recognized in the stock market. Since the dependent variables in our regression models are *gross* accounting and market rates of return, and not true cash flows, firms which utilize greater tax sheltering will report lower gross-profit rates. The ratio of depreciation and amortization to capital stock D/K measures this confounding effect, and its regression coefficient is expected to have a negative sign when the dependent variable is accounting rates of return. The efficient-market hypothesis implies that the regression coefficient should be zero when the dependent variable is a market-based rate of return.

A firm's actual inventory consists of two components: desired or planned inventories and unplanned or unanticipated inventories. When sales do not match expectations, inventories increase (when sales are slower than expected), and conversely. The firm adjusts inventories until the desired level with respect to actual sales is achieved. If the optimal or desired ratio of inventory to sales is constant, then increases in this ratio reflect unplanned inventory accumulations because sales are lower than anticipated, and vice versa. The implication is that the regression coefficient for the inventory/sales ratio I/S will be negative because when sales are less than anticipated, profits will be below the expected level. Also, firms with lower inventory/sales ratios are making higher profits vis-à-vis other firms because they are utilizing capital more productively. This will explain a positive coefficient.

A direct measure of the productivity of capital is turnover, or the ratio of sales to assets S/K. If a firm can increase this ratio, it is utilizing its entire capital stock, primarily a fixed expense, more efficiently, and one would expect profits and rates of return to increase.

Since the data are collected over a series of years (1960–1974) which were turbulent with respect to recession, inflation, and the dramatic OPEC-engendered oil-price increase in 1973 and associated rapid escalation in prices of most nonrenewable resources, some adjustment must be made for these significant shocks to corporate policies and rates of return. We introduce dummy variables (taking on the value 0 or 1) for each separate year (D60 to D74) to account for these shocks which are unaccounted for by the variables just described.

Of course, the variables of interest in the regression model are effective tax rates—the effective tax with respect to capital stock or book value ($t = T/K$) and the effective tax with respect to pretax profits ($t' = T/\Pi$). The dependent variables are the internal rate of return (ROR), which is calculated as the rate of interest that equates pretax profits Π discounted over fifteen years to the book value of the firm; the return on equity (ROE), which is the pretax profits Π divided by the market value of common and preferred stock outstanding; and the return on equity plus debt (RODE), which is the pretax profits Π plus interest payments divided by the market value of common and preferred stock plus interest-bearing debt outstanding. The market-based rates of return do not require discounting, for the investor in the marketplace has already used profits or earnings to estimate future earnings and discounts these to determine the worth of a share of stock.

Results

The estimates of the estimated regression coefficients for the shifting parameters S_1, S_2, and S_3 discussed above for firms within seven industry categories are shown in table 4–3 and for eighteen selected firms are shown in table 4–4. The regression procedure produces an estimated coefficient for each independent variable in the equation, tax rate and the other explanatory variables. Statistics are also produced which indicate the reliability of those coefficients and of the regression as a whole. These statistics are reported with the regression coefficients in tables 4–5 and 4–6. The standard error of the coefficient is the estimated standard deviation of the distribution of probable regresssion coefficients based on sampling error. A useful rule of thumb for judging the coefficients' reliability is that if the standard error is less than one-half the size of its corresponding regression coefficient, then the coefficient is significantly different from zero and is, therefore, of statistical importance. Standard errors are reported in parentheses below each regression coefficient in the tables. The standard error of the estimate is a measure of variation in the dependent variable which the regression equation does not explain. The adjusted coefficient of determination $\bar{R}^2$ (as adjusted for the number of variables in the regression equation) measures the percentage of variance in the dependent variable "explained" by the independent variables.

Table 4–3
Estimated Regression Coefficients by Industry

Industry	S_1	S_2 for ROE	S_2 for RODE	S_3 for ROE	S_3 for RODE
Miscellaneous metals (sixteen firms)	2.08*	1.8*	1.6*	.04*	.03*
Copper ores (fifteen firms)	1.67*	1.08*	.90*	−.08	.06
Lead and zinc (eleven firms)	1.12*	.76*	.60*	.08*	.04
Gold (nine firms)	2.74*	.43*	.43	.06*	.06*
Bituminous coal (ten firms)	1.94*	3.35*	3.03*	.25*	.21*
Gas and oil production and exploration (eleven firms)	1.98*	.82*	.57	.005	.006
Miscellaneous nonmetals (ten firms)	2.11*	1.32*	1.38*	.02	.03

*Statistically significant at the .05 level.

Table 4–4
Estimated Regression Coefficients by Firm

Industry	S_1	S_2 for ROE	S_2 for RODE	S_3 for ROE	S_3 for RODE
Miscellaneous Metals					
Amax (molybdenum, aluminum, iron, coal)	1.20*	.98*	.95*	.14	.15*
Foote Mineral (Ferrous alloys and lithium compounds)	1.95*	2.02*	1.50*	−.14*	−.16*
Hecla Mining (silver, lead, and zinc)	1.43*	.23*	.25	.03*	.03*
United Nuclear (nuclear operating and uranium mines)	−1.14	−2.03	−.26	.015	.0012
Copper					
Callahan Mining (silver and copper)	2.20*	1.02*	1.02*	.02	.02
Newmont Mining (magma copper, metals, oil and gas)	1.92*	1.52*	1.52*	.07	.04
Lead and Zinc					
Cominco (lead and zinc)	1.40*	.32*	.34*	.15*	.13*
Gulf Resources and Chemicals (metals, coal, lithium, lead, zinc, and silver)	3.54*	.49	.27	.04	.13
Hudson Bay Mining and Smelting (lead and zinc)	1.61*	.71	.67	.11*	.10*
St. Joe Minerals Corp. (lead, zinc, coal, oil and gas)	2.23*	.69*	.70*	.33	.16
Gold					
Campbell Red Lake Mine (gold)	2.95*	.18	.17	−.83*	−.42*
Dome Mines Ltd. (gold)	1.85*	.33	.33	.14	.14
Pato Consolidated Gold Dredging (gold, South America)	1.97*	2.37*	2.33*	.05	.05
Coal					
Pittston Co. (coal)	1.69*	.68*	.56*	.12*	.07*
Crude-Oil Products					
American Petrofina (oil and petrochemical)	1.48*	4.22*	3.63*	.36*	.31*
Felmont Oil (oil, gas, and ammonia)	.58	−1.76*	−1.81*	−.02*	−.02*
Miscellaneous Mining					
Arundel Corp. (mining, heavy construction, building material)	1.96*	2.71*	2.74*	−.03	.18
Freeport Minerals Co. (sulfur, oil and gas)	1.55*	1.06*	1.04*	.27*	−.07

*Statistically significant at the .05 level.

If $\bar{R}^2 = 1.0$, then all variation (100 percent) is explained by the independent variables.

Notice in tables 4–5 and 4–6 that except in the case of the effective tax rate, not all explanatory variables are included in each regression equation. When a variable added little or nothing to the explanatory power of the regression equation ($\bar{R}^2$), it was not included. The purpose of including explanatory variables other than effective tax rates is to statistically remove

Table 4–5
Estimated Regression Equations by Industry

Industry	Dependent Variable	t	t'	D/E	I/S
Miscellaneous metals (sixteen firms)	ROE		.0354 (.016)		
	RODE		.027 (.014)	−.033 (.012)	
	ROR	2.080 (.141)			−.013 (.003)
	ROE	1.765 (.258)		.027 (.013)	
	RODE	1.553 (.199)			
Copper Ores (fifteen firms)	ROE		−.082 (.047)	.152 (.039)	
	RODE		.064 (.043)	.065 (.036)	
	ROR	1.665 (.328)		−.014 (.005)	
	ROE	1.075 (.309)		.174 (.037)	
	RODE	.903 (.284)		.083 (.034)	
Lead and zinc (eleven firms)	ROE		.075 (.038)	−.090 (.024)	
	RODE		.038 (.030)	−.063 (.018)	
	ROR	1.122 (.250)		−.123 (.019)	−.023 (.007)
	ROE	.760 (.334)		−.075 (.024)	
	RODE	.602 (.256)		−.051 (.019)	
Gold Mining (nine firms)	ROE		.058 (.024)		.0096 (.0029)
	RODE		.063 (.025)		.014 (.003)
	ROR	2.742 (.154)			
	ROE	.427 (.242)			
	RODE	.427 (.243)			
Bituminous coal (ten firms)	ROE		.250 (.058)	−.036 (.017)	
	RODE		.209 (.039)	−.030 (.011)	
	ROR	1.943 (.167)		.010 (.004)	

S/K	D/K	Var	Year Dummies					Adjusted R^2	Standard Error
	−.758 (.405)	.412 (1.671)	D73	.922 (.026)	D68	−.043 (.026)		.16	.076
		−.930 (1.267)	D73	.085 (.022)				.17	.066
		.500 (.741)	D73	.062 (.013)	D69	−.024 (.013)		.73	.037
		−.206 (1.266)	D73	.150 (.023)				.36	.066
		−.150 (1.064)	D73	.125 (.019)	D68	−.028 (.019)		.41	.055
		−14.316 (5.089)	D65	−.058 (.030)	D73	.456 (.025)		.37	.057
		−14.834 (4.650)	D73	.050 (.023)	D65	−.053 (.027)		.31	.052
		4.079 (4.832)	D73	.102 (.024)				.45	.054
		−13.986 (4.679)	D73	.070 (.024)	D65	−.057 (.028)		.46	.053
		−14.430 (4.309)	D73	.071 (.022)	D65	−.052 (.026)		.39	.049
.167 (.036)		−1.000 (1.414)	D73	.115 (.031)				.43	.066
.165 (.028)		−1.242 (1.104)	D73	.091 (.025)				.47	.051
.246 (.030)		−1.832 (1.137)						.72	.052
.137 (.039)		−.496 (1.445)	D73	.145 (.033)				.44	.065
.138 (.030)		−.718 (1.106)	D73	.114 (.025)				.50	.050
−.077 (.034)		−5.171 (.642)	D70	−.102 (.026)				.63	.059
−.798 (.035)		−5.108 (.653)	D70	−.101 (.026)				.65	.060
	.212 (.017)	−.599 (.591)	D73	.176 (.033)				.89	.055
−.134 (.051)	.048 (.016)	−5.468 (.640)	D70	−.106 (.026)				.62	.059
−.136 (.052)	.077 (.016)	−5.500 (.644)	D70	−.107 (.026)				.59	.062
−.058 (.033)		−4.014 (.919)	D70	.109 (.035)				.52	.080
−.054 (.022)		−1.796 (.613)	D73	.053 (.023)				.49	0.54
		−.220 (.209)	D73	.036 (.008)	D70	.013 (.008)		.80	.017

Table 4–5 (continued)

Industry	Dependent Variable	t	t'	D/E	I/S
	ROE	3.353 (.749)			
	RODE	3.025 (.494)		−.021 (.011)	
Crude-oil producers and oil-gas field exploration (eleven firms)	ROE		.0054 (.016)	.105 (.038)	
	RODE		.0016 (.014)	.100 (.034)	
	ROR	1.980 (.401)			.042 (.103)
	ROE	.816 (.308)		.108 (.037)	
	RODE	.574 (.280)		.103 (.033)	
Miscellaneous Nonmetal mining (ten firms)	ROE		.023 (.074)	.248 (.054)	
	RODE		.028 (.070)		
	ROR	2.106 (.335)			−.010 (.005)
	ROE	1.319 (.465)		.255 (.049)	
	RODE	1.376 (.429)			

factors affecting rates of return which might distort the regression coefficient of interest.

For convenience, the estimates for the tax-shifting parameters S_1 (the effective tax rate with respect to book value on the internal rate of return), S_2, and S_3 (the effective tax rate with respect to book value on the market rate of return and on the equity-plus-debt market rate of return) are extracted from tables 4–5 and 4–6 for the seven resource industries and for eighteen selected firms. These figures are shown in tables 4–3 and 4–4. An asterisk designates estimates which are statistically different from zero at the .05 probability-of-error level. The Musgrave-Krzyzaniak shifting measure S_1 ranges from 1.12 to 2.11 in the industry regressions and from .58 to 3.54 for the individual firms, and it is statistically significant in all but two cases.

Recall that the Musgrave-Krzyzaniak measure $S_1 = 1.0$ indicates full shifting and $S_1 = 0.0$ implies that no shifting takes place. The industry results

S/K	D/K	Var		Year Dummies			Ad-justed R^2	Standard Error
		−2.999	D71	.119	D73	.083	.48	.084
		(1.011)		(.036)		(.038)		
−.059		−.961	D73	.089			.54	.051
(.021)		(.619)		(.023)				
		−2.508	D60	−.133	D73	.040	.21	.082
		(.829)		(.042)		(.024)		
		−2.559	D60	−.133	D73	.035	.25	.074
		(.744)		(.038)		(.021)		
	1.828	−.206	D74	.0096			.66	.105
	(.171)	(1.106)		(.003)				
		−2.167	D60	−.129	D73	.052	.27	.079
		(.809)		(.041)		(.023)		
		−2.309	D60	−.130	D73	.044	.29	.072
		(.736)		(.037)		(.021)		
		3.188	D73	.072	D68	.077	.73	.048
		(2.263)		(.035)		(.038)		
		4.444	D73	.092	D68	−.0799	.47	.046
		(2.062)		(.024)		(.036)		
		−4.874	D73	.082			.52	.031
		(1.059)		(.018)				
		1.791	D73	.098	D68	−.067	.78	.044
		(1.531)		(.033)		(.032)		
		3.169	D73	.122	D68	−.069	.59	.040
		(1.285)		(.023)		(.029)		

shown in table 4–3 for S_1 strongly support the contention that an average firm in the seven extractive industries shifts the corporate income tax by more than 100 percent (from 112 percent in lead and zinc to 211 percent in miscellaneous metals). The specification of the theoretical and empirical model makes it impossible to ascertain what portion of the tax is shifted forward to consumers and what portion is shifted backward onto purchased inputs.

The alternative shifting measures lend partial support to the position that shifting takes place. The effective tax rate measured with respect to book value is used in shifting measure S_2, which is estimated for market rate of return on equity (ROE) and market rate of return on equity plus debt (RODE). This measure is statistically significant in all but one case and strongly indicates shifting for bituminous coal, miscellaneous metals, and miscellaneous nonmetals. A significant degree of shifting is indicated in the

Table 4–6
Estimated Regression Equations by Firm

Firm	Dependent Variable	t	t'	RODE	D/E	I/S
Amax	ROR		−.0030 (.0478)		−.149 (.015)	
	ROE		.143 (.074)			.024 (.009)
	RODE		.149 (.047)			−.103 (.005)
	ROR	1.204 (.311)			−.131 (.026)	
	ROE	.975 (.532)				
	RODE	.950 (.479)				
	ROR			1.034 (.427)		
Foote Mineral Company	ROR		−.041 (.022)			
	ROE		−.139 (.025)		.207 (.036)	
	RODE		−.157 (.023)		.195 (.029)	
	ROR	1.947 (.323)				
	ROE	2.018 (.489)				
	RODE	1.498 (.389)				
	ROR			.125 (.244)		
Hecla Mining Co.	ROR		.036 (.006)			
	ROE		.026 (.005)			
	RODE		.026 (.005)			
	ROR	1.429 (.139)				
	ROE	.225 (.174)				
	RODE	.249 (.169)				
	ROR			1.488 (.648)		
United Nuclear Corp.	ROR		.0015 (.004)			
	ROE		.015 (.012)			
	RODE		.0012 (.007)			

S/K	D/K	Year Dummies						Adjusted R^2	Standard Error
		D60	.063 (.0099)	D66	.029 (.009)	D61	.026 (.0096)	.96	.008
		D60	.083 (.015)	D66	.032 (.012)			.75	.011
		D60	.077 (.009)	D66	.032 (.008)	D70	.028 (.008)	.90	.007
		D62	.037 (.013)					.92	.012
		D60	.062 (.016)	D74	.0049 (.0017)	D66	.0296 (.011)	.80	.010
								.27	.035
		D65	.088 (.022)	D66	.059 (.022)			.71	.020
		D72	.078 (.023)	D70	.042 (.024)			.86	.021
		D71	.042 (.021)					.82	.019
.136 (.041)		D68	.036 (.017)					.84	.015
		D73	.166 (.027)	D70	.108 (.025)			.81	.024
		D73	.145 (.021)	D70	.072 (.0198)			.81	.019
								.06	.039
.573 (.039)		D63	−.0599 (.021)					.96	.020
		D74	.006 (.001)	D65	.052 (.015)			.83	.015
		D74	−.006 (.001)	D65	.053 (.015)			.84	.014
.324 (.042)		D73 9	.048 (.014)					.99	.012
		D74	−.006 (.001)	D62	−.070 (.028)			.67	.020
		D74	−.006 (.001)	D62	−.675 (.027)			.69	.020
								.25	.083
		D63	.045 (.009)	D65	.029 (.009)			.72	.008
		D61	−.135 (.024)	D62	.991 (.024)			.81	.023
		D61	−.134 (.016)	D74	.003 (.001)			.86	.013

Table 4–6 *(continued)*

Firm	Dependent Variable	t	t'	RODE	D/E	I/S
	ROR	−1.137 (1.374)				
	ROE	−2.026 (4.047)				
	RODE	4.261 (2.302)				
	ROR			.211 (.121)		
Callahan Mining	ROR		.335 (.049)			
	ROE		.023 (.041)			
	RODE		.023 (.041)			
	ROR	2.198 (.091)				
	ROE	1.018 (.088)				
	RODE	1.018 (.088)				
	ROR			1.853 (.170)		
Newmont Mining Corp.	ROR		−.398 (.244)			
	ROE		.065 (.083)		−.156 (.064)	
	RODE		.036 (.067)		−.173 (.051)	
	ROR	1.918 (.554)				−.092 (.017)
	ROE	1.520 (.431)				
	RODE	1.522 (.283)				
	ROR			1.010 (.441)		
Cominco Ltd.	ROR		.243 (.086)		−.554 (.064)	
	ROE		.153 (.048)			
	RODE		.128 (.044)			
	ROR	1.380 (.179)				
	ROE	.321 (.158)				
	RODE	.340 (.137)				
	ROR			1.767 (.912)		

S/K	D/K		Year Dummies			Adjusted R^2	Standard Error
		D63	.044 (.009)	D74	.028 (.009)	.74	.008
		D61	.135 (.026)	D72	.099 (.026)	.79	.024
		D61	.139 (.014)	D74	−.004 (.001)	.90	.011
						.14	.015
		D71	−.134 (.034)	D67	−.088 (.033)	.80	.032
.286 (.044)	−.914 (.285)	D61	−.070 (.016)			.88	.013
.286 (.044)	−.914 (.285)	D61	−.070 (.016)			.88	.013
		D73	.205 (.010)	D71	−.026 (.019)	.98	.009
		D73	.0999 (.010)	D62	.034 (.009)	.94	.009
		D73	.0999 (.010)	D62	.034 (.009)	.94	.009
						.90	.023
		D66	.098 (.054)	D70	.115 (.066)	.26	.051
.428 (.080)		D66	.032 (.017)			.84	.016
.390 (.065)		D66	.035 (.014)			.86	.013
.733 (.140)		D71	.112 (.043)			.78	.028
.121 (.068)		D73	.095 (.030)			.85	.015
		D73	.113 (.014)	D66	.033 (.013)	.87	.012
						.26	.051
.277 (.085)		D65	.059 (.026)			.90	.023
.197 (.041)		D66	.0498 (.014)			.75	.012
.158 (.037)		D66	.047 (.013)			.72	.011
.571 (.044)	−3.002 (.768)	D67	−.115 (.018)			.98	.011
		D73	.079 (.017)	D66	.041 (.015)	.64	.015
		D73	.064 (.015)	D66	.0397 (.013)	.63	.013
						.17	.069

Table 4–6 *(continued)*

Firm	Dependent Variable	t	t'	RODE	D/E	I/S
Gulf Resources and Chemicals	ROR		.439 (.105)		−.141 (.022)	
	ROE		.039 (.206)			
	RODE		.125 (.072)			
	ROR	3.541 (.281)				
	ROE	.488 (.843)				
	RODE	.268 (.502)				
	ROR			.458 (.424)		
Hudson Bay Mining and Smelting	ROR		.195 (.020)			
	ROE		.105 (.019)			
	RODE		.100 (.019)			
	ROR	1.612 (.378)				−.032 (.0075)
	ROE	.708 (.364)				
	RODE	.674 (.360)				
	ROR			1.127 (.329)		
St. Joe Minerals Corp.	ROR		.111 (.121)		−.424 (.090)	
	ROE		.325 (.113)			
	RODE		.159 (.087)		−.171 (.066)	
	ROR	2.215 (.298)			−.238 (.079)	
	ROE	.689 (.193)				
	RODE	.696 (.162)				
	ROR			1.603 (.232)		
Campbell Red Lake Mines	ROR		−1.864 (.143)			
	ROE		−.729 (.029)			
	RODE		−.423 (.177)			

S/K	D/K		Year Dummies			Adjusted R^2	Standard Error
		D71	.290 (.073)	D68	−.125 (.045)	.79	.069
		D60	−.386 (.094)	D71	.316 (.107)	.77	.071
		D60	−.224 (.041)	D62	.169 (.037)	.85	.036
		D63	.057 (.025)	D73	.059 (.026)	.93	.024
		D60	−.383 (.077)	D71	−.318 (.076)	.77	.070
		D60	−.248 (.045)	D62	.168 (.502)	.82	.040
						.17	.081
		D72	−.126 (.021)	D70	−.067 (.021)	.92	.020
		D73	.109 (.019)	D69	.043 (.019)	.83	.019
		D73	.097 (.019)			.82	.018
		D73	.116 (.020)			.94	.016
		D73	.131 (.028)	D71	−.068 (.029)	.77	.022
		D73	.117 (.028)	D71	−.065 (.029)	.74	.022
						.45	.051
.342 (.093)		D71	−.103 (.024)			.93	.021
.177 (.083)		D67	.043 (.023)			.81	.021
.168 (.065)		D68	−.035 (.017)			.88	.016
		D73	.111 (.020)			.95	.018
		D74	.0073 (.0013)	D66	.055 (.019)	.85	.018
		D74	.0068 (.001)	D66	.050 (.016)	.88	.015
						.78	.037
		D71	−.134 (.046)			.93	.045
		D63	.026 (.010)			.98	.009
	.056 (.015)	D63	.023 (.009)			.99	.008

Table 4–6 *(continued)*

Firm	Dependent Variable	t	t'	RODE	D/E	I/S
	ROR	2.949 (.237)				
	ROE	.178 (.163)				
	RODE	.171 (.162)				
	ROR			1.767 (.157)		
Dome Mines Ltd.	ROR		.268 (.092)			
	ROE		.136 (.068)			
	RODE		.136 (.068)			
	ROR	1.853 (.320)				
	ROE	.332 (.327)				
	RODE	.332 (.327)				
	ROR			.751 (.724)		
Pato Consolidated Gold Dredging	ROR		.058 (.022)			
	ROE		.045 (.016)			
	RODE		.045 (.016)			
	ROR	1.966 (.439)				
	ROE	2.371 (.343)				
	RODE	2.330 (.3395)				
	ROR			.975 (.443)		
Pittston Co.	ROR		.167 (.051)		−.0988 (.020)	
	ROE		.123 (.028)		.124 (.013)	
	RODE		.068 (.022)		.056 (.011)	
	ROR	1.693 (.126)				
	ROE	.688 (.135)			.157 (.011)	
	RODE	.558 (.112)			.086 (.008)	.034 (.014)
	ROR			−.428 (.495)		

S/K	D/K	Year Dummies				Adjusted R^2	Standard Error
		D73	.984 (.034)			.99	.015
.069 (.006)		D69	.019 (.011)			.97	.010
.098 (.006)		D69	.019 (.011)			.99	.010
						.91	.053
.332 (.074)		D73	.081 (.013)	D74	.005 (.001)	.95	.010
.154 (.067)		D68	−.021 (.011)			.63	.010
.154 (.067)		D68	.021 (.011)			.63	.010
		D73	.216 (.018)			.92	.012
2.205 (.681)		D73	.040 (.017)			.56	.011
2.205 (.681)		D73	.040 (.017)			.56	.011
						.006	.043
		D68	.288 (.052)	D70	−.118 (.052)	.76	.050
		D61	.089 (.038)	D70	−.084 (.038)	.59	.037
		D61	.090 (.038)	D70	−.083 (.038)	.59	.036
		D68	.348 (.043)	D73	.183 (.046)	.85	.040
−4.119 (.690)		D68	.087 (.029)			.79	.026
−4.190 (.684)		D68	.087 (.029)			.79	.026
						.22	.091
		D70	.0386 (.017)			.85	.012
		D68	−.041 (.011)	D61	−.027 (.009)	.93	.008
		D68	−.023 (.009)	D69	−.016 (.007)	.85	.007
		D74	.0037 (.0005)			.95	.007
		D68	−.046 (.010)	D61	−.023 (.008)	.94	.007
		D68	−.029 (.007)			.90	.006
						.02	.032

Table 4–6 (continued)

Firm	Dependent Variable	t	t'	RODE	D/E	I/S
American Petrofina	ROR		.045 (.044)		.024 (.022)	
	ROE		.359 (.035)		−.149 (.025)	.102 (.034)
	RODE		.314 (.035)		−.128 (.027)	.096 (.034)
	ROR	1.480 (.419)			−.125 (.0198)	
	ROE	4.218 (.377)		(.023)	−.177 (.032)	.140
	RODE	3.639 (.570)				
	ROR			.673 (.283)		
Felmont Oil Co.	ROR		−.0058 (.002)			.061 (.031)
	ROE		−.108 (.005)		−.507 (.122)	
	RODE		−.107 (.005)		−.507 (.140)	
	ROR	.575 (.297)				
	ROE	−1.755 (.613)				.135 (.036)
	RODE	−1.806 (.640)				.132 (.037)
	ROR			.953 (.119)		
Arundel Corp.	ROR		.159 (.132)			
	ROE		−.028 (.198)		.339 (.046)	
	RODE		.180 (.218)			
	ROR	1.951 (.241)				
	ROE	2.711 (.252)		(.121)	.376	
	RODE	2.736 (.336)				
	ROR			.399 (.121)		
Freeport Minerals Co.	ROR		.138 (.061)			
	ROE		.265 (.082)			
	RODE		−.066 (.029)		1.837 (.458)	
	ROR	1.554 (.896)				
	ROE	1.058 (.369)				.019 (.007)
	RODE	1.040 (.361)				.019 (.006)
	ROR			.406 (.629)		

S/K	D/K		Year Dummies			Adjusted R^2	Standard Error
	.780 (.361)	D71	−.026 (.012)			.91	.012
		D70	.026 (.012)			.92	.011
		D64	.026 (.014)			.89	.011
		D71	−.025 (.012)			.91	.011
		D71	−.022 (.011)			.94	.010
		D74	.003 (.001)			.78	.016
						.28	.033
.537 (.029)	−.343 (.109)	D73	.052 (.009)			.98	.008
.342 (.060)		D73	.062 (.018)	D68	−.039 (.018)	.88	.017
.342 (.069)		D73	.067 (.021)			.84	.020
.514 (.037)	−.358 (.107)	D73	.078 (.012)	D61	.034 (.009)	.97	.008
.663 (.088)		D62	.058 (.024)			.81	.022
.664 (.092)		D62	.057 (.026)			.79	.023
						.83	.022
.029 (.013)		D68	.084 (.041)			.36	.032
		D68	−.157 (.060)			.85	.045
		D73	−.209 (.067)			.41	.052
		D73	.048 (.018)			.83	.016
	3.277 (.661)	D73	.146 (.119)			.98	.016
	2.067 (.660)	D73	.278 (.031)			.89	.023
						.43	.030
.395 (.044)	−4.076 (.703)	D64	−.031 (.018)			.88	.016
.383 (.038)	−5.255 (.847)	D64	−.038 (.016)	D69	.048 (.024)	.91	.014
		D73	.0699 (.010)	D67	−.037 (.009)	.83	.008
.151 (.077)		D73	.064 (.033)			.64	.027
		D73	.086 (.015)	D64	.055 (.013)	.75	.010
		D73	.086 (.014)	D69	.054 (.013)	.76	.010
						.05	.047

other four industries, for all estimates exceed the no-shifting values expected for S_2 (0.0 or .10 to .40, depending on assumptions made concerning capitalization).

The third shifting measure S_3 measures the effective tax rate with respect to pretax profits and is also statistically significant and positive in most cases measured for ROE and RODE. If we judge by relative magnitudes among industries, it indicates that significant tax shifting (perhaps over 100 percent) takes place in the coal industry and partial shifting occurs in all but the gas and oil production and exploration and copper industries (the no-shifting value expected for S_3 is 0.0 or .005 to .015, depending on the assumptions made concerning capitalization). If the three shifting measures are considered simultaneously, the conclusion of rejecting the null hypothesis that no tax shifting takes place is strongly upheld in the coal, lead and zinc, gold, and miscellaneous metals, and nonmetals industries. Shifting measure S_1 alone indicates shifting in every industry. It appears that less shifting takes place in the copper and gas and oil exploration and production industries, although the results are not uniform among the different measures. The evidence does not support the hypothesis of no shifting, and it appears that shifting varies by industry. These findings are somewhat counterintuitive, particularly in the case of coal, which many consider to be a highly competitive industry and consequently firms lack market power, and oil and gas, which many consider to be oligopolistic and subsequently firms may possess market power.

The large international oil companies profited enormously from the 1973–1974 and 1979–1980 OPEC price increases. This implies that they were not maximizing monopoly profits before the OPEC decisions (and thus were not omniscient and all-powerful). Also it should be noted that their tax advantages have been so extensive that they probably have not been concerned with shifting taxes onto others, at least until the advent of mandated change in accounting practices for foreign operations, the windfall-profits tax, and deletion of percentage depletion. The oil and gas companies in this sample do not include the gigantic seven sisters of international oil and, as such, are more likely to be competitive.

The bulk of coal produced in the United States is sold as boiler fuel for electric power plants. Large mine development is most frequently undertaken after a long-term contract is signed with one or more utilities. The contract is used to secure long-term financing so that the mine can be developed. Provisions for passing through of tax and other costs increases are being written into these contracts with increasing frequency. Utilities do little to resist these provisions, since they are able to pass on cost increases either immediately when fuel-adjustment clauses are permitted by state regulatory commissions or more slowly as rate hearings take place. Thus, coal is in an opportune position because of this institutional structure to pass tax increases forward to consumers. Furthermore, if coal production is a competitive industry, then theory leads one to suspect that its supply curve is relatively elastic. This implies that most of the tax will be passed on to others because there are no economic profits.

The estimated shifting measures for firms (fifteen years of data for each) (tables 4–3 and 4–6) most likely represent more short-run than long-run situations. The data are only from time series, and when year-to-year changes for an individual firm are sampled, it is less likely that a firm can adjust toward desired capital stock, pricing policy, and so on.

The five shifting estimates for eighteen individual firms are abstracted from the regression equation for each firm (shown in table 4–4). The first fact which is immediately apparent is that there is more variation in the shifting measures among firms than there is among industries. This pattern is to be expected because the fortunes and accounting practices of individual firms fluctuate more than the averages for all firms in the industry. Remember that the Musgrave-Krzyzaniak shifting measure of $S_1 = 1.0$ means that full shifting has occurred. The estimates for S_1 all are statistically significant and are greater than 1.0 except for United Nuclear and Felmont Oil. Although these estimates are somewhat higher than the short-run Musgrave-Krzyzaniak findings for other industries, generally they confirm Musgrave's and Krzyzaniak's conclusions. The market-based measures of tax shifting S_2 and S_3 confirm the conclusion that tax shifting takes place. Again, recall that if investors discount the effect of tax liabilities on the market value of the firm accurately and quickly, then an S_2 ranging from .1 to .4 reveals no tax shifting. Virtually all studies of the behavior of the stock market confirm the efficient-market hypothesis. The implication is that all relevant information is capitalized by the market rapidly and accurately. By employing this criterion, the results indicate that the corporate income tax is shifted by all eighteen firms except for United Nuclear, Campbell Red Lake Mine, Gulf Resources and Chemical, Hecla Mining, Dome Mines, and Felmont Oil.

Now look at the third measure of tax shifting, S_3. Again recall that only the relative magnitude of this measure is interpretable and a value of .05 to .015 indicates no tax shifting. Again, most firms are able to shift all or a significant portion of their corporate tax burdens except for Foote Mineral, Hecla Mining, United Nuclear, Callahan Mining, Campbell Red Lake Mine, Felmont Oil, and Arundel Corp.

What, then, do these estimates of corporate tax shifting for industries (a mixed long-run and short-run situation) and individual firms (a short-run situation because each firm is tracked over fourteen years for year-to-year changes) reveal? The Musgrave-Krzyzaniak findings of positive shifting are confirmed for most of the firms; the alternative measures for individual firms S_2 and S_3 confirm these findings in most cases. Therefore, we have strong but incomplete support for their findings of short-run shifting. Our cross-sectional data of firms over time as classified into seven industries lend strong support that long-run tax shifting is taking place in the coal, miscellaneous metals, and miscellaneous nonmetals industries and somewhat weaker support for shifting in the other four industries. This methodology is unable to discern the extent to which forward shifting places the tax burden on consumers and to which backward shifting puts the tax burden on inputs other than equity capital purchased by the extractive firm.

Now we take a detailed look at the estimated regression equations that provided the shifting measures. The results for the seven industries are shown in table 4–5, and those for the eighteen individual firms are contained in table 4–6. There are five regression equations for each industry. Each provides one of the shifting measures. There is an extra equation for each firm, listed last, which is now analyzed.

The final equation estimated for each firm in table 4–6 is a two-variable regression which relates the pretax rate of return based on book value (ROR) to the market-determined rate of return with respect to equity value plus interest-bearing debt (RODE). If investors holding shares in each firm agreed that accountants who calculate book value are accurately depicting the value of the firm, then the regression coefficient would equal 1.0 and the two return measures would vary closely together ($\bar{R}^2$ would be close to 1.0). The estimated regression coefficients shown in table 4–6 (the last row labeled ROR for each firm) vary greatly. This implies that the accountants' estimated book value and that estimated by investors often do not agree. The fact that the estimated regression coefficients show little consistency from firm to firm suports the observation that accounting methods in the extractive industries are not standardized. This observation was confirmed in a debate among stock market analysts. *Business Week* and *Fortune* staff writers disagreed over the future prospects for Amax corporation because it had chosen to amortize rather than currently charge interest expense for capital investment in a major new molybdenum mine being developed in Colorado.

As previously explained, dependent variables are rates of return based on book value or on the value of the firm as determined by investor behavior in stock markets: ROR is the rate of return which equates book value and capitalized future earnings (the *internal rate of return*), ROE is calculated as the ratio of current earnings to the market value of total stock outstanding, and RODE is calculated as the ratio of earnings plus interest payments to the value of total stock outstanding plus interest-bearing debt. The independent variables are listed across the top of each table. Two variables measure the effective tax rate: t is the effective rate with respect to the book value of the firm, and t' is the rate with respect to the current earnings of the firm. Their coefficients (S_1 and S_2 for the effective tax rate with respect to book value t and S_3 for the effective tax rate with respect to current earnings t') are the measures of tax shifting. The standard errors are given in parentheses below each regression coefficient. The tax-rate coefficients are almost invariably of the sign anticipated and are statistically significant.

The explanatory variables described previously are shown in the next set of columns. The variable *var* measures risk and is computed for only the industry regression equations. Because its computation requires data on firms' performance over a series of years, the variable is fixed for each individual firm and would mean nothing in a regression equation for one firm (table 4–6). The variation in the market rate of return on equity (ROE) is anticipated as having a positive effect on rate of return because it measures risk. The coefficient for the debt/equity ratio D/E is expected to be either positive or negative; the inventory/sales ratio I/S is expected to be negative; the sales/asset ratio S/K is expected to be positive; and the depletion/amortization ratio D/K is expected to be positive or negative. The next set of columns lists the regression coefficients for dummy variables that remove the positive or negative effects of unusual yearly occurrences which deviate from normal patterns of earnings. These are selected for statistical significance and can represent any year from 1960 (D60) through 1974 (D74). A dummy variable takes on the value 1.0 when the year it represents occurs and 0.0 otherwise.

The second-to-last column gives the adjusted coefficient of determination or explanatory power of the regression equation. If $\bar{R}^2$ is multiplied by 100.0, it gives the percentage of the variation in the dependent variable explained by the independent variables in the regression equation. Note that only those explanatory variables which yielded a significant contribution to this explanatory power were included in the equations. Those variables left out are revealed by blank spots in the tables. Although the estimated values of $\bar{R}^2$ vary a lot, they show that the regression equations explain a good deal of the variation in rates of return and are "pretty good" for data from individual firms. Another statistic measuring the accuracy of the regression equation is shown in the final column—the standard error of the estimate. It is an estimate of the unexplained variation in the dependent variable.

The coefficient for the variance of the rate of return (*var*) is expected to possess a positive sign, which indicates that investors require higher rates of return when a firm's earnings are volatile. The data did not bear this presumption out, for the coefficient is either statistically insignificant (table 4–5) or negative when significant. However, the coefficient of the debt/equity ratio D/E was frequently negative, which was also unexpected and may have confounded the measurement of risk. However, for individual firms, either the coefficient of the debt/equity ratio D/E was not included in the equation because of its low explanatory power, or when it was included, it displayed negative and positive signs with roughly the same frequency. This disparate pattern was also true of the inventories/sales ratio I/S. This result is understandable for the rapid price increase for gold resulting from the phased devaluation of U.S. dollar in 1971 and 1972, and the OPEC-induced oil-price increases benefited the gold and oil industries. The equations for the gold and oil industries and the equations for individual oil firms reveal a positive coefficient, but the coefficient is negative for the remaining industries. The turnover, or sales/capital, ratio S/K measures how efficiently the firm is using both natural and human-made capital and is generally positive. However, the gold and coal industries (table 4–5) display a strong and significant negative sign, indicating that firms which increased their reserves and held aboveground inventories (in both industries, reserves are a significant portion of capital) relative to sales again benefited from price increases which occurred in the 1970s. The gold and oil industries and individual firms, with the exception of Pato Consolidated Gold Dredging, also possess a positive sign for the ratio of depletion and amortization expenses to capital (D/K) for the same reason. When D/K is included, the sign is negative in all other industry (table 4–5) and firm (table 4–6) regression equations (except for Arundel Corporation).

The annual dummy variable (D60 through D74) indicates particularly good or bad years for various industries and firms. Because of price increases in virtually all nonrenewable resources in 1973 accompanying the OPEC oil-price increases and investors' expectations of possible shortages and exhaustion of low-cost resources, a positive coefficient occurs frequently for the 1973 dummy variable. The other annual dummy variables generally single out poor years. Taken together, all the "explanatory" variables explain from 15 to 70 percent of the variation in rates of return. Obviously, innumerable factors affect rates of return and, by implication, market prices for stocks which we did not take into account because of their unpredictability or the lack of available data. As mentioned previously, if the rate of return for an individual firm could be forecast accurately, then those willing to go into the bowels of this regression equation could become rich in short order. However, the market quickly captures and capitalizes information, both measurable and immeasurable, so that the best *statistical* estimate of future stock prices is that they are random and thus unpredictable.

Burden of State Taxes

Data released by corporations publicly permit an examination of shifting of the federal corporate income tax. However, these same data cannot be used to test for shifting of state corporate income and other taxes because these corporations, almost without exception, operate in more than one state. The data, therefore, cannot be segregated by state, and another approach is necessary to test for shifting of state taxes. One data source that is segregated by state is competitive coal leases for extraction from federal lands. We analyze the amount paid for these leases in competitive bidding in eight states. An alternative way to test for tax shifting and other tax-induced effects is to simulate how resource markets respond to alternative tax policies. This approach is undertaken for the domestic coal and copper markets in chapter 6.

The Mine Leasing Act of 1920 established procedures for the sale of certain mineral rights located on federal lands. It is distinct from the General Mining Law of 1872, which in effect permits exploration for hard-rock minerals to take place on federal lands. Once a claim is discovered and reserves are shown to be economically exploitable, the claim staker is awarded, under the 1872 law, title in fee simply after investing a minimal amount in the claim (Anderson 1976). The 1920 Mine Leasing Act, on the other hand, establishes procedures for the federal government to sell mineral rights (not all rights to the land and mineral body, as in the 1872 General Mining Law) at competitive market prices. The Mine Leasing Act applies to fuels (coal, gas, and oil). Under provisions of the Freedom of Information Act, we obtained data on 170 competitive coal leases made between 1951 and 1971 in Colorado, Montana, New Mexico, North Dakota, Oklahoma, Utah, Washington, and Wyoming from the U.S. Bureau of Land Management.

The theory of the mine discussed in chapter 2 shows that extracting a nonrenewable resource gives rise to user costs which represent forgone future profits when the resource is mined now rather than at some future date. It was also shown, by using the simplest model of the firm, that user costs increase at the rate of interest. The discussion of economic rent in chapter 3 indicates that user costs for mines making no economic profits, or from an accountant's and management point of view making just sufficient profits that it is worthwhile to keep the mine in operation rather than shutting down and abandoning it, start out at near-zero and grow at the rate of interest. Thus, sites having superior ore grades, access to transportation, proximity to marketing centers, and other advantages earn economic rent (see figure 3–3). Those sites enjoying cost advantages will continue to be operated when institutions and individuals expropriate part of or all these economic rents.

The power to tax is a coercive and persuasive one, and the potential for securing economic rents to bolster government revenues is often irresistible. Others seeking to use their legal and market powers to take a share of rents are unions, mineral landowners, and transportation systems. The federal government is the largest single landowner in the United States and has large holdings in the Western states as well as owning the continental shelf, where large deposits of domestic exhaustible resources (oil and natural gas) are to be found. As owner of mineral rights, the government may seek to maximize economic return for these rights when they are sold under the Mineral

Leasing Act. Coal leases are sold only after discovery has taken place and certain information about the size and quality of a potential reserve is known. When leases to these mineral rights are sold, bidders are theoretically willing to pay up to the expected economic rent (for more detail see Gaffney 1967, editor's appendix). If the lease is structured as a royalty wherein the lessee pays the lessor a percentage of the total revenues (a gross lease) or the accounting profits (a net or profits lease), then both parties share risks that the site may not pan out. Most leases in the private sector and those made for state-owned land are negotiated as gross leases, but the federal government prefers another approach. Most federal leases combine a fixed, nonnegotiable royalty paid as the resource is extracted with an initial lump-sum payment which is most frequently bid for competitively and which thus captures the federal government's share of *differential* economic rents. This is called a *bonus bid*. In this situation, the bidder and the developer of the resource bear the majority of the risks, and the government bears little risk.

In order to ensure that these so-called bonus bids are competitive, government bureaucrats analyze each lease and establish minimum acceptable bids. Several environments exist for bidding: oral bids, sealed bids, and a mixture of the two. In some cases, preferential bids are allowed in which only certain parties may bid (such as small firms and individuals as opposed to corporations). The theoretical value of the maximum bid in a bonus-bid situation is the firm's estimated net present value of future economic rents. Obviously, if other groups and institutions have already extracted some of the economic rent, then bids will be lower to reflect this condition. We analyze data on 170 competitive coal leases made in eight Western states from 1951 to 1971 (when the de facto and temporary federal leasing moratorium went into effect). We test the premise that differences in state tax rates affect costs and therefore are capitalized into bonus bids. However, if extractive firms are able to shift taxes to consumers or other groups, then state-to-state differences in bid patterns should reflect only differences in geologic quality and accessibility. The hypothesis is that state borders should make no differences in bonus bids unless state taxes significantly affect economic rents. If no state-to-state differences are found, it implies that taxes are shifted.

Coal bonus bids are made on a per-acre basis. The royalty provisions over the period sampled remained relatively constant. A base charge of $0.25 per acre was made for the first year after signing the lease, $0.50 in the second through fifth years, and $1 in the sixth through twentieth years. The royalty rates varied in these data, but not to a significant extent. The most frequent arrangement was a royalty of $0.15 per ton for underground mining and $0.175 for surface mining which increased to $0.20 in the eleventh through twentieth years of the lease.

The bonus bid per acre was first analyzed for evidence that geologic-quality variables, reported by the Bureau of Land Management, were statistically significant. These data include the Btu (British thermal unit) per pound of coal, seam thickness, percentage of sulfur content, percentage of ash, percentage volatile, an expected percentage recovery of the total reserve, and the size of total reserves. A potentially important variable was not reported by the Bureau of Land Management—the ratio of overburden to seam thickness—and this factor could not be tested by us. Other variables—miles to nearest transportation and miles to nearest consumption center—were also tested for their impact.

As the number of bidders at a sealed or oral auction increases, the size of the winning bid also increases (Mead 1967). Collusion is more difficult as the number of conspirators increases. When faced with competition, a bidder is forced to bid close to or at the maxium rather than playing a holding-back strategy game.

The exercise of testing for significance of all variables describing geologic quality, number of bidders, and distance to transportation and markets excluding the effect of state borders proved futile. In a lengthy series of statistical tests, the expected signs of respective regression coefficients were counter to what one would rationally expect, or the coefficients, when of the "correct" sign, were not statistically significant.

The second test was to simultaneously check for the effect of state borders, reflecting differential state taxes which may or may not be shifted, the effect of other location and geologic variables affecting costs, and type of bidding procedure. The results were also negative. The result of one specific statistical test is shown in table 4–7. The bonus bid per acre was regressed on the year the lease was signed, to reflect growth in coal demand, the number of bidders, and dummy variables (equal to 0 or 1) for the year 1965 and for each of the states. Note that statistical procedure requires that one state (Wyoming was chosen) not be assigned a dummy variable, for its effect is absorbed in the constant of the regression. It is also important to note that the statistical significance of quality and cost variables, which are not included in this equation, was also tested in other regressions. Seam thickness, number of acres, percentage of sulfur, Btu, moisture, volatility, and ash content as well as miles to possible consumption, miles to transportation, and, quite surprisingly, mine-mouth price for coal extracted in the West for each respective year were of the "wrong" sign and mostly statistically insignificant. This is why they are left out of the reported regression equation (table 4–7).

The regression coefficients may be evaluated if the benchmark for statistical significance of the coefficient being twice its standard error (reported in parentheses) is recalled. Only two coefficients shown in table 4–7 pass this test—the year of lease and the 1965 dummy variable. Failing to

Table 4–7
Regression Estimate for the Effect of State Borders and Quality Factors on Bonus Bid per Acre

	Regression Coefficient	Standard Error
Year of lease	1.36	(.68)
Number of bidders	1.82	(1.95)
1965 dummy variable	29.46	(10.73)
Colorado	15.15	(11.04)
Montana	11.90	(14.24)
New Mexico	1.90	(17.95)
North Dakota	2.79	(15.23)
Oklahoma	3.74	(14.38)
Utah	6.76	(10.04)
Washington	44.78	(42.66)
Wyoming	N/A	
Adjusted R^2	.06	

meet this benchmark means that the remainder of the coefficients are not different from zero in a probability or statistical sense and thus have no discernible effect on the level of bonus bids. It comes as no surprise that the regression "explains" very little of the variation in the bonus bids (only 6 percent, as shown at the bottom of table 4–7). What do these negative results mean? The implication for the hypothesis we are testing is that state border (tax) differentials on bonus bids (economic rents) have no effect. This extremely weak negative evidence indicates that the taxes are shifted and therefore do not affect economic rents as evidenced in bonus bids.

However, one other explanation is possible—that tax differentials are so insignificant that their effect does not show up in the regression equations. Recent changes in state tax rates are considered in the next chapter, but a bit of advance information may help out here. The year 1975 may well become known as the "annus mirabilus" for the coal-rich northern Plains states. It marked the advent of Montana's 30 percent maximum severance tax rate on surface coal. Since then, New Mexico, Colorado, and North Dakota have increased their coal taxes. Wyoming is a relatively heavy taxer as well. (For further discussion see Church 1978*b*.) However, in the period under consideration (1951–1971), taxes were not high. The key question is how great tax differentials were during this period, which is a question we are unable to answer with any degree of accuracy. However, the review of state tax laws conducted to describe exhaustively the relevant bases and rates in 1977 included historical searches. These data were not formally collated and evaluated, but they indicate that tax differentials did exist during this time. This indication lends support to tentative conclusions that state tax differentials are not reflected in bonus bids and that they are shifted.

Table 4–8
Regression Coefficients for Bonus Bid per Acre

Year	Number of Leases	Increase in Bid per Acre for Each Additional Bidder (Regression Coefficient)	R^2
1961	19	$ 1.00 (.18)	.63
1962	17	18.51 (1.82)	.86
1963	12	4.73 (1.40)	.48
1964	15	4.87 (1.40)	.44
1965	19	95.32 (45.13)	.16
1966	18	4.24 (2.05)	.16
1967	11	6.26 (9.85)	.05

The coal-lease data have another story to tell. Although the number of bidders was not a significant variable in the regression equation reported in table 4–8, conclusive findings by other researchers stimulated a further investigation of this matter. The bonus bid per acre was regressed on the number of bidders in separate years (when a sufficient number of leases to make for statistically valid results occurred). The number of leases, the regression coefficients and standard errors for the number of bidders, and the $\bar{R}^2$ are shown in table 4–8 for 1961 through 1967. These segregated data lend strong support to the contention that bid amounts increase as the number of bidders increase. Since all the coefficients (except for 1967) are positive and pass the benchmark test, the explanatory power of this variable is strong. Resource quality and state dummy variables were also tested in these year-by-year equations, but the results were the same as those found previously: There are no statistical significance. Next look at what tax rates were in 1977 and how the complex array of tax provisions can be converted to effective tax rates so that tax climates in various states may be compared.

Appendix 4A

Assume that all units of measurement are normalized so that initial prices equal unity,

$$P_L = P_R = P_K = P_X = 1$$

and that P_K is chosen as the numeraire.

Tax on Output

Assume $T_K = T_R = T_L = 0$.

Totally differentiate equations 4.1 through 4.6, where E_X denotes elasticity of demand for X and E_L is the elasticity of supply of labor. Also A_{ij} denotes the partial elasticity of substitution of factor i for factor j weighted by the proportion of factor j in total cost, and f_L, f_R, f_K are respective factor shares:

$$\frac{dX}{X} = f_L \frac{dL}{L} + f_K \frac{dK}{K} + f_R \frac{dR}{R} \tag{A4.1}$$

$$\frac{dX}{X} = E_X \, dP_X + E_X \, dT_X \tag{A4.2}$$

$$\frac{dK}{K} = A_{KL} \, dW + A_{KR} \, dP_R + \frac{dX}{X} \tag{A4.3}$$

$$\frac{dL}{L} = A_{LL} \, dW + A_{LR} \, dP_R + \frac{dX}{X} \tag{A4.4}$$

$$\frac{dL}{L} = E_L \, dW \tag{A4.5}$$

$$dT_X + dP_X = f_L \, dW + f_R \, dP_R \tag{A4.6}$$

$$\frac{dR}{R} = A_{RL} \, dW + A_{RR} \, dP_R + \frac{dX}{X} \qquad \text{for land} \tag{A4.7}$$

Substituting in equations A4.1 and A4.4 minus A4.7 and A4.3 minus A4.7 with substitutions gives:

$$\begin{pmatrix} f_L E_L + f_K(A_{KL} - A_{RL}) - E_X f_L & f_K(A_{KR} - A_{RR}) - E_X f_R \\ E_L - A_{LL} + A_{RL} & A_{RR} - A_{LR} \end{pmatrix}$$

$$\times \begin{pmatrix} \dfrac{dW}{dT_X} \\[2ex] \dfrac{dP_R}{dT_X} \end{pmatrix} = \begin{pmatrix} E_X \\[2ex] 0 \end{pmatrix}$$

$$(A4.1a)$$

$$(A4.4a)$$

Solving gives.

$$D = [f_L E_L + f_K(A_{KL} - A_{RL}) - E_X f_L](A_{RR} - A_{LR})$$
$$\quad - (E_L - A_{LL} - A_{RL})[f_K(A_{KK} - A_{RR}) - E_X f_R]$$

$$\frac{dW}{dT_X} = \frac{E_X(A_{RR} - A_{LR})}{D} \leq 0$$

$$\frac{dP_R}{dT_X} = \frac{- E_X(E_L - A_{LL} + A_{RL})}{D} \leq 0$$

Tax on Mineral Property Rights

An analogous process performed on equations 4.1 to 4.7 (see text of chapter) and A4.1 to A4.7 and all other taxes equal zero gives:

$$\begin{pmatrix} E_X f_L - f_L E_L - f_K(A_{KK} - A_{RL}) & E_X f_R - f_K(A_{KK} - A_{RR}) \\ E_L - A_{LL} + A_{RL} & A_{RR} - A_{LR} \end{pmatrix}$$

$$\times \begin{pmatrix} \dfrac{dW}{dT_R} \\[2ex] \dfrac{dT}{dT_R} \end{pmatrix} = \begin{pmatrix} f_K(A_{KR} - A_{RR}) - E_X f_R \\[2ex] A_{RR} - A_{LR} \end{pmatrix}$$

Solving gives

$$D = [E_x f_L - f_L E_L - f_K(A_{KK} - A_{RL})](A_{LR} - A_{RR})$$
$$- (A_{LL} - A_{RL} - E_L)[E_x f_R - f_K(A_{KR} - A_{RR})]$$

$$\frac{dW}{dT_R} = \frac{(A_{LR} - A_{RR})[f_K(A_{KR} - A_{RR}) - E_x f_R]}{D}$$

$$- \frac{(A_{RR} - A_{LR})[E_x f_R - f_K(A_{KR} - A_{RR})]}{D}$$

$$\frac{dP_R}{dT_R} = \frac{(A_{RR} - A_{LR})[E_x f_L - f_L E_L - f_K(A_{KL} - A_{RL})]}{D}$$

$$- \frac{(A_{LL} - A_{RL} - E_L)[f_K(A_{KR} - A_{RR}) - E_x f_R]}{D}$$

Tax on Capital

A tax on capital is solved in an analogous manner:

$$D = [f_L E_L + f_K(A_{RL} - A_{RL}) - E_x f_L](A_{LR} - A_{RR})$$
$$- (A_{LL} - A_{RL} - E_L)[f_K(A_{KR} - A_{RR}) - E_x f_R]$$

$$\frac{dW}{dT_K} = \frac{[E_x f_K - f_K(A_{KK} - A_{RK})](A_{LR} - A_{RR})}{D}$$

$$- \frac{(-A_{RK} + A_{LK})[f_K(A_{KR} - A_{RR}) - E_x f_R]}{D}$$

$$\frac{dP_R}{dT_K} = \frac{[f_L E_L + f_K(A_{KR} - A_{RL}) - E_x f_L[(A_{RK} - A_{LK})}{D}$$

$$- \frac{(A_{LL} - A_{RL} - E_L)[E_x f_K - f_K(A_{KK} - A_{RK})]}{D}$$

Tax on Labor

And a tax on labor gives

$$D = [E_x f_L - f_L E_L - f_K(A_{KL} - A_{RL})](A_{LR} - A_{RR})$$
$$- (A_{LL} - A_{RL} - E_L)[E_x f_L - f_K(A_{KR} - A_{RR})]$$

$$\frac{dW}{dT_L} = \frac{[f_K(A_{KL} - A_{RL}) - E_X f_L](A_{LR} - A_{KR})}{D}$$

$$- \frac{(A_{KL} - A_{LL})[E_X f_L - f_K(A_{KR} - A_{RR})]}{D}$$

$$\frac{dP_R}{dT_L} = \frac{[E_X f_L - f_L E_X - f_K(A_{KL} - A_{RL})](A_{KL} - A_{RL})}{D}$$

$$- \frac{(A_{LL} - A_{RL} - E_L)[f_K(A_{KL} - A_{RL}) - E_X f_L]}{D}$$

5

Nominal and Effective Tax Rates

Nonrenewable-resource tax policies carried out by national, state, and local governments are diverse in the tax bases utilized, nominal tax rates legislated, and the manner in which taxes are administered and enforced. In order to narrow our scope, we focus on national, state, and local taxes in the United States. However, tax practices vary among nations. For example, Gillis and Bucovetsky (1978) made a comparison of tax policies in Bolivia, Chile, and Indonesia and their effect on exploration and investment. DeYoung (1978) of the U.S. Geological Survey makes a similar review of tax policies in the United States, Canada, and Australia.

In this chapter, we first survey estimates of the effective U.S. corporate income tax rate and then take up the matter of state taxes on nonrenewable-resource extraction. Taxes falling on copper and coal extraction are reviewed in detail and then converted to an effective tax rate output, so that complex state tax structures may be compared.

In the case of U.S. federal tax policy, the primary tax affecting the extractive industries is the corporate income tax. However, a number of federal taxes affect particular resource industries. A summary of the federal tax structure on the coal (table 5–10) and copper (table 5–27) industries is included here to illustrate these differences. Resources falling under the Mine Leasing Act of 1920 generate royalty income for the federal government, part of which is shared with state governments.

The tax treatment of exploration and development expenses and depletion is reviewed in chapter 3. It is shown that the tax-shelter effect of the percentage-depletion allowance is governed by the corporate tax rate and limited by the maximum of 50 percent of net income. Percentage depletion reduces the effective corporate income tax rate (t) by providing what is, in effect, a subsidy on output or a negative severance tax. The subsidy was shown to equal t times the depletion allowance rate divided by $1.0 - t$, where t is the nominal tax rate (.46 for the highest bracket and .48 prior to 1978). The percentage-depletion rates are established by law for each industry and are reported in table 5–1. Note that the percentage-depletion option was eliminated for the major oil corporations in 1975 and was reduced from 27.5 to 22 percent for the independent producers. The subsidy is in effect until the limitation of 50 percent of net income is reached, and after this point the shelter is ineffective.

Cost depletion must be used by the major gas and oil producers and is an option for other extractive firms. Cost depletion per unit of output is computed by dividing the invested capital in developing the reserve at discovery by the estimated total recoverable reserves and allocating this cost

153

Table 5–1
Percentage-Depletion Rates

Mineral Product	Rate (%)
Oil and gas (independent producers)	22
	20 (1981)
	18 (1982)
	16 (1983)
	15 (1984)
Sulfur	23
Bauxite	23
Lead and zinc	23
Uranium	22
Iron	15
Copper	15
Limestone and phosphate rock	15
Precious metals	15
Bituminous and anthracite coal	10
Stone, sand, and gravel	5

per unit as the resource is extracted. A third alternative for value at discovery is to estimate total royalty, rental, and bonus payments which will be paid over the life of the reserve. This cost or value per unit of reserves is multiplied times annual production to give the total cost-depletion deduction. Cost depletion is less attractive than percentage depletion because the value of the reserve and thus the tax deduction are fixed at the date of discovery in the former, whereas a percentage of gross current market value is used in the latter. When resource prices are rising, percentage depletion is more attractive, particularly when coupled with the tax treatment of exploration and development costs.

The effect of expensing exploration and development costs, although firms may amortize these expenses, is a second tax advantage granted to extractive firms. Its effect is to defer tax liabilities and thus grant the firm an interest-free loan from the government. The treatment of intangible drilling expenses was tightened up in the 1976 Tax Reform Act, although the expensing provision is still available. Capital-gains provisions in the corporate income tax also can be used by resource firms to reduce taxes. The label *tax preference* or *tax expenditure* is used by government agencies to refer to tax provisions which reduce effective tax rates.

In 1963 the U.S. Treasury analyzed the 1960 tax returns of domestic corporations declaring depletion deductions and found that percentage depletion was declared by roughly three-quarters of these, that the limitation of 50 percent of net income became effective for about 20 percent of this group, and that cost depletion accounted for one-fortieth of the firms declaring depletion deductions. A more recent study by the IRS of 1971 corporate tax returns reveals the importance of all the tax preferences to extractive firms. These tax benefits are reported in table 5–2 as a percentage of net income. The returns show that extractive industries are able to shelter roughly one-half of their profits and metal mining was able to shelter more than 100 percent, compared to 10.3 percent for the average of all industries. As a result, effective corporate income taxes fell well below the nominal rate of 48 percent for all industries. It is easy to see that extractive industries benefit to a greater extent than other industries, but the 50 percent net-income limitation for depletion keeps the effective tax rates from falling precipitously below normal tax rates.

These data underestimate the full extent of tax-preference items. Much exploratory oil and gas drilling is conducted by limited partnerships which pass on tax savings to wealthy investors who deduct them from their individual income taxes and shelter other sources of personal income. The effects of limited partnerships and subchapter S corporations, which pay no corporate income taxes but pass all income (positive or negative) on to their shareholders who are then solely liable for personal income taxes, are unknown. However, the effect is most likely substantial. For example, oil and gas tax shelters normally provide losses in the first year of 80 to 100 percent of an investor's initial investment. For those with a 60 percent marginal tax rate, a 100 percent loss the first year means that the investor gets back 60 percent of the initial investment. The losses diminish over time, and successful exploration yields positive income in the future.

Brannon (1975a) estimates that the 22 percent depletion rate on oil and gas, the 10 percent rate on coal, and the 22 percent rate on uranium, when coupled with the maximum depletion limitation of 50 percent of net income

Table 5–2
Effect of Tax Provisions

Industry	Ratio of Total Tax-Preference Items[a] to Net Income	Income Tax as Percentage of Net Income before Foreign Tax and Some Investment Tax Credits
All industries	10.3	43.7
Mining	62.5	35.2
Metal mining	107.2	Not disclosed
Coal mining	57.1	39.2
Crude petroleum and natural gas	57.9	33.1
Nonmetallic mining (ex fuds)	43.9	39.9

Source: U.S. Department of Treasury, Internal Revenue Service, *Statistics of Income 1971 Corporation Income Tax Returns* (Washington: GPO, 1976).

[a]Items include capital-gains provisions, amortization depreciation, depletion, deferral, accelerated depreciation, certified pollution-control facility, reserves for losses on bad debt (financial inst), depletion, capital gains.

and the expensing of exploration and development costs, result in total tax benefits of 21, 4.4, and 9.3 percent, respectively, for the oil and gas, coal, and uranium extractive firms in 1975. He also estimates the total tax benefits as a percentage of delivered price of energy for utility consumers as 12.9 percent for oil, 11.7 percent for natural gas, 3.4 percent for coal, 1.4 percent for gas conversion from coal, 2.8 percent for uranium, and 4.5 percent for shale oil. These differential tax benefits and resulting net-price differentials distort economic choices and, in the critical case of energy, have lulled consumers into believing energy is cheaper than its true social cost. Government price controls on natural gas and "old" oil also have magnified these price and resource distortions. If consumers were faced with prices which reflect the true costs of energy, it is likely that more conservation and energy efficiency would be achieved. This is particularly important because investment decisions made today affect the efficiency of energy use for the next five to fifty years. This occurs because vehicles, buildings, and plant and equipment possess useful lives which vary over this range, and once they are constructed, it is far moe difficult and expensive to modify their technology in order to improve energy efficiency.

The effect of special tax provisions (note that percentage depletion varies by industry) and others, such as the treatment of coal royalties as capital gains and the investment tax credit on the effective corporate income tax rate are substantial. Siegfried (1974) uses 1963 IRS statistics on corporate income and a somewhat different methodology from the studies cited above to estimate the effective corporate tax rates by industry. When depletion and investment tax credit are considered, his results are shown in table 5–3.

The most recent study of effective federal corporate income tax rates was

Table 5–3
Effective Corporate Income Tax Rates by Industry, 1963
(percent)

Industry	Effective Tax Rate with Depletion	Effective Tax Rate with Depletion and Investment Tax Credit
Iron ore	1.8	0.0
Copper, lead, zinc, gold, silver ore	32.8	31.1
Miscellaneous metal mining	−20.9	−21.1
Coal mining	22.6	
Crude petroleum, natural gas, and liquids	24.3	18.2
Miscellaneous metallic minerals except fuels	18.9	24.1
All corporate tax returns	41.4	46.2

Source: John J. Siegfried, "Effective Average U.S. Corporation Income Tax Rates," *National Tax Journal* 27 (June 1974). Reprinted with permission.

conducted by the U.S. Treasury. They estimated the effective tax rates for extractive firms from individual corporate tax returns for 1972 (U.S. Department of the Treasury 1978). Domestic corporate revenues and taxes were separated from worldwide revenues and taxes in order to give a picture of U.S. operations. This procedure is especially important in the oil industry because of the seven multinational giants. Efforts were made to eliminate spurious and misleading data from the sample, such as data from corporations not subject to tax [subchapter S corporations and domestic international sales corporations (DISC), both of which file returns but are taxed through other legal entities], those corporations without income, and corporate entities which have tax exemptions. Loss carry-forwards were added to taxable income in order to give a more accurate picture of tax liabilities for the single year of 1972. The effective tax rate on U.S.-source income broken down for all corporations and corporations with under $1 billion in assets is shown in table 5–4. Extractive industries are ranked among the corporations paying the lowest tax rates—coal mining (eighteenth in rank), petroleum and natural gas (seventeenth in rank), mining not classified elsewhere (miscellaneous metal mining and nonmetallic minerals) (sixteenth in rank), nonferrous primary metals (twelfth in rank), and ferrous primary metals (eighth in rank). The industry categories include mining activities as well as processeing and manufacturing the basic metal product. Table 5–4 reveals that only the banking industry enjoys greater tax benefits than coal mining and domestic petroleum and natural-gas production. It should be noted that when corporations with assets under $1 million are compared, the differentials in effective tax rates among extractive firms and other industries are more modest.

These estimated large tax differentials are supported by data compiled from Form 10-K reports made by publicly held corporations to the Securities and Exchange Commission (SEC). Anderson (1977) computes the effective tax rate as the sum of federal income taxes, state income taxes, and foreign income taxes with respect to corporate income (the income definition excludes state and local severance, franchise, and property taxes). His estimate of effective domestic tax rates is made after subtracting foreign taxes and foreign income. These estimates are reported in table 5–5 for twelve mining companies.

Anderson also collected data from Form 10-K reports on fifteen representative manufacturing concerns. He concludes that the mining industry's tax burden is on an average only two-thirds that for manufacturing. If all state and local taxes were included as tax liabilities, Anderson (1977) estimates that this differential benefit would be less—on the order of 15 percent.

Lower effective tax rates for extractive firms produce two interrelated effects. One is to lower these firms' cost of capital, which causes additional

Table 5–4
Effective Tax Rates on U.S.-Source Income

Industry	All Corporations		Corporations with under $1 Million of Assets	
	Industry Rank	Effective Tax Rate (%)	Effective Tax Rate (%)	Corporations Included (%)
Manufacturing not classified elsewhere	1	42.0	32.3	86.2
Paper and allied products	2	38.4	32.4	79.5
Credit dealers, brokers, insurance agents	3	38.3	29.0	95.1
Wholesale and retail trade	4	38.0	32.0	94.9
Communications	5	36.1	27.6	79.6
Electric, gas, and sanitary services	6	35.3	28.0	89.8
Lumber and wood products (nonfurniture)	7	34.6	32.2	87.5
Primary metals: ferrous	8	33.7	33.2	75.8
Contract construction	9	33.4	28.4	95.2
Services	10	31.6	26.5	97.6
Transportation	11	30.1	26.6	94.3
Primary metals: nonferrous	12	29.4	25.4	72.6
Real estate	13	28.9	26.2	94.7
Agriculture, forestry, and fisheries	14	28.1	23.4	94.4
Unclassifiable businesses	15	27.7	26.1	98.8
Mining not classified elsewhere	16	25.6	22.4	76.9
Petroleum and natural gas	17	24.7	23.8	85.5
Coal mining	18	19.4	24.7	84.3
Banking	19	18.6	26.8	6.7

Source: U.S. Department of the Treasury, Office of Tax Analysis, *Effective Income Tax Rates Paid by United States Corporations in 1972* (Washington: GPO, 1978).

Note: Included are all corporations with income and corporations with under $1 million in assets, by industry, in 1972.

Table 5–5
Taxation of the Mining Industry, 1973
(percent)

Firm	Tax Burden			
	Effective Domestic Tax Rate	*Effective Rate Reported to SEC*	*Effective World Tax Rate*	*Federal Income Tax Rate*
Amax	25.6	28.8	37.4	20.1
American Smelting and Refining	25.6	17.2	29.6	13.0
Anaconda	32.6	24.0	36.3	18.6
Diamond Crystal Salt	32.8	42.6	33.8	29.0
Foote Mineral	47.9	25.0	47.9	21.7
Homestake Mining	31.2	29.0	31.2	28.9
Freeport Minerals	+	20.0	+	39.2
Kennecott Copper[a]	+	26.1	45.3	+
Phelps Dodge	43.2	37.3	43.6	35.5
Pittston	42.2	21.5	42.2	14.0
St. Joe Minerals	+	34.3	+	30.6
Westmoreland Coal	31.3	0.	31.3	0.0
Average	34.7	25.5	37.9	22.8

Source: Robert C. Anderson, "Taxes, Exploration and Resources," in *Nonrenewable Resource Taxation in the Western States*, Lincoln Institute of Land Policy Monograph (Cambridge, Mass., 1977). Reprinted with permission.
[a]The two depletion figures for Kennecott refer to minerals and coal, respectively; total depletion benefit is 22.4.
+ denotes unavailable.

investment flowing into the industry—this argument can be traced to Harberger (1955) and is explained in chapter 4. The other effect is that added investment, which is induced by a lower effective tax rate, results in higher output and lower prices which, although advantageous to consumers, results in inefficient use of resources. Let us use a simple numerical example contrived by Anderson (1977). Suppose that tax advantages reduce the effective tax rate for extractive firms to one-half that faced by manufacturing firms. In 1969 the average corporation earned 14.8 percent return on invested capital, which represents a 7.7 percent after-tax return at a 48 percent corporate income tax rate. If extractive firms pay half this rate, they have to earn only 10.1 percent before taxes, compared with 14.8 percent for manufacturing to earn the same after-tax return. This, in effect, lowers the cost of capital to firms extracting nonrenewable resources. If every $1 of capital produces $2 of output yearly, a "ballpark" number representative of the U.S. economy, then the supply curve of the extractive firm is shifted 9.4 percent by the tax advantage—that is, the price of output is lower, depending on the responsiveness of the demand curve, than it would have been without tax preferences.

In a more sophisticated rendition of this model, Harberger (1955, pp. 448–449) estimates how much an investor would be willing to pay for $1 pretax investment in six extractive industries. The investor is willing to pay a premium because she or he is comparing after-tax returns in the resource industries which enjoy tax benefits to an after-tax dollar from returns in other industries which are assumed to enjoy no such advantages. The results are based on conditions prevailing in 1955: petroleum, $1.95; sulfur, $2.12; iron, $2.13; copper, $1.96; lead and zinc, $2.27; and coal, $2.30.

Agria (1969) evaluates the special provisions already discussed in addition to including the leasing, developing, and sale of gas and oil leases before production to be treated as capital-gains income. She estimates that these provisions, along with the disadvantage of state severance taxes on extraction (which she estimates are small), result in 1.5 multiple advantage for investment in gas and oil production and a 1.8 multiple in coal compared to activities having no such tax advantages. Harberger's assault on the depletion allowance as discriminatory and fostering economic inefficiency as a result of distorted investment decisions did not go unnoticed. Agria (1969), Steiner (1959), and a host of other economists have supported his position that the tax preferences result in nonneutralities and should be eliminated.

McDonald (1962, 1963, 1967, 1970) has made a career of defending the depletion provision by looking at how these tax provisions affect consumer prices. His defense is basically a tax-shifting argument. He holds that if the corporate income tax is fully passed on to consumers (that is, if firms increase price by the full amount of the net income tax, which incidently the empirical model presented in chapter 3 does not support consistently), then a corporate income tax rate uniform on all industries results in nonuniform tax shifting. Thus, prices to consumers increase by different proportions for different industries. If this is the case, the uniform tax rate is nonneutral. He argues that neutral treatment requires a tax benefit for the extractive industries because they are more capital-intensive (require more capital per unit of output) than manufacturing enterprises. Capital-intensive industries must earn more profit per dollar of sale to generate equal return on capital and thus are burdened by heavier effective tax rates, measured by changes in consumer prices (assuming all equity financing) when corporate tax rates are equal. Full forward shifting of the corporate income tax means prices of output are distorted when capital intensities in the production process are not uniform. McDonald estimates that the rate of return for domestic firms was 10.3 percent in manufacturing versus 14.5 percent in oil and gas production in the 1950s, with this difference representing a risk premium. The respective capital intensities of .523 and 1.43 are used to compute a "neutralizing exemption" equal to 14.5 percent for oil and gas production. This income exemption is necessary in order for both sectors to be earning an equivalent risk-adjusted rate of return. McDonald has acknowledged that the 14.5 percent deduction from gross revenue is less than the 23 percent effective depletion allowance that he estimates as existing in the 1960s. However, at other times he has estimated the required adjusting deduction to be 21.5 percent and thus argues that the depletion allowance existing before 1969 was proper.

McDonald's work has been subjected to extensive criticism by Eldridge (1950, 1962), Steiner (1959), Wright (1976), and others. These criticisms led McDonald to recant his position that percentage depletion is fully justified to argue for a reduced depletion rate (a 14.5 percent rate) and to acknowledge numerous restrictions on his method. McDonald (1967, 1970) also recommends a percentage-depletion rate dependent on time and not on the rate of extraction and notes that landowners (lessors) may enjoy most of the benefits of these special tax provisions so that economic rent thereby escapes taxation. Gaffney (1967) argues that the tax benefits of depletion allowances accrue only to successful firms and thus are disadvantageous to marginal firms. He agrees that the depletion provision allows the economic rents on superior lands to escape taxation. McDonald ignores distortions in investment induced by tax preferences which Harberger and others have emphasized. Since full tax shifting in the long run occurs in only some of the extractive industries and does not necessarily fall on consumers only, McDonald's position that percentage depletion is required for tax neutrality remains a relic gathering dust on the shelves of academia.

Another way of looking at the possible benefits of corporate tax advantages is to estimate how much additional output or reserves are produced in response to the forgone tax revenues engendered by the reallocation of investment into the extractive sector. The widely criticized CONSAD report (1969) concludes that $1.48 billion of taxes forgone in the 1960s are a result of special tax provisions produced an increase of 150 million barrels in oil supplies at a cost of $9.70 per barrel. Cox and Wright (1975) analyze limitations of the CONSAD report and evaluate the effectiveness of tax advantages by posing this question: How much would it cost to sign contracts directly with firms to produce the same output as the depletion and expensing of exploration and development expense tax treatment provides? For percentage depletion it would be $75 million cheaper to contract directly for additional production, but in the case of intangible drilling expense, contracting would cost from $100 million to $125 million more than the tax preference.

Stiglitz (1975a) is also concerned about the nonneutrality of federal corporate income taxes and the resulting distortion in investment. His analysis refutes McDonald. He derives and optimal differential tax for oil versus manufacturing industries based on the objective of minimizing the efficiency loss, "deadweight loss," engendered by percentage-depletion tax treatment. He estimates the supply elasticity of oil as 1.0 (a 1 percent increase in the price of oil increases production 1 percent) and price elasticity of supply to lie between 2.0 and infinity for manufacturing. The demand elasticity (the percentage fall in consumption divided by a percentage increase in price) for oil is also low (estimated at .5) compared to manufacturing (estimated at 2.0 to 5.0). Oil's capital intensity, as discussed above, is high (.2 to .45) as opposed to that found in manufacturing (.2 to .3). Based on these rough estimates, Stiglitz concludes that oil industry's tax rate should be 3.0 to 82.5 times the manufacturing tax rate in order to minimize tax distortions. Wright (1976) reviews both Stiglitz's estimate of optimal tax rates and McDonald's models of tax distortions of output prices and develops his own hybrid model. He concludes that discrimination does exist and results in approximately 20 to 30 percent more oil and gas investment than would otherwise be the case.

Taxation and Regulation

Taxes are but one distorting influence to investment and production of natural resources. The other major source is direct regulation. And the interaction of tax and regulatory policies makes the picture more complex than either does in isolation. Since this book is about taxation and not about regulation, these interactions are only touched on.

One example has been alluded to previously—the Federal Power Commission's regulation of natural gas transported interstate. In the 1950s through the 1970s the avowed policy was to keep gas prices low in order to benefit consumers and consuming states. However, the rising costs of all sources of energy in the 1970s made it more lucrative for producing companies to keep gas in unregulated intrastate markets and made it less profitable to search for new supplies. These economic realities were transformed into political reality in the Natural Gas Policy Act of 1978. That act established a complex set of rules which allow for the gradual deregulation of natural-gas prices, sold both intra- and interstate, under the auspices of the Federal Energy Regulatory Commission (FERC). The price of "new" (post-April 1977) onshore and offshore natural gas is to rise at a rate roughly equivalent to the rate of inflation. *New gas* is defined as coming from offshore leases beginning after April 1977 and from onshore wells

1. 2.5 miles or more from the nearest marker well
2. Any new well at least 1,000 feet deeper than a marker well closer than 2.5 miles
3. New onshore reservoirs

4. Stripper well production (less than 60,000 cubic feet per day)
5. Alaska pipeline gas

Full price decontrol of new gas will become effective in 1985. A further option is full price decontrol in 1979 for *high-cost gas*, which includes:

1. Wells begun after February 1977 and production obtained from deeper than 18,000 feet
2. Geopressured brine
3. Occluded gas from coal seams
4. Devonian shale
5. Other high-cost conditions on which the FERC rules

In early 1980 this provision meant that high-cost gas sold for $4 per 1,000 cubic feet (Mcf) as opposed to $2.30 for new regulated gas. The prospects stimulated investment in deep wells. Amoco Production Co. earmarked more than one-third of its 1980 exploration budget for deep wells, and Conoco Inc. budgeted about $100 million. If a producer elects to accept the high-cost gas price, he or she must give up all credits, exemptions, and deductions under the federal income tax. State severance taxes are allowed to be added to the price for all categories of gas and passed through to consumers. The phased decontrol process is complex, and old gas will remain controlled. However, the legislation is a needed effort to remove regulations which hinder production and distort resource use.

In 1980 Congress enacted a phased price decontrol for oil production and simultaneously decided to tax away much of the "windfall profits" engendered by the act. The tax is expected to raise $228 billion between 1980 and 1990. Oil is defined into three catgories, each having its own tax rate:

1. Wells producing prior to 1979 are called *tier one*, and the difference between the base price (an average of $12.81 per barrel) and current market price is taxed at 70 or 50 percent for the first 1,000 barrels per day for independents.
2. Stripper wells (low-production wells) and wells in the national petroleum reserves are *tier two*, and the difference between the base price ($15.20 per barrel) and current market price is taxed at 60 or 30 percent for the first 1,000 barrels per day for independents.
3. New (post-1978) oil, heavy oil, and tertiary recovery oil is called *tier three*, and the difference between the base price ($16.55 per barrel) and current market price is taxed at 30 percent for all production.

In all cases, the base price is increased at a rate commensurate with inflation. Alaskan oil, and oil produced from state and Indian owned lands are exempt. Furthermore, state severance taxes enacted prior to March 31, 1979, are deductible from this tax (up to a 15 percent rate) as long as the tax is uniform on the entire price of each barrel of oil. The windfall-profits tax is deductible from the corporate income tax, which lowers its effective rate, particularly for the independents who continue to use percentage depletion. The major oil companies are expected to supply about 90 percent of the tax revenue. Price decontrol and the provision for severance-tax deductibility and exemption of production on state and Indian lands are the factors which account for the boom in state revenues reported in chapter 1.

Expected windfall-profits tax includes an array of tax credits which are as complex and controversial as the tax itself. The intent of these credits is to award to nonconventional and renewable sources of energy some of the tax benefits which depletable resources have long enjoyed. The revenue losses from 1980–1990 are estimated at $5.1 billion. Solar advocates and others have maintained that price regulation and tax preferences for oil and gas have been sufficient to top the economic balance in favor of conventional sources and energy gluttony and away from conservation and nonconventional sources. The tax credits enacted in 1980 vary from 10 to 15 percent for biomass equipment to convert waste into fuel, cogeneration, coal gasification, geothermal and ocean thermal, solar, wind, small-scale hydroelectric, oil shale, alcohol engine fuel, and intercity bus equipment. The act includes expanded tax credits for homeowners for insulation, conservation, solar, wind, and geothermal energy equipment.

It is clear than when extractive industries enjoy special tax advantages, investment decisions are distorted. Nonconventional-energy advocates lobbied successfully for similar special treatment. Virgin nonrenewable resources are cheaper under the existing tax code than they might otherwise be, which makes recycling less attractive and introduces further distortions (Anderson and Spiegelman, 1977). The preferable way to correct all tax-induced misallocations is to pursue an objecive of tax neutrality. The elimination of percentage-depletion provisions for large petroleum producers in 1969 represents a triumph for economists preaching tax neutrality. The U.S. Treasury instituted limitations in 1976 on the expensing of exploration and development costs for oil producers which reduce this tax distortion. Price deregulation is a step in the right direction. The question is unresolved as to what the effective tax rate will become after these changes have worked their way through the economy. In order to answer the question of what a neutral tax rate is for extractive industries vis-à-vis manufacturing, the question posed in chapter 4 must be resolved: What is the final incidence (burden) of the corporate income tax for industry categories? Only then will it be possible to compute a theoretically and empirically correct depletion allowance and tax rates which are neutral. This question will surely be addressed by researchers as data revealing the result of recent changes in federal tax laws becomes available. We now consider nominal and effective tax rates imposed by state and local governments.

State Taxation

States and local governments in which nonrenewable resources have been discovered and extracted levy taxes directed specifically at resource firms as well as applying taxes that affect all business enterprise. Discriminatory taxes are those where the tax base is the nonrenewable resource or where a special tax rate is applied to extractive firms. These taxes include severance, property, and net-proceeds taxes. We ignore taxes levied on payrolls since they are primarily nondiscriminatory (small distortions are introduced because extractive firms utilize less labor-intensive technologies than manufacturing) and are not considered in this book. The effect of state corporate income taxes on extractive firms resembles that of the federal tax. The effect of state tax differentials is considered only in the detailed analysis of taxes falling on coal and copper producers.

Unfortunately, not a great deal is known about effective tax rates levied by states possessing nonrenewable resources, and only limited published data are available on nominal tax rates and bases. The complexity of state tax laws means that a direct comparison of nominal tax rates is virtually impossible. Since taxes enter both fixed and variable costs in many complex ways, the firm generally compares the after-tax profits expected from specific projects when evaluating alternative locations for possible development. Analysts perform detailed calculations of all extraction, processing, and tax costs so that management of resource firms possesses detailed knowledge of tax-rate effects on net accounting rates of return. Regrettably, their data on historical costs in existing facilities and expected future costs in proposed projects and tax effects are not available. One result is that policy analysts and state legislators are usually forced to accept on faith the analyses of the tax structure made available by industry lobbyists.

An alternative source of information on existing nominal and effective tax rates is state and local government data on tax collections. Summaries of existing tax laws are available from the individual states directly and are compiled by private firms such as the Commerce Clearing House. Major defects in these data are that the descriptions of tax laws are incomplete and the legislation itself is not specific enough to know how the tax is actually administered. Little information is available concerning actual tax administration. Tax rates and administrative practices of local units of government are not compiled exhaustively by anyone.

A University of New Mexico research team under my direction, collected tax statutes from all coal- and copper-producing states. The statutes were reviewed by second- and third-year law students and compiled into summarized form presented later in this chapter. We asked the basic questions: What is the tax base and what is the tax rate? The tax statutes, court rulings, and administrative practice proved to be so obscure or unknown and dependent on laws stretching back for several decades that mating published information with professional opinion was required to answer these two basic questions adequately. This required telephone calls to tax bureaucrats. We learned that even they had trouble understanding and administering the laws and court rulings and that the actual practice of tax administration often deviated significantly from what appeared to be the meaning and intent of the tax statute. It is clear that wide discretion is left to administrative and enforcement agencies in carrying out state tax laws and in enforcing the many provisions. The conclusion is that collecting information from state tax statutes does not necessarily provide good data about actual tax liabilities and thus is inadequate for estimating the effective tax rates on the extractive industries.

I also attempted to gather data on actual collections for state taxes on extractive activities segregated by resource and tax. The attempt proved futile, for the majority of states queried did not maintain information on tax collections broken down by industry, or else they considered the information to be privileged. Had this information been available, it would have been a straightforward matter to divide actual tax collections by the appropriate tax base to determine effective tax rates. In lieu of this procedure, data on nominal tax rates were transformed to estimates of effective tax rates for the coal and copper industries by simulating costs and technology. This exercise is reported later in this chapter after nominal tax rates are described.

The taxes levied by states whose rates or structures apply differentially to extractive firms may be classified into corporate income taxes, property taxes, and taxes on output. We first review these from a broad perspective and then focus on taxes levied on coal and copper producers.

Nominal State Tax Rates

Income Taxes. Most states levy a tax on corporate net income. However, the methods of assigning the in-state portion of income for corporations operating in more than one state and how the federal income tax base is utilized make this tax base vary among states. In some cases, states utilize the federal base but do not allow the percentage-depletion provision (this provision does introduce discrimination) or a deduction for federal income taxes paid (both of which would reduce the base substantially). The practice of six Western states in this respect is shown in table 5–6. There is a long history of litigation over the methods used by states to allocate national

Table 5–6
Corporate Income Taxation in Six Western States

State	Maximum Rate (%)	Federal Income Tax Used as Tax Base?	Allow Deduction of Minerals Depletion?	Federal Income Tax Deductible?
Arizona	10.5	No	No	Yes
Colorado	5.0	Yes	Yes	No
Idaho	6.5	Yes	Yes	No
Montana	6.25	Yes	Yes	Yes (limited)
New Mexico	5.0	Yes	Yes	No
Utah	6.0	No	No	Yes

Source: G. Laing, "An Analysis of the Effects of State Taxation on the Mining Industry in the Rocky Mountain States" (Master's thesis, Colorado School of Mines, 1976). Reprinted with permission.

corporate net income to each state. The techniques are varied and subject to interpretation under the commerce clause of the Constitution. In 1980 the Supreme Court held that a reasonable and consistent method is acceptable even though it is chosen to maximize tax revenues to the state.

Property Taxes. While the property tax falls on virtually all real property, methods of assessment and tax rates are unique to each separate taxing jurisdiction. The practices, as reported by Laing (1976), of eight Western states for valuing various resources in situ (mineral estate) and improvements necessary for extraction are shown in table 5–7. Three methods of valuation for the mineral in situ are found within these eight states: capitalized earnings (which are based on capitalized projected accounting profits), a multiple of gross sales revenues, and a multiple of gross revenues minus certain operating expenses (net proceeds). The notation *unit valuation* refers to the practice of valuing an entire operation such as a mine, pipeline, or railroad and then allocating it in piecemeal fashion. Portions of estimated value are assigned to taxing jurisdictions (counties, school districts, municipalities, and special districts) by a formula based on factors such as miles of track, location of sales, and so forth. An intriguing problem occurs in Colorado, for example, where an Amax molybdenum mine is located in one county but a tunnel constructed for ore extraction and handling waste material and tailings is located in another county. The practice in many Western states is for state governments to assess mines, or at least to provide advisory appraisals to local jurisdictions. This practice is less prevalent in the Midwest and East where "self-assessment" and negotiated assessments frequently occur. The third assessing method is to appraise the replacement cost less depreciation (RCLD) of mining equipment and other identifiable improvements.

Appraisals carried out by state and local taxing jurisdictions are often the result of "self-assessment" or negotiation. This practice stems from the fact that tax assessors are unwilling or unable to assess extractive sites because their staffs often lack expertise and because the mineowners and mine operators tend to have considerable political influence. For example, Gaventa (1975) looked into property taxation of coal mines in eastern Appalachia. Five counties in the northeastern portion of Tennessee account for about 80 percent of the state's coal reserves, and within these counties nine corporations own 80 percent of the coal. Land with coal reserves is worth from $50 to $150 per acre, but Gaventa found assessments in this area to average $25 to $30 per acre even though state law calls for assessments to be at 90 percent of market value for land (mineral rights) and improvements. The net result is that the total assessed value for the nine corporations is less than 4 percent of the total assessed property value in this area. Gaventa found many of the same practices in the other Appalachian coal states of Kentucky, West Virginia, and Virginia.

Certain states attempt to value resources (mineral property rights) as accurately as possible by using the capitalized-income-stream approach. For example, the Arizona Department of Revenue conducts appraisals utilizing the Hoskold capitalization formula based on extensive data reported by each firm for each mine site. Arizona applies a 60 percent assessment rate to extractive industries but lower rates to other commercial enterprises and residential land and improvements. Because of the many uncertainties associated with estimating future net-income streams to determine the property-tax base and because the tax is a fixed cost during the lean years and for three years after a mine shutdown, the Arizona property tax has been subject to a barrage of court battles.

While the Arizona courts have held that the Hoskold method of capitalizing income shall be followed, the recent litigation has centered on the accuracy of estimates of future prices, mine life and reserves, and estimated costs. Because of problems in administering the property-tax base established as the net present value of a site, most states have substituted simpler methods. It has also been reported to me that Colorado appraisers do not actually appraise based on capitalized income as reported in table 5–7. A gross-revenues or net-proceeds tax basis, as reported for seven of the eight states in table 5–7, in effect converts the property tax to an output-based tax. The data in table 5–7 also reveal that resources are treated to different assessment practices defined in tax legislation or incorporated into procedural manuals issued by state tax departments.

Anderson (1977) examined firms which report property taxes as a separate expense in Form 10-K reports made by corporations in 1973 to the SEC. These reports include information on the sources and uses of funds by

Table 5-7
Property Taxation of Western Mining Operations

State	Mineral	Valuation Method — Mineral Estate	Valuation Method — Improvements	Basis	Assessment Rate (%)
Arizona	All	Capitalized earnings unit valuation		Market value	60
Colorado	Coal, Industrial minerals	Capitalized earnings unit valuation		Market value	30
	Metal mines	Gross or net proceeds	Appraisal	Previous year	100
Idaho	All	Net proceeds	Appraisal	Book value	30
				Previous year	100
				RCLD	Variable
Montana	All except coal	Net Proceeds	Appraisal	5-year average	100
	Surface coal	Gross proceeds		Previous year	18
	Underground coal	Gross proceeds		Previous year	13.3
Nevada	All	Net proceeds	Appraisal	Previous year	12 / 35
New Mexico	All except potash	Net proceeds	Appraisal	5-year average	35 / 100
	Potash	Gross proceeds	Appraisal	50% of previous year	33⅓
Utah	Metal mines	Net proceeds	Appraisal	3-year average × 2	33⅓ / 33⅓
	All other	Net proceeds	Unit valuation	5-year average	100 / 30
Wyoming	All	Gross proceeds	Appraisal	Previous year	30 / 100 / 25

Source: G. Laing, "An Analysis of the Effects of State Taxation on the Mining Industry in the Rocky Mountain States" (Master's thesis, Colorado School of Mines, 1976). Reprinted with permission.

the corporation, although frequently property tax and other taxes are lumped into general categories of cost of goods sold or miscellaneous fixed expenses.

Anderson randomly selected a number of mining, metal mining and fabricating, petroleum, and manufacturing firms reporting these data in 1973. He computed property taxes paid as a percentage of capital (assets) of the firm. Based on these data, shown in table 5–8, one must conclude that mining and petroleum are taxed less heavily per dollar of assets than the selected manufacturing firms. The tax rate for mining and petroleum firms is only 61 percent of the tax rate for manufacturing firms. Since the capital intensity associated with mining activity is considerably higher than in manufacturing, one must conclude, based on these rather sketchy data, that extractive firms

Table 5–8
Property Tax Rates for Various Industries, 1973 Data
(percent)

Mining		
Amax		0.71
ASARCO		3.00
Diamond Crystal Salt		2.28
Foote Mineral		2.02
Freeport Minerals		1.85
Kennecott		1.34
Phelps Dodge		1.53
Pittstown		1.68
Westmoreland Coal		0.96
	Average	1.71
Metal Mining and Fabricating		
Aluminum Co. of America		1.34
Bethlehem Steel		2.10
Republic Steel		2.69
Reynolds Metals		1.35
	Average	1.87
Petroleum		
Atlantic Home Products		2.92
Copperweld		3.12
Corning Glass		2.90
Dow Jones		3.23
DuPont		3.18
Fort Howard Paper		2.24
General Mills		3.06
General Tire and Rubber		3.00
Goodyear		1.58
Westinghouse		2.09
	Average	2.73

Source: R.C. Anderson, "Taxes, Exploration and Reserves," in *Non-Renewable Resource Taxation in the Western States* (Cambridge, Mass.: Lincoln Institute of Land Policy, 1977). Reprinted with permission.

are not discriminated against by property-tax statutes or by tax administrators.

Taxes on Output. Because of the difficulties in administering property taxes and problems in achieving differential effective rates among different resources which some jurisdictions desire, a growing tendency has been to substitute severance taxes for property taxes. For example, Louisiana exacted a 3.3 percent severance tax on gas and oil extraction in 1972 to substitute for a projected increase in general property taxes. The obvious advantages of the severance tax are that it can be discriminatory with respect to specific resources and that it can be a prodigious revenue producer. Output-based tax structures for eight Western states are shown in table 5–9. It is apparent that the tax is levied on physical output by weight, physical output classified by quality (coal by Btu in Montana), or the value of output. The tax rates vary widely among states, and rates also vary among resources. It is apparent from this survey that states tax according to what they can extract from resource firms. Output taxes are, however, preferred by the industry to property taxes. One advantage is that tax payments coincide with extraction and avoid the fixed-cost nature of a property tax, which is particularly burdensome during periods of production at levels below the planned capacity (due to business cycles and other unforeseen problems) or when shutdown occurs. Also, with income taxes and taxes on output, the government shares risk in that tax payments are contingent on profits or output whereas a fixed tax is not. Agria (1969) estimates that oil producers bear 80 percent of the burden of state severance taxes, but because of regionalized markets and high transportation costs, coal producers bear only 50 percent of the burden.

It should also be noted that states and local governments receive revenues from lease royalties and payments from public lands. Furthermore, resources falling under the Mine Leasing Act of 1920 on federal lands provide revenues to states. In 1976 the state share was raised to 50 percent from 37.5 percent of all lease payments.

Mining activity in Colorado has been low relative to other Western states recently, and its tax laws today reflect past booms of silver and gold mining. Metal mines, now primarily molybdenum, are assessed for the property tax at 50 percent of gross proceeds or 100 percent of net proceeds, depending on whichever amount is considered greater by the state, and not as capitalized income reported previously in table 5–7. Coal, asphalt, and other nonmetal mining are assessed at the county level, and the procedures are variable. Self-assessment by the mining firms is frequent, although the law specifies that the method is based on capitalized net earnings. Underground mining equipment is not assessed because of the influence of the Union Pacific Railroad, which owned underground mines that produced coal for its steam engines.

Table 5–9
Summary of Western State Severance Taxes on Mineral Production

State and Commodity	Tax	Rate and Basis
Arizona All minerals	Transaction privilege tax	2.5% of gross value
Colorado Coal	Coal tonnage	$0.007 per ton
Idaho All minerals	Mine license	2% of value
Montana	Metalliferous Mine license	Up to 1.438% of gross value

Coal	Coal severance	
Btu/lb	Surface (per ton)	Underground (per ton)
Under 7,000	$0.12 or 20% of value	$0.05 or 3% of value
7,000–8,000	$0.22 or 30% of value	$0.08 or 4% of value
8,000–9,000	$0.34 or 30% of value	$0.10 or 4% of value
Over 9,000	$0.40 or 30% of value	$0.12 or 4% of value
Micaceous minerals	Micaceous minerals License	$0.05 per ton
All minerals	Resource indemnity trust	0.5% of gross value over $5,000
Cement and gypsum	Cement and gypsum license	$0.04 per 376 lb of cement $0.05 per ton of gypsum

Nevada	No mineral severance tax	
New Mexico All minerals	Resource excise (resource, processor, or service)	0.75% of taxable value
Potash	Resource excise	0.125% of taxable value
Molybdenum	Resource excise	0.125% of taxable value
Copper	Severance	0.5% of gross value
Potash	Severance	2.5% of gross value
Uranium	Severance	1% of gross value
Coal	Severance	0.5% of gross value
Pumice, gypsum, sand	Severance	0.125% of gross value
Clay, nonmetallics, gold, silver, lead, zinc, thorium molybdenum, rare earths, and other metals	Severance	0.125% of gross value
Utah Gold, silver, copper, lead, iron, zinc, tungsten, uranium, or other valuable metal	Mining occupations	1% of gross value
Wyoming Gold, silver, other precious metals, soda saline, uranium, or other valuable metal	Severance	2% of gross value
Coal, trona, oil shale Other fuel	Severance	4% of gross value

Source: G. Laing, "An Analysis of the Effects of State Taxation on the Mining Industry in the Rocky Mountain States" (Master's thesis, Colorado School of Mines, 1976). Reprinted with permission.

Colorado has experienced difficulties in assessing captive mines belonging to utilities and other integrated industries, as have other states, because an objective sale price is absent. Colorado Fuel and Iron Corporation, in order to secure tax advantages, maintains that all its profits are due to steel production and not its coal production. Utilities are also known to allocate profits between captive mines and power-generation sales in an effort to avoid taxes and to avoid constraints placed on profits by regulatory bodies.

Idaho bases its property-tax valuation on the previous years' profit, and county-level assessments are monitored by the state. However, air-pollution equipment is exempted from the property tax, a frequent practice in many states. In one instance where residuals from the stack gases are converted to sulfuric acid, the exemption does not apply. Montana bases property-tax valuation on the previous year's net proceeds after the five-year-average method was declared unconstitutional. Nevada also bases its assessments on 100 percent of net proceeds (deductions are specified by state statute) from the previous year and on the appraised value of equipment. Pollution-equipment exemptions are allowed. Although improvements have appreciated in market value at up to a 35 percent annual rate, assessments have been increased only modestly. The property tax is also levied on leasehold interest by capitalizing the net income to the lessor.

New Mexico also assesses mineral and certain industrial properties at the state level. Property-tax assessments are based on 100 percent of the net proceeds except for a 50 percent rate on gross proceeds for potash and a 75 percent rate on improvements for uranium. The previous years' net proceeds or an average of the past five years is used in copper-mine assessments. State spokespersons concede that these procedures are flexible and actual valuations reflect political power and self-assessment practices. A coal severance surtax was enacted in 1978, and the severance tax rates on uranium, gas, and oil were boosted in 1980.

Utah recently replaced property-tax assessments based on reserves remaining in situ with a method based on recent net-income flows. Presently, a guideline of twice the last three years' net proceeds is used as a proxy for capitalizing future income streams from metal mines. A 100 percent assessment rate is applied to this computation while a 30 percent rate applies to improvements. Utilities carry a 25 percent assessment rate, and residential construction is given a lower 20 percent rate.

In Wyoming, minerals account for nearly one-half of the entire property-tax base. These properties are assessed at both state and local levels based on 100 percent of the previous year's production valued at the mine mouth. Air, water, and land (reclamation) pollution exemptions are allowed but have proved to be difficult to administer partially because income from processed pollution is netted out of these exemptions.

Taxes on Coal and Copper. The data on state taxes on firms extracting nonrenewable resources are neither up to date nor exhaustive. Our detailed study of coal and copper taxation is summarized in an extensive set of tables which are one of the most exhaustive and authoritative reviews available. A similar survey was conducted by Stinson (1977). However, we believe our data to be more complete, and thus the data are presented in entirety. Tables 5–11 through 5–26 cover taxation of coal by the sixteen states which produce virtually all the U.S. coal. Tables 5–28 through 5–36 cover the nine states which either now produce or may produce copper. Federal taxes are considered in table 5–10 (coal) and table 5–27 (copper). All significant state taxes affecting the cost of investing in and operating a mine/mill (coal) and mine/mill/smelter (copper) operation are included. "Insignificant" state taxes that are excluded are believed to yield less than $5,000 per site annually.

The state taxes cover sales and use taxes on equipment and supplies, property taxes (which fall on both equipment and the mine itself and are assessed by a number of differing methodologies), severance and other output-based taxes, and corporate-income-based taxes. Taxes on labor (payroll taxes) and other miscellaneous taxes are ignored because of the uniformity of their rates with other business enterprises or their marginal importance. No in-depth perusal of state tax statutes, administrative regulations and enforcement practices, or court rulings is required to show that the tax bases and rates are both diverse and complex. The tables are organized in the first column by tax bases defined by output, profit, physical capital, and land along with a reference to the relevant enabling statute. A brief description of each legislated tax base and its associated rate is also noted in the second and third columns. Converting these complex nominal taxes to estimates of effective tax rates with respect to either total revenue or net revenue necessitates models of the technology of the industry. A modest literature (primarily for Western states) exists and is reported in the fourth column of each table. Our estimates of effective tax rates based on output are reported later. A final column cites 1977–1978 changes in the statutes and also contains remarks about the determinations of various tax bases.

Estimating Effective Tax Rates

Converting nominal tax rates to estimates of effective tax rates becomes an involved task when one realizes that extraction costs and transportation costs vary from site to site and state to state and that the nominal tax bases and rates are themselves complex. The result is a paucity of information on effective tax rates. A direct estimating approach of dividing actual tax revenues by the value of total output for each resource is infeasible because

Table 5-10
1977 Federal Taxes on Coal

Tax Type: Economic Definition	Tax Base: Legal Definition	Nominal Tax Rate	Effective Rate from the Literature	Remarks
Output				
Reclamation fee (30 USC 1232)	Tonnage produced or value of coal at mine site	Surface: $0.35/ton Underground: $0.15/ton or 10% of value, whichever is less		Not to exceed 2% of f.o.b. mine-mouth price; does not apply to lignite coal
Black-lung benefits P.L. No. 95-227, 2, 92 Stat. 11 (1978)]	Tonnage produced	Surface: $0.25/ton Underground: $0.50/ton		
Profit				
Corporate income (IRC 61, 63)	Gross income minus allowable deductions. In addition ot the usual deductions, mining corporations may include amortization of pollution-control facilities, research, and experimental expenditures; a depletion allowance (cost or 10% of gross income, whichever is greater); development expenditures; and current exploration expenditures (must be recaptured when mine reaches producing state).	0-$25,000 20% $25,000–$50,000 22% $50,000–over 48%	26.7% of world-wide net income[a] 19.4% of U.S.-source net income[a]	Tax credits include investment tax credit; credit for tax on foreign subsidiaries; new-jobs tax credit; and tax credit for work-incentive-program expenses.
Capital gains (IRC 1201)	Net capital gains	30% of net capital gains if less than amount of regular corporate income tax		This provision is limited to certain classes of assets, usually those not part of the normal business of the corporation. Coal royalties qualify when a corporation or an individual does not conduct mining operations. IRS S 6-31-C.

Table 5-10 *(continued)*

Tax Type: Economic Definition	Tax Base: Legal Definition	Nominal Tax Rate	Effective Rate from the Literature	Remarks
Minimum tax on tax-preference items (IRC 56, 57)	Tax-preference items less the greater of $10,000 or full amount of regular income tax	15%		Tax-preference items include capital gains (if capital-gains tax is used); accelerated depreciation on real property (amounted in excess of straight-line method); excess of percentage depletion over cost depletion; excess of accelerated amortization of pollution-control devices over normal amortization; and excess of itemized deductions over 60% of adjusted gross income. Note that the 1978 tax revision eliminates capital gains from this provision.
Accumulated earning (IRC 532, 533)	Accumulated taxable income	27.5% ≤ $100,000 38.5% > $100,000		Applies to corporations whose earnings and profits accumulate beyond reasonable needs of the business (The purpose being to avoid individual ordinary income tax for shareholders)

aU.S. Department of Treasury, *Effective Income Tax Rates Paid by United States Corporations in 1972* (Washington: GPO, May 1978). These rates apply to corporations with under $1 million of assets (84.3% of coal-mining corporations).

Table 5–11
1977 Alabama Taxes on Coal

Tax Type: Economic Definition	Tax Base: Legal Definition	Nominal Tax Rate	Effective Rate from the Literature	Remarks
Output				
Severance (40-13-6)	Tons severed, reported by firms	$0.20/ton + $0.135/ton[a] ($0.50/ton + $0.135/ton in DeKalb Col) + $0.20/ton shipped through Mobile docks[a]		Some of $0.135/ton may be refunded; currently about 90% is refunded. Much less than half of production goes through Mobile.[a]
Profit				
Corporate income (40-18-31-35)	Net income (UDITPA)	5%		Same as federal tax base, except no coal-depletion allowance[b]
Franchise (40-14-40-46)	Outstanding stock, surplus, undivided profits, bonds, or notes	0.3%		Foreign corporations taxed on capital stock employed in state
Physical Capital				
Property (40-8-1-2)	State: 25% of market value of personal property; County: 15% to 25% of market value of personal property[a]	State: 0.65%[a] County: 2.1% average		Mining machinery exempt
Sales and use (40-23-1)	Gross receipts	1.5% on mine machinery		Other items taxd at 4%
Land				
Property (40-8-1-2)	State: 25% of market value of real property; county: 15% to 25% of market value of real property[a]	State: 0.65%[a] County: 2.1% average		Mineral rights self-assessed, generally valued at $5/acre

[a]Phone call to Mr. Stough, Property Tax Division, Alabama Department of Revenue.
[b]Phone call to Melvin Johnson, License Division, Alabama Department of Revenue.
[c]Phone call to Dan DeVaughn, Sales Tax Division, Alabama Department of Revenue.

Table 5-12
1977 Arizona Taxes on Coal

Tax Type: Economic Definition	Tax Base: Legal Definition	Nominal Tax Rate	Effective Rate from the Literature	Remarks
Output				
Education excise (42-1361)	Gross proceeds from all sales in and and out of state	Combined rate of 2.5%[a] on all sales plus 4% on in-state retain sales	2.5% of revenue[b]	2.5% is, in effect, a severance tax. As of June 1, 1978, combined rate is 2% for 2 years[c]
Special excise (42-1271)	Gross proceeds from all sales in and out of state	Combined rate of 2.5%[a] on all sales plus 4% on in-state retail sales	2.5% of revenue[b]	4% is retail sales tax.
Transaction privilege (42-1309, 1310)	Gross proceeds from all sales in and out of state	Combined rate of 2.5%[a] on all sales plus 4% on in-state retail sales	2.5% of revenue[b]	
Profit				
Corporate income (43-101, 199)	Adjusted federal taxable income less federal income tax attributable to state	Graduated up to 10.5% of income over $6,000	5.5% of taxable income[d] 0.99% of revenue[b]	Federal income tax deduction is under litigation.[c] 100% of domestic dividends exempt. Only percentage depletion allowed (10%) for mines established after 1953.

Physical Capital Property (42-124, 136, 201, 227)	60% of fair market value[e]	Average 11.51%[a] 3.4% of revenue[b]	Rate varies by locality. 11.51% is statewide average for all mining property (real, personal, minerals).[a] Fair market value based on a capitalized-net-income approach or comparative sales approach.[d] Tax reform of assessment ratios which vary by property class and district are expected in 1979[c]
Use (42-1401, 1361)	Purchase price	4%	Most mining equipment exempt
Sales (42-1361, 1309, 1310)	Sales price	4%	Most mining equipment exempt

[a]Phone calls on May 12, 1978 and June 13, 1978 to Donald Ross, Arizona Bureau of Revenue, and Joe Langlois, Arizona Department of Revenue.

[b]D.W. Gentry, "Financial Modeling of Mining Ventures—The Effects of State Mine Taxation," in *Non-Renewable Resource Taxation in the Western States*, Lincoln Institute of Land Policy Monograph 77-2 (Cambridge, Mass., 1977).

[c]Julie French, Magma Corporation.

[d]"Taxation of Producing Mineral Interests," memorandum from Joe Langlois, Arizona Department of Revenue, February 6, 1978.

[e]Letter from A.D. Coumides, Arizona Mining Association.

Table 5-13
1977 Colorado Taxes on Coal

Tax Type: Economic Definition	Tax Base: Legal Definition	Nominal Tax Rate	Effective Rate from the Literature	Remarks
Output Coal tonnage (34-23-101)	Tonnage at point of severance	$0.07/ton		Repealed effective January 1, 1978. Replaced by a severance tax of $0.60/ton adjusted for quarterly changes in whole price index of all commodities.[a] Self-assessed.
Property (39-6-107)	Stockpiles of coal assessed at 5% of fair value[b,c]	6%[d]	See rates under Profit, property tax.	Fair value is approximately sales price less transportation cost[c]
Profit Corporation income (39-22-304, 39-22-301)	Net income	5%	0.6% of gross value[e] $0.053 ton[e] 1.15% of revenue[f]	Federal taxable income with adjustments
Property (39-6-111)	Capitalized net income assessed at 30%[b]	6%[d]	0.3% of gross value[e] $0.025/ton[e] 0.92% of revenue[f] (Note: these figures are for total property taxes.)	Appraisers use estimated ore reserves, average annual production, economic royalty rate, and capitalization rate to determine capitalized net income.[g] Minimum capitalization rate is 11.5%.[b] Maximum royalty rate is 12.5% of current selling price.[b]
Physical Capital Property Real property (buildings)	Value of improvements is based on location and desirability, functional use, current replacement cost less depreciation, comparison with similar property, and earning or productive capacity. Assessed at 30%.[g]	6%[d]	See rates under Profit, property tax	Improvements are valued at book value using straight-line depreciation and no salvage value.[g] Mine improvements within a mine excavation are exempt[b]

Personal property (equipment)	Same as above.	6%	See rates under Profit, property tax.	Mining company lists and values all equipment[g]
Supplies	Actual value of amount on hand at assessment date assessed at 30%[b]	6%	See rates under Profit, property tax.	Mining company determines this amount[e,g]
Sales and use (39-26-102, 39-26-203)	Sales price	3% state levy[h] 4% maximum county and city levy[h]		Sales and purchases of coal are exempt.
Land Property Surface land	Land on which minerals are depleted is valued at selling price of similar land assessed at 30%.[b]	6%[d]	See rates under Profit, property tax.	
	Land on which minerals are not depleted is valued at present net worth when land is depleted.[b]	6%	See rates under Profit, property tax.	See remarks under Profit, property tax, for current capitalization rate.

[a]Mark C. Resta, "Natural Resource Taxation in the Four Corner States" (Master's thesis, University of New Mexico, 1978).

[b]"Natural Resources 6000," from "Circulator of Recommendations to County Assessors," Division of Property Taxes, Colorado.

[c]Phone call on May 11, 1978 to Jack O'Donnell, Division of Property Taxation, Colorado.

[d]Phone call on April 26, 1978 to John Williams, Division of Property Taxation, Colorado.

[e]"Comparison of Tax Burden on the Minerals Extraction Industry—Colorado and Selected States," Memorandum to Senator T. Bishop from the Legislative Council Staff, February 20, 1976. Note these rates are for 1975.

[f]D.W. Gentry, "Financial Modeling of Mining Ventures—The Effects of State Mine Taxation," in *Non-Renewable Resource Taxation in the Western States*, Lincoln Institute of Land Policy Monograph 77-2 (Cambridge, Mass., 1977). Note that these rates are for a hypothetical mine in its tenth year of cash flow in 1976.

[g]Glen John Steward Laing, "An Analysis of the Effects of State Taxation on the Mining Industry in the Rocky Mountain States" (Master's thesis, Colorado School of Mines, 1976).

[h]Phone call on March 29, 1978 to Mr. Wiche, Colorado.

Table 5-14
1977 Illinois Taxes on Coal

Tax Type: Economic Definition	Tax Base: Legal Definition	Nominal Tax Rate	Effective Rate from the Literature	Remarks
Output				
Sales (439-440)	Gross receipts from intrastate sales	State: 4%[a] Local: 1% (county of mine mouth)		
Profit				
Corporate income (Ch. 120, Sec. 2)	Net income	4%		Federal taxable income, with adjustments. Minor changes pending.
Franchise (157-131)	Stated capital plus paid in surplus attributable to state	0.05%		Minimum: $25 Maximum: $1 million
Physical Capital				
Property (499-611)	33⅓% of replacement cost less depreciation	Downstate average[a] 5.38%		Tax rate varies with locality. Personal property tax expires 1979, but probably will be extended.
Sales and use (439-440)	Gross receipts	State: 4%[a] Local: 1%		Applies to inputs bought in and out of state
Land				
Property (499-611)	33⅓% of fair cash value of real property	Downstate average[a] 5.6%		Tax rate varies with locality. Assessed by county. Assessment method varies, but market value appears to be general approach.

[a]Phone call to Dale Young, Illinois Department of Revenue.

Table 5-15
1977 Indiana Taxes on Coal

Tax Type: Economic Definition	Tax Base: Legal Definition	Nominal Tax Rate	Effective Rate from the Literature	Remarks
Output Sales (6-2-1-37)	Gross receipts from retail sales	4%		Sales to utilities and manufacturers exempt
Profit Corporate income (6-2-1-6-3-1)	(a) Adjusted gross income from Indiana sources	3%[a]		(a) $b + (a$ or c, whichever is greater) Taxable income under IRC plus state income and local property taxes
	(b) Net income	3%		(b) Adjusted gross income less the greater of adjusted-gross-income tax or gross-income tax.
	(c) Gross income less $1,000	0.375%		(c) Rate decreases 0.0125% each year
Physical Capital Property (6-1-46-1)	33⅓% of cost less depreciation	State: 0.01%[b] Incorporated areas: 2% Unincorporated areas: 1.25%		Political subdivisions may levy emergency taxes over these statutory limits.
Sales (6-2-1-37)	Gross receipts	4%		Sales of electricity, gas, and water for use in mining exempt. Sales of materials directly consumed exempt.[c]
Land Property (6-1-46-1)	33⅓% of true cash value	State: 0.01% Incorporated: 2% Unincorporated: 1.25%		True cash value: Coal: $60/acre Surface land: $450/acre times a factor based on land use

[a]Phone call to Spence Walton, Corporate Tax Division, Indiana Department of Revenue.

[b]Phone call to Gordon McIntyre, Property Evaluation Unit, Division of the (Indiana) State Board of Tax Commissioners.

[c]Phone call to Frank Sanders, Sales Tax Division, Indiana Department of Revenue.

Table 5-16
1977 Kentucky Taxes on Coal

Tax Type: Economic Definition	Tax Base: Legal Definition	Nominal Tax Rate	Effective Rate from the Literature	Remarks
Output				
Severance (143.010, 020)	Gross value of coal severed	4.5%		
Sales (139.200)	Gross receipts on intrastate sales	5%		Sales to electric utilities and residential users exempt. Sales to industrial users to extent that fuel cost exceeds 3% of cost of production exempt
Profit				
Corporate income (141.010, 040)	Net income	4% of first $25,000, 5.8% of rest		Base is federal taxable income plus taxes deducted from same.[a]
Franchise (136.070)	Total capital in and out of state (stock, surplus, etc.)	0.07%		
Physical Capital				
Property (132.190, 420)	≥ 90% market value: original cost less depreciation (30% of cost minimum)[b]	0.45% state[c] 0.6% county		Pollution-control equipment taxed at 0.15% of fair cash value by state, exempt from county tax
Sales and use (139.200, 139.310)	Selling price	5%		Machinery for new and expanded industry exempt
Land				
Property (132.190, 420)	≥ 90% market value[b]	0.45% state 0.6% county		Mineral rights on nonproducing land, valued at $50–$150/acre, soon will not be taxed except for large holdings. Value of minerals on producing land affects market value of land.[b]

[a] Phone call to Paul Tanner, Kentucky Bureau of Revenue.
[b] Phone call to Jim Calhoun, Kentucky Bureau of Revenue.
[c] Phone call to R. Rojas, Kentucky.

Table 5-17
1977 Montana Taxes on Coal

Tax Type: Economic Definition	Tax Base: Legal Definition	Nominal Tax Rate	Effective Rate from the Literature	Remarks
Output Coal severance (84-1313-1313)	Contract sales price at mine mouth on output over 20,000 tons/year[a]	< 7,000 Btu/lb 20%[a] > 7,000 Btu/lb 30%	25% of revenue[b]	For underground mines, rates are 3% and 4%. Also a graduated tax ranging from $0.12 to $0.40 ($0.05 to $0.12 for underground per ton. Method yielding greatest amount is used—currently this is percentage method. Severance tax currently under litigation.[a]
Property (84-3013)	Previous year's output times average price at mine mouth for county of production less production taxes times 45%[a] (33⅓% for underground)	Average 20.1%[c] (1976)		Rate varies by county.
Mineral mining (84-7003-7006)	Gross value at mine mouth	$25 + 0.5% of gross value over $5,000		
Profit Corporation license (84-1501-15023)	Net income from state sources	6.75%		Net income is state portion of federal taxable income, with adjustments.
Physical Capital Property (84-301)	12% of original installed cost[a]	Average 20.1%[c] (1976)		Rate varies by county.
Land Property (84-301)	12% of market value of land price paid to government on patented mines and mining claims.[a] Market value on reserved rights of entry.	Average 20.1%[c] (1976)		Rate varies by county.

[a]Phone call to John Clark, Research Division, Department of Revenue, Montana.

[b]D.W. Gentry, "Financial Modeling of Mining Ventrues—The Effects of State Mine Taxation," in *Non-Renewable Resource Taxation in the Western States*, Lincoln Institute of Land Policy Monograph 77-2 (Cambridge, Mass., 1977).

[c]Letter of January 3, 1977 from Kenneth K. Morrison, Property Assessment Division, Department of Revenue, Montana.

Table 5-18
1977 New Mexico Taxes on Coal

Tax Type: Economic Definition	Tax Base: Legal Definition	Nominal Tax Rate	Effective Rate from the Literature	Remarks
Output				
Severance (72-18-6)	Severance of coal at mine mouth	$0.38/ton on steam coal[a] $0.18/ton on metallurgical coal[a]		Effective July 1, 1977. Prior to that date, severance tax was 0.5% of sales value less deductions for putting product in marketable form not to exceed 50% of gross value[a]
Severance surtax	Severance-tax payment[a]	Percentage rise in consumer price index from calendar year just prior to year in which severance rates are computed.[a]		Effective July 1, 1978[b]
Resource excise (resources tax, processor tax, service tax)	Sales price less royalties paid to state or United States, and sales of coal to state, United States, and certain nonprofit organizations[a]	0.75%[a]		
Conservation	Same as resource excise tax[a]	0.18%[a]		This base was effective as of July 1, 1977. Prior to that date, base was same as severance-tax base prior to July 1, 1977.[a] (See above remarks under severance tax.)
Gross receipts (72-16A-4)	Gross receipts	4% state levy 0.25% maximum city levy		Exported coal is exempt (approximately 15% of output). Tax on coal consumed in state is paid for by user.
Profit				
Corporation income (72-15A-2, 6)	Net income[b]	5%	0.48% of revenue[c] 0.3% of gross value $0.106/ton[d]	Net income is federal tax base, adjusted.
Property (72-29-12)	Mineral property is valued at 300% of annual net-production value. The annual net-production value is market	2.813% (average levy of total local taxes for three	4.36% of revenue[c] (includes all property taxes)	Net value of production may be last year's or average of five preceding years' net value. Once one of these methods is

value less royalties paid to state, United States, or Indian tribe; costs of extraction, milling, treating, reducing, transporting, and selling the product; and depreciation costs.[a] Assessed at 33⅓%.	major coal-producing counties)	1.1% of gross value, 5% per ton[d]	chosen, it must be used thereafter.[a] Appraised by the state property appraisal department.[a]
Physical Capital Property (79-29-12) Improvements, equipment, materials, supplies, and other properties used in connection with mine are assessed at 33⅓% of original value less straight-line depreciation.[a,f]	2.813%[e] (average levy of total local taxes for three major coal-producing counties)	4.36% of revenue[c] (includes all property taxes) 1.1% of gross value, 5% per ton[d]	Appraised by the state property appraisal department[a]
Gross receipts (72-16A-4) Gross receipts[b]	4% state levy[b] 0.25% maximum city levy[b]		Effective July 1, 1978, state levy was decreased to 3.75% and maximum city levy was increased to 0.5%.[b] Paid on all plant and equipment and operating and maintenance supplies. Out-of-state purchases taxed by compensating-use tax.
Land Property (79-29-12) If mine is on fee title land, it is assessed on per-acre basis times an assessment ratio of 33⅓%.[g]	2.813%[e] (average levy of total local taxes for three major coal-producing counties)	4.36%[c] of revenue (includes all property taxes) 1.1% of gross value, $0.05/ton[d]	Most coal mining is done on federal or Indian lands so there is no assessed value[g]

[a] Mark C. Resta, "Natural Resource Taxation in the Four Corners States" (Master's thesis, University of New Mexico, 1978).

[b] Commerce Clearing House, Inc., State Tax Guide, Washington, D.C. 1978.

[c] D.W. Gentry, "Financial Modeling of Mining Ventures—The Effects of State Mine Taxation," in Non-Renewable Resource Taxation in the Western States, Lincoln Institute of Land Policy Monograph 77-2 (Cambridge, Mass., 1977). Note that these rates are based on a hypothetical mine in its tenth year of cash flow in 1976.

[d] "Comparison of Tax Burden on Minerals Extraction Industry—Colorado and Selected States—1974," Memorandum to Senator T. Bishop from the Legislative Council Staff, February 20, 1976.

[e] Phone call on May 12, 1978 to Mrs. Hill, Property Taxation Division, New Mexico.

[f] Phone call on March 27, 1978 to Billy Martin, Property Taxation Division, New Mexico.

[g] Phone call on May 15, 1978 to Paul McDonald, Property Tax Division, New Mexico.

Table 5-19
1977 North Dakota Taxes on Coal

Tax Type: Economic Definition	Tax Base: Legal Definition	Nominal Tax Rate	Effective Rate from the Literature	Remarks
Output Severance (57-61-01)	Short tons	$0.65/ton plus $0.01/ton for every one-point rise in wholesale price index over June 1977 level		Currently about $0.74/ton. Expires 1979, likely to be renewed.[a]
Sales and use (57-39.2, 40.2)	Retail sales or purchase price	3%		
Profit Income (57-38-11)	Net income	3% of first $3,000 4% $3,000–8,000 5% $8,001–$15,000 6% over $15,000		Net income same as federal taxable income
Corporate privilege (47-38-66)	Net income over $2,000	1%		Net income same as federal taxable income
Physical capital Property (57-02-24, 28)	8.4% of market value of improvements on land[a]	19.9% average[a]		No tax on personal property
Sales and use (57-39.2, 40.2)	Retail sales or purchase price	3%		
Land Property (57-02-24, 28)	8.4% market value of land[a]	19.9% average[a]		Mineral deposits not included in assessment. Land valued as commercial property.

[a]Phone call to Bill Cudworth and Marcie Dickerson, North Dakota Bureau of Revenue.

Table 5–20
1977 Ohio Taxes on Coal

Tax Type: Economic Definition	Tax Base: Legal Definition	Nominal Tax Rate	Effective Rate from the Literature	Remarks
Output				
Severance (5749.02)	Tonnage severed at mine mouth	$0.04/ton		This tax is self-assessed subject to periodic audits. [a]
Profit				
Corporation franchise (5733.05)	25% of the first $25,000 of adjusted federal net income plus 8% of any income greater than $25,000 or 0.5% of value of issued and outstanding shares of stock[b]	See Tax Base.		
Physical Capital				
Property (5711.22)	Equipment and machinery are valued at book value less depreciation times an assessment rate of 48%.[c]	4.39%[c] in 1976		Owner determines tax base.[c] Assessment rate will decrease gradually to 35% in 1985.[c]
Sales and use (5739.02, 5741.02)	Retail sales price	4% state levy 0.5% maximum county levy[c]		Coal sales exempt. Equipment and machinery used directly in mining exempt[c].
Land				
Property Mine (mineral rights)	Market value of minerals in place is based on sales of similar properties. If there are no sales of similar properties, leases and physical characteristics such as seam thickness and overburden are considered. The assessed value is the 35% of market value.[c,d]	4.39% in 1976		No specific formula is used to determine market value.[d]

Table 5-20 (*continued*)

Tax Type: Economic Definition	Tax Base: Legal Definition	Nominal Tax Rate	Effective Rate from the Literature	Remarks
Surface land	If surface rights are owned with mineral rights, value is determined in value of mine. If surface rights are owned separately, market value is based on sales of similar land assessed at 35%.[d]	4.39% in 1976		By law, a physical reassessment must take place every 6 years and market values must be updated every 3 years.[c]

[a]Council of State Governments, *State Coal Severance Taxes and Distribution of Revenue* (Lexington, Ky.: Council of State Governments, September 1976).

[b]"1978 Ohio Corporation Franchise Tax Report Instructions," State of Ohio, Department of Taxation; and Ohio Statute 5733.05

[c]Phone call on May 4, 1978 to Ron Nucha, Tax Research Division, Ohio. Note that the nominal tax was calculated from actual county revenues divided by the tax base.

[d]Phone call on May 5, 1978 to Bob Kinney, Tax Equalization Division, Ohio.

Table 5-21
1977 Pennsylvania Taxes on Coal

Tax Type: Economic Definition	Tax Base: Legal Definition	Nominal Tax Rate	Effective Rate from the Literature	Remarks
Output				
Property				
Greene County (Major underground coal-mine producer in 1975)	Coal mined during the years is assessed at $4,000 to $8,000 per acre times assessment ratio of 20%.[a]	7.31%[a] (county plus average school-district levy)		Determined by individual counties. It is estimated this procedure places an average value of $1.30/ton on coal for assessment purposes.[a] Also see remarks under Land, property tax, Greene County.
Indiana County (Second major underground coal mine producer in 1975)	Coal planned to be mined based on 10-year projection by the operator is assessed at $50/acre times assessment ratio of 35%.[b]	7.51%[b] (county plus average school-district levy)		Note that average seam thickness is less than in Greene County, which helps account for difference in assessed values.[b] Also see remarks under Land, property tax, Indiana County.
Profit				
Corporation income (72-3420C)	Net income	10.5%[c]		Net income is federal tax base adjusted (state income tax must be added back).
Capital stock (72-7601, 7602)	Actual value in cash of corporation's capital stock at end of year	1%		Self-assessed subject to adjustment by Department of Revenue
Physical Capital				
Property				
Greene County	Market value of machinery, equipment, and buildings which do not support machinery.[a] Market value is based on cost, income, and selling price assessed at 20%.	7.31%[a] (county plus average school-district levy)		Determined by individual counties. In 1978, shell structures of buildings housing equipment used directly in mining may be assessed.[b] It is estimated that in 1977, 20% to 30% of machines, equipment, and buildings at mine could be assessed.[a] Also, see remarks under Land, property tax, Greene County.

Table 5–21 *(continued)*

Tax Type: Economic Definition	Tax Base: Legal Definition	Nominal Tax Rate	Effective Rate from the Literature	Remarks
Indiana County	Same as in Greene County but 35%.[b]	7.51%[b] (county plus average school-district levy)		See first remark above. Also, see remarks under Land, property tax, Indiana County.
Clearfield County (Major surface-mining county in 1975)	Same as in Green County but assessed at 40%.[d]	7.21%[d] (county plus average school-district levy)		See first remark above. Also, see remarks under Land, property tax, Clearfield County.
Sales Tax (72-7204)	Retail sales	6%		Retail sales of coal are exempt. Mining equipment used directly in mining is exempt.[e]
Land Property Greene County	Coal not yet mined is assessed at $400 to $800 per acre times assessment ratio of 20%.[a]	7.31%[a] (county plus average school-district levy)		Determined by individual counties. Mineral rights are assumed to have no market value until they are sold to coal operator.[a]
	Surface land is appraised at 20% of market value.[a]	7.31% (county plus average school-district levy)		Last reassessment was done in 1976.[a] The actual assessment/market-value ratio based on 1976 market values as determined by the State Tax Equalization Board is 25.2%.[f]

County	Description	Rate
Indiana County	Coal that an operator has the mineral rights to but will not mine based on a 10-year projection by the operator is assessed at $30/acre times assessment ratio of 35%.[b]	7.51%[b]
	Acreage that has been mined out but is still used by coal operator is assessed at $15/acre times assessment ratio of 35%.[b]	7.51%[b]
	Surface land is appraised at 35% of market value.[b]	7.51%[b]
	Market value is determined by private appraisal company. The last reassessment was done in 1968.[b] The actual assessment/market-value ratio based on 1976 market values as determined by the State Tax Equalization Board is 28.7%.[f]	
Clearfield County	Coal mineral rights are assessed at $10/acre whether depleted or not times assessment ratio of 40%.[d]	7.21%[d] (county plus average school-district levy)
	Surface land is assessed at 50% of market value.[d]	7.21%[d] (county plus average school-district levy)
	Last reassessment was done in 1960.[d] The actual assessment/market-value ratio based on 1976 market values as determined by the State Tax Equalization Board is 28.2%.[f]	

[a] Phone call on May 3, 1978 to John Cole, Coal Assessor, Greene County.

[b] Phone call on May 3, 1978 to David S. Wilson, Chief Assessor, Indiana County.

[c] Phone call on April 24, 1978 to Mr. Douglas, Corporation Taxes, Pennsylvania.

[d] Phone call on May 3, 1978 to Mr. West, Chief Assessor, Clearfield County.

[e] Phone call on April 24, 1978 to Wayne Dietrich, Pennsylvania.

[f] Pennsylvania State Tax Equalization Board, "1976 Market Values of Taxable Real Property," June 30, 1977.

Table 5-22
1977 Tennessee Taxes on Coal

Tax Type: Economic Definition	Tax Base: Legal Definition	Nominal Tax Rate	Effective Rate from the Literature	Remarks
Output Coal severance (67-5901-5902)	Tons severed	$0.20/ton		Self-reported
Sales (67-3002)	Sales price on intrastate sales	4.5%		Coal used directly (part of product) exempt
Profit Corporation excise (67-2704)	Net income (UDITPA)	6%		Federal taxable income plus operating-loss deduction, Tennessee excise tax, and other adjustments.[a]
Franchise (67-2903)	Stock, surplus, undivided profits at book value	0.15%		Foreign corporations taxed on amount of capital stock employed in state.
Physical Capital Property (67-401-734)	30% of actual value[b]	Average 3.5%[b]		Self-evaluated. No state tax. County rates vary. Pollution-control equipment exempt.[b] The 4.5% rate is due to be reduced to 3% after June 30, 1978, but will probably remain at 4.5% for one more year. Utilities and fuel used directly in process exempt.
Sales and use	Sales price	1% on mining machinery 4.5% on machinery to remove over-burden		
Land Property (67-401-734)	40% of actual value[b]	Average 3.5%[b]		Actual value based on discounted present value of recoverable reserves.

[a]Phone call to Allen Curtis, Corporate Income Tax Division, Tennessee Department of Revenue.
[b]Phone call to George Tidwell, Property Assessment Division, Tennessee Department of Revenue.

Table 5-23
1977 Utah Taxes on Coal

Tax Type: Economic Definition	Tax Base: Legal Definition	Nominal Tax Rate	Effective Rate from the Literature	Remarks
Profit				
Corporation income (59-13-3, 5, 6)	Net income less federal income tax	4%[a]	0.65% of revenue[b]	Net income is approximately same as federal net income. The rate has decreased from 6% in 1976.[b]
Property	23% of capitalized net income[c]	6.64%[c]	3.31% of revenue[b]	Capitalized net income is determined from average annual income of five previous years capitalized at 20%. The base is self-assessed subject to audit.[c]
Physical Capital				
Sales and use (59-15-4, 59-16-3)	Purchase price	4% state 0.75% county limit		Coal sales are exempt.[a] Pollution-control equipment is exempt.
Prepaid sales and use (63-51-3, 59-15-4, 59-16-3)	Estimates of construction costs in developing or utilizing a natural resource	4% state 0.75% county limit		The purpose of this method of paying sales and use taxes is to appropriate funds needed for schools and highways in areas of natural resource development.

[a]Phone call of March 29, 1978 to Mr. Osika, State Tax Commission.

[b]D.W. Gentry, "Financial Modeling of Mining Ventures—The Effects of State Mine Taxation," in *Non-Renewable Resource Taxation in the Western States*, Lincoln Institute of Land Policy Monograph 77-2 (Cambridge, Mass., 1977). Note these figures are based on a hypothetical coal mine in its tenth year of cash flow in 1976.

Table 5-24
1977 Virginia Taxes on Coal

Tax Type: Economic Definition	Tax Base: Legal Definition	Nominal Tax Rate	Effective Rate from the Literature	Remarks
Output Property				Determined by individual counties based on some overall state regulations.
Buchanan County Major underground coal-producing county in 1975) Mineral license (58-774)	Gross receipts as severed[a]	1%		Self-assessed.[a]
Profit Corporation income (58-151.031)	Adjusted federal net income	6%		Virginia state income taxes are not deductible.
Physical Capital Property Buchanan County	Machinery, tools, motor vehicles, and delivery equipment are valued at 10% of cost. This is done by a state assessor.[a]	5.5%[a]		Beginning in 1978, value of machinery and equipment will be self-assessed.[a] See remark under Land, property tax.

Sales and use (58-441.14)	Sales price	3% state levy	Sales of coal are exempt.[b] Machinery and equipment used by mines are exempt, but equipment used by contractors is not exempt.[b]
Land Buchanan County Surface land	Market value of surface land. If land is sold between reassessments, land value is adjusted to 60% of price recorded on deed.[a]	5.5%[a]	Last reassessment was in 1972. Next reassessment will be in the fall of 1978.[a] Virginia is transferring to a 100%-of-fair-market-value program by 1982.[c]
Minerals	Active mine acreage is assessed at $400 to $16,00 per acre depending on seam thickness.[a] Acreage which has been depleted is assessed at $15 to $30 per acre.[a]	5.5%[a]	

[a]Phone call on May 10, 1978 to Ruby Rise, Assessors Office, Buchanan County.

[b]Phone call on May 10, 1978 to Mr. West, Sales and Use Tax Division, State Taxation Department, Virginia.

[c]Phone call to Real Estate Appraisal and Mapping, State Taxation Department, Virginia.

Table 5–25
1977 West Virginia Taxes on Coal

Tax Type: Economic Definition	Tax Base: Legal Definition	Nominal Tax Rate	Effective Rate from the Literature	Remarks
Output Privilege (11-13-2, 3, 21)	Gross proceeds from sale are used whenever possible. If not available or if felt to be an inaccurate indicator of value, gross proceeds from subsequent sale, sales of similar products, or cost plus markup is used.[a]	3.5% (base rate) 0.35% (additional rate)		There is a $50 exemption per business person. Prepaid freight on outgoing products. West Virginia excise, and certain federal excise taxes may be deducted.[a]
Profit Carrier	If coal mine also transports goods, it must pay 3.3% on intrastate gross income derived from transport and/or 6.6% on prorated interstate net income.[a]	See Tax Base.		
Corporation net income (11-24-4, 9)	Net income. The smallest amount of privilege and/or carrier tax; corporation net-income tax without credits; or net-income tax imposed on corporation subject to privilege and/or carrier tax.	6%		Net income is federal tax base plus state income taxes less 37.5% of excess of net long-term capital gains over net short-term capital losses, with other adjustments.

Physical Capital Property (11-3-1, 7)	Replacement cost less an initial 30% depreciation allowance and thereafter straight-line depreciation assessed at average county rate of 63.9%.[a]	2.5%[a] (average county levy for class III property)	Pollution-control facilities are valued at salvage value (15% of cost).[a] Nominal varies from 1.5% to 3.28%.[a]
Sales and use (11-15A-2)	Purchase price of tangible personal property furnished or delivered for use within state	3%	Most coal sales and equipment and machinery used directly or consumed in mining are exempt.[a]
Land Property	Value of coal in ground is based on all the following factors: current royalty rates, fee sales, Btu content, and seam thickness. Average assessed value per acre for four of the major coal-mining counties is $172.00.[b]	2.5%[a] (average county levy for class III property)	State determines an appraisal value which can be adjusted by counties with state approval. Then counties assess this value at 50% to 100%. State is currently reappraising coal properties.[a] State's share of tax is ⅜ of 1%.[a]

[a]Phone calls on May 15, 1978 and June 16, 1978 to Fred Sapp, Research Division, State Tax Department, West Virginia.
[b]Same as above. These figures are for Wyoming, Boone, Logan, and Raleigh counties.

Table 5-26
1977 Wyoming Taxes on Coal

Tax Rate: Economic Definition	Tax Base: Legal Definition	Nominal Tax Rate	Effective Rate from the Literature	Remarks
Output Excise (39-227)	Market value	7% + 1.5%		1.5% expires year after taxes collected equal $250 million.
Severance (39-277)	Market value	1.6%		Expires year after taxes collected equal $160 million.
Sales (29-6-404, 504)	Gross receipts from intrastate retail sales[a]	State: 3% 12 counties: 1%[a]		Fuel sold to persons engaged in manufacturing, processing, agriculture, or transportation exempt.[a]
Property (39-272, 277)	Market value of previous year's output	State: 0.6%[b] Average county: 6.3%[b]		Price excludes any crushing, screening, or other processing and costs.[b]
Physical Capital Property (37-272, 277)	10%–12% of cost less depreciation	State: 0.6%[b] Average county: 6.3%[b]		Pollution-control equipment exempt.
Sales and use (39-6-404, 504)	Purchase price	State: 3% 12 counties: 1%[a]		Power and fuel used exempt.[a] County levy is on retail sales only.

[a] Phone call to Don Bright, Wyoming.
[b] Phone call to Mr. Bauer, Property and Excise Tax Division, Wyoming.

Table 5-27
1977 Federal Taxes on Copper

Tax Type: Economic Definition	Tax Base: Legal Definition	Nominal Tax Rate	Effective Rate from the Literature	Remarks
Profit				
Corporate income (IRC 61, 63)	Gross income minus allowable deductions. In addition to usual deductions, mining corporations may include amortization of pollution-control facilities; research and experimental expenditures; depletion allowance (cost or 15% of gross income, whichever is greater); development expenditures; and current exploration expenditures (must be recaptured when mine reaches producing stage).	0–$25,000 20% $25,000–$50,000 22% $50,000–over 48%	32.4% of world-wide net income[a] 29.4% of U.S.-source net income[a]	Tax credits include investment tax credit; credit for tax on foreign subsidiaries; new-jobs tax credits; and tax credit for work-incentive-program expenses.
Capital gains (IRC 1201)	Net capital gains	30% of net capital gains if less than amount of regular corporate income tax		This provision is limited to certain classes of property, usually those not part of normal business of corporation.
Minimum tax on tax-preference items (IRC 56, 57)	Tax-preference items less greater of $10,000 or full amount of regular income tax	15%		Tax-preference items include capital gains (if capital-gains tax is used); accelerated depreciation on real property (amount in excess of straight-line method); excess of percentage depletion over cost depletion; excess of accelerated amortization of pollution-control devices over normal amortization; and excess of itemized deductions over 60% of adjusted gross income. Note that the 1978 tax revision eliminates capital gains from this provision.

Table 5-27 *(continued)*

Tax Type: Economic Definition	Tax Base: Legal Definition	Nominal Tax Rate	Effective Rate from the Literature	Remarks
Accumulated earning (IRC 532, 533)	Accumulated taxable income	27.5% ≤ $100,000 38.5 > $100,000		Applies to corporations whose earnings and profits accumulate beyond reasonable needs of the business (the purpose being to avoid individual ordinary income tax for shareholders)

[a]U.S. Department of Treasury, *Effective Income Tax Rates Paid by United States Corporations in 1972* (Washington: GPO, May 1978). Note that these rates are an average for the nonferrous primary-metals industry.

Table 5-28
1977 Arizona Taxes on Copper

Tax Type: Economic Definition	Tax Base: Legal Definition	Nominal Tax Rate	Effective Rate from the Literature	Remarks
Output				
Education excise (42-1361)	Gross proceeds from all sales in and out of state less freight charges and purchase price of concentrates. All sales are calculated as if cathodes.[a]	Combined rate of 2.5%[b] on all sales plus 4% on in-state retail sales	2.5% of revenue[c]	2.5% if it is, in effect, a severance tax. As of June 1, 1978, combined rate is 2% for 2 years.[a]
Special excise (42-1271)	Gross proceeds from all sales in and out of state less freight charges and purchase price of concentrates. All sales are calculated as if cathodes.[a]	Combined rate of 2.5%[b] on all sales plus 4% on in-state retail sales.	2.5% of revenue	4% retail sales tax
Transaction privilege (42-1309, 1310)	Gross proceeds from all sales in and out of state less freight charges and purchase price of concentrates. All sales are calculated as if cathodes.[a]	Combined rate of 2.5%[b] on all sales plus 4% on in-state retail sales	2.5% of revenue	
Profit				
Corporate income (43-101, 199)	Adjusted federal taxable income less federal income tax attributable to state	Graduated up to 10.5% of income over $6,000	5.5% of taxable income[d] 0.99% of revenue[c]	Federal income tax deduction is under litigation.[a] 100% of domestic dividends exempt. Percentage depletion allowed (15%) only for mines established after 1953.

Table 5-28 (continued)

Tax Type: Economic Definition	Tax Base: Legal Definition	Nominal Tax Rate	Effective Rate from the Literature	Remarks
Physical Capital Property (42-124, 136, 201, 227)	60% of fair market value[e]	Average 11.51%[b]	3.4% of revenue[c]	Rate varies by locality. 11.51% is state-wide average for all mining property (real, personal, minerals.)[b] Fair market value based on capitalized-net-income approach or comparative-sales approach.[d] Tax reform of assessment ratios which vary by property class and district is expected in 1979.[a]
Use (42-1401, 1361)	Purchase price	4%		Most mining equipment exempt
Sales (42-1361, 1309, 1310)	Sales price	4%		Most mining equipment exempt

[a]Julie French, Magma Corporation.

[b]Phone calls on May 12, 1978 and June 13, 1978 to Donald Ross, Arizona Bureau of Revenue, and Joe Langlois, Arizona Bureau of Revenue.

[c]D.W. Gentry, "Financial Modeling of Mining Ventures—The Effects of State Mine Taxation," in *Non-Renewable Resource Taxation in the Western States*, Lincoln Institute of Land Policy Monograph 77-2 (Cambridge, Mass., 1977).

[d]"Taxation of Producing Mineral Interests," Memorandum from Joe Langlois, Arizona Department of Revenue, February 6, 1978.

[e]Letter from A.D. Coumides, Arizona Mining Association.

Table 5-29
1977 Michigan Taxes on Copper

Tax Type: Economic Definition	Tax Base: Legal Definition	Nominal Tax Rate	Effective Rate from the Literature	Remarks
Profit				
Single business (208-31)	Adjusted federal net income. State income tax must be added back.	2.35%		Federal tax base plus total compensation for labor plus capital expenditures expensed in first year plus state income taxes and other minor adjustments.
Property (208-31)	Based on what mine would sell for (basically capitalized-net-income approach) assessed at 50%[a]	3.89%[b] (average of actual and potential copper-mining counties)		Mining company submits reserve estimates, cost, and selling price to the state geologist who then calculates value of mine. This is done annually. There is no additional assessment on equipment, buildings, or surface land (if owned by mine) as they are considered in mine value.[a]
Physical Capital				
Sales and use	Retail sales price[c]	4%		Most copper sales and machinery and equipment used directly in mining are exempt.[b]

[a]Phone call on May 15, 1978 to Bob Reed, Geological Survey, Department of Natural Resources, Michigan.
[b]Phone call on May 12, 1978 to Rick Willits, Office of Tax Revenue and Analysis, Michigan.
[c]Phone call on May 12, 1978 to Rick Willits, Office of Tax Revenue and Analysis, Michigan.

Table 5-30
1977 Montana Taxes on Copper

Tax Type: Economic Definition	Tax Base: Legal Definition	Nominal Tax Rate	Effective Rate from the Literature	Remarks
Output				
Metalliferous-mines license (84-2002-2005)	Gross value of smelter output	<$100,000 0.15% $100,000– 0.575% $250,000– $250,000– 0.86% $400,000– $400,000– 1.15% $50,000 >$500,000 1.438%		Based on annual average value as published in *Engineering and Mining Journal.*[a]
Mineral mining (84-7003-7006)	Gross value of smelter output	$25 +0.5% of gross value over $5,000		
Profit				
Corporation license (84-1501-1502)	Net income from state sources	6.75%		Net income is state portion of federal taxable income, with adjustments.
Property (84-301)	3% of gross value of smelter output	Average 20.1%[b] (1976)		Rate varies by county. Gross value determined as in metalliferous-mines tax.
Physical Capital				
Property (84-301)	12% of original installed cost[a]	Average 20.1%[b] (1976)		Rate varies by county. New developments are assessed at 7% for 5 years. There are specific exemptions for pollution control.[c]
Land				
Property (84-301)	12% of market value of land.[a] Price paid to government on patented mines and mining claims. Market value on reserved rights of entry.	Average 20.1%[b] (1976)		Rate varies by county. See above remarks.

[a]Phone call to John Clark, Research Division, Department of Revenue, Montana.

[b]Letter of January 3, 1977 from Kenneth K. Morrison, Property Assessment Division, Department of Revenue, Montana.

[c]Patricia Foley, Massachusetts Institute of Technology.

Table 5-31
1977 Nevada Taxes on Copper

Tax Type: Economic Definition	Tax Base: Legal Definition	Nominal Tax Rate	Effective Rate from the Literature	Remarks
Profit				
Net proceeds of mines (362.120, 362.270)	Gross income less costs of production, transportation, marketing, etc.	4% average 5% maximum		Net proceeds of mines tax imposed on net value of output and value of personal property (capital)
Physical Capital				
Net proceeds of mines	35% of market value	4% average 5% maximum[a]		
Sales and use (372.105, 372.185)	Retail sales or purchase price	2% state 1% county		Utility purchases exempt
Land				
Property (361.045, 362.040)	35% of market value	4% average 5% maximum	2.8% of revenue[b]	Property tax is assessed only on real property of patented mines on which less than $100 in improvements has been made in past years.

[a]Phone call to Erline Hines, Nevada Department of Revenue.

[b]D.W. Gentry, "Financial Modeling of Mining Ventures—The Effects of State Mine Taxation," in *Non-Renewable Resource Taxation in the Western States*, Lincoln Institute of Land Policy Monograph 77-2 (Cambridge, Mass., 1977).

Table 5–32
1977 New Mexico Taxes on Copper

Tax Type: Economic Definition	Tax Base: Legal Definition	Nominal Tax Rate	Effective Rate from the Literature	Remarks
Output Severance	If product has posted or market price at point of production. The base is this price less expenses incurred in hoisting, crushing, and loading necessary for putting product in marketable form not to exceed 50% of gross sales value; and rental and royalty payments belonging to state and United States[a] If product must be processed before sale, the base is proceeds from first sale less processing costs and freight charges.[a]	0.5%[a]		
Resource excise (resource tax, processor tax, service tax)	Sales price less royalties paid to state or United States, and sales of copper to state, United States, and certain nonprofit organizations[a]	0.75%[a]		
Profit Corporation income (72-15A-6)	Net income	5%		Net income is federal tax base, adjusted.

Property	Mineral property is valued at 300% of annual net-production value. The annual net-production value is market value less royalties paid to state, United States, or Indian tribe; costs of extraction, milling, treating, reducing, transporting, and selling the product; and depreciation costs.[a] Assessed at 33⅓%.	2.38%[b] (Grant County)	Net value of production may be last year's or average of five preceding years' net value. Once one method is chosen, it must be used thereafter.[a] Appraised by the state property appraisal department.[a]
Physical Capital Property	Improvements, equipment, materials, supplies, and other properties used in connection with mine are assessed at 33⅓% of original value less straight-line depreciation.[a,c]	2.38%[b] (Grant County)	Appraised by the state property appraisal department.[a]
Gross receipts	Gross receipts[d]	4% state levy[d] 0.25% maximum city levy[d]	Effective July 1, 1978, state levy was decreased to 3.75% and maximum city levy was increased to 0.5%.[d] Paid on all plant and equipment and operating and maintenance supplies. Out-of-state purchases taxed by compensating-use tax.
Land Property	If mine is on fee title land, it is assessed on per-acre basis times assessment ratio of 33⅓%.[e]	2.38%[b] (Grant County)	An average per-acre assessment is $30.[e]

[a]Mark C. Resta, "Natural Resource Taxation in the Four Corners States (Master's thesis, University of New Mexico, 1978).

[b]Phone call on May 12, 1978 to Mrs. Hill, Property Tax Division, New Mexico.

[c]Phone call on March 27, 1978 to Billy Martin, Property Tax Division, New Mexico.

[d]Commerce Clearing House, Inc., State Tax Guide, Washington D.C., 1978.

[e]Phone call on May 15, 1978 to Paul McDonald, Property Tax Division, New Mexico.

Table 5-33
1977 Tennessee Taxes on Copper

Tax Type: Economic Definition	Tax Base: Legal Definition	Nominal Tax Rate	Effective Rate from the Literature	Remarks
Output				
Sales (67-3002)	Sales price on intrastate sales	4.5%		4.5% rate due to be reduced to 3% after June 20, 1978, but probably will remain at 4.5% for one more year.[a]
Profit				
Corporation excise (67-2704)	Net income (UDITPA)	6%		Federal taxable income, with adjustments
Franchise (67-2903)	Stock, surplus, undistributed profits at book value	0.15%		Foreign corporations taxed on amount of capital stock employed in state.
Physical Capital				
Property (67-401-734)	30% of actual value	Average 3.5%[a]		No state tax. County rates vary. Pollution-control equipment exempt.
Sales (67-3002)	Sales price	1% on machinery[a] 4.5% on machinery to remove overburden[a]		Water, gas, electricity, fuel oil, coal used directly in process exempt.[a]
Land				
Property (67-401-734)	40% of actual value	Average 3.5%		

[a]Phone call to George Tidwell, Director of Property Assessment Division, Tennessee.

Table 5–34
1977 Texas Taxes on Copper

Tax Type: Economic Definition	Tax Base: Legal Definition	Nominal Tax Rate	Effective Rate form the Literature	Remarks
Profit Franchise	Capital and paid-in surplus attributable to in-state sales	0.425%		If property tax assessed by county is greater, this amount is used. This is true for ASARCO.[b] Minimum tax for large corporations is $55.
Physical Capital Property (7145, 7245)	School: 75% of market value State and county: 27% of market value[c]	1.74% 1.25%[c]		Varies by county. Assessment rates and tax rates for highest county.
Sales and use (20.01, 20.03)	Retail sales or purchase price	4% state 1% local		Utility purchases exempt
Land Property (7145, 7245)	School: 75% of market value State and county: 27% of market value[c]	1.74% 1.25%[c]		Varies by county. Assessment rates and tax rates for highest county.

[a]Phone call to Reuben Valdez, Franchise Tax Division, Texas Department of Revenue.
[b]Patricia Foley, Massachusetts Institute of Technology.
[c]Phone call to John Dillon, Ad Valorem Tax Division, Texas Department of Revenue.

Table 5–35
1977 Utah Taxes on Copper

Tax Type: Economic Definition	Tax Base: Legal Definition	Nominal Tax Rate	Effective Rate from the Literature	Remarks
Output Occupation (59-5-67)	Gross value (sales price) less transportation costs, reduction costs if necessary to sale, and $50,000	1%	0.741% of gross value[a] 0.5% of gross income[b]	Reduction costs can include cost of assaying, sampling, refining, and transportation.
Profit Corporation income (59-13-3, 5, 6)	Net income less federal income tax	4%[c]	3.1% of net income[b]	Net income is approximately same as federal net income. Rate has decreased from 6% in 1976.[d]
Property (59-5-57)	Twice average net annual proceeds from three previous years	6.64%[d]		Nominal rate is average county levy.[d] Base is self-assessed and then adjusted by property tax department.[e]

Physical Capital			
Property	20% of fair cash value as determined by sliding-scale percentage of original cost[f]	6.6%[d]	Nominal rate is average county levy.[d] Base is self-assessed and then adjusted by property tax department.[e]
Sales and use (59-15-4, 59-16-3)	Purchase price	4% state 0.75% county limit	Copper sales are exempt.[c] Pollution-control equipment is exempt.
Prepaid sales and use (63-51-3, 59-15-4, 59-16-3)	Estimates of construction costs in developing or utilizing natural resource	4% state 0.75% county limit	Purpose of this method of paying sales and use taxes is to appropriate funds needed for schools and highways in areas of natural-resource development.
Land			
Property	$5/acre	6.64%[d]	

[a]Mark C. Resta, "Natural Resource Taxation in the Four Corner States" (Master's thesis, University of New Mexico, 1978). This is an average for 1968–1973.

[b]"Colorado's Proposed Tax Sparks Interest in Other States Mining Taxes," *Engineering and Mining Journal*, June 1975, pp. 32–37.

[c]Phone call of March 29, 1978 to Mr. Osika, State Tax Commission.

[d]Phone call of April 24, 1978 to Mr. Osika, State Tax Commission.

[e]Glenn J.S. Laing, "An Analysis of the Effects of State Taxation of the Mining Industry in the Rocky Mountain States" (Master's thesis, Colorado School of Mines, 1976).

[f]Phone call of May 1, 1978 to Mr. Osika, State Tax Commission.

Table 5-36
1977 Wisconsin Taxes on Copper

Tax Type: Economic Definition	Tax Base: Legal Definition	Nominal Tax Rate	Effective Rate from the Literature	Remarks
Profit				
Metalliferous-minerals occupation (70-375)	Average of net proceeds from preceding 3 years	Ranges from 6% for net proceeds less than $4 million to 20% for net proceeds over $30 million		Effective July 1, 1977. Tax base is self-determined. This tax replaces copper-production tax of 1.5% on market value of metals recovered. Copper price used is average for year.
Corporation Income (71-03, 71-04, 71-09(2a)]	Wisconsin taxable income, which is gross income with standard deductions for expenses and percentage-depletion allowance ranging from 15% on first $100,000 of gross income from sales of ore to 3% on income over $300,000.	Ranges from 2.1% on first $1,000 of taxable income to 7.4% on income over $6,000		Depletion allowance may not exceed 50% of net income as computed without allowance. Depletion allowance is being changed to cost-depletion allowance, which is being phased in over next 10 years.[a]
Physical Capital				
Sales (77-52)	Retail sales	4% state levy 0.5% maximum county levy		Most copper sales would be exempt.[b] Equipment and machinery used directly in mining would be exempt.[b]
Land				
Property	Surface land is valued by certified assessor at full value of private sale. Buildings and improvements are valued at depreciated value. Assessment ratio of 10% to 90% is then applied depending on location.[a]	1.779% of market value before assessment ratio is applied is an average of total levys in relevant towns[b] (1976).		Mining equipment is exempt from property taxes.[b] Generally property is reassessed when sold or every ten years.[a] Minerals in place are not assessed, but study is now being done on how to do this.[a]

[a]Phone call to Phil Bradbury, Department of Revenue Research and Analysis, Wisconsin.
[b]Phone call on May 12, 1978 to Al Davis, Department of Revenue, Wisconsin.

states are unable to provide segregated information. Individual tax returns are confidential information. Industry strives to maintain confidentiality of its internal data on costs, revenues, and taxes by mine site. An exhaustive examination of Form 10-K reports filed by forty-five extractive firms with the SEC confirmed that publicly available data on taxes paid by each firm are not sufficiently detailed for use in estimating effective state taxes. The critical deficiencies are that virtually all firms operate in multiple states and that many firms have divisions and subsidiaries which are not engaged in resource extraction and processing. Although some data segregated by major operations are available, they are not consistent from firm to firm or year to year. Thus, this avenue of approach had to be abandoned.

The remaining approach for transforming nominal tax rates to estimates of effective tax rates is to simulate firm and industry technology and behavior. An initial approach was taken by the U.S. Bureau of Mines (1972) in the 1960s in an attempt to estimate extraction costs, which would then be utilized to estimate economic reserves. This approach was employed by several investigators in the 1970s to estimate effective tax rates.

The method followed is to create a typical mine described in varying levels of detail with respect to exploration, development and capital costs (amortized over mine life), and operating costs. State and federal taxes on inputs, profits, and output are applied to this typical mine, and effective rates are estimated with respect to either the value of output or profits. In some studies, only taxes on output were considered, so that this method is vastly simplified. In other studies, taxes on inputs and corporate income taxes were considered. In these studies a detailed technological and cost model is constructed so that the complexities of various state tax codes could be accurately simulated.

The best-known efforts to estimate the effective state tax rates were made by Boyle (1977), Bronder (1976b), Laing (1976), Gentry (1977), Bradley (1979), and Stinson and Voelker (1978) for coal, uranium, and metal mining in Western states. A selection of their estimates of effective tax rates, expressed in percentage of the value of output or in dollars per physical unit of output (coal), is reported in table 5–37. Unfortunately, the studies are inconsistent. Different tax bases were estimated, and the nominal tax structures were assumed for different years. The researchers (except for Boyle) employ a hypothetical mine in order to make these estimates. This approach, sometimes known as the "mine on wheels," has a large number of methodological problems, which are discussed shortly. However, it may be concluded from table 5–37 that effective rates vary considerably among the states; and the patterns for the most part are consistent for the five researchers. This finding implies that, to the extent that taxes do not fall on economic rents, differential state taxes create distortions in the allocation of resources. The level of the resulting inefficiencies depends on the size of these differentials and the degree of resource reallocation.

Table 5-37
Effective Tax Rate on Output

State	Coal					Uranium	Metal Mining
	State and Local Taxes FY 1976 on Value of Output (%) (Boyle 1977)	Property, Severance, and Income Taxes on Value of Output (%) (Laing 1976)	Tax Rates ($/Ton)			State and Local Taxes on Value of Output (%) (Boyle 1977)	Property, Severance, and Income Taxes on Value of Output (%) (Gentry 1977)
			State and Local (Bronder, 1976b)		State and Local (Loomis 1975)		
			Surface	Underground			
Arizona	9.6	5.8	0.46	1.48	0.49		5.5
Colorado	1.4	2.1	0.16	0.45	0.09	5.3	3.2
Kansas	1.3						
Montana	25.3	28.0	1.48	1.55	1.51		4.5
New Mexico	2.7	6.1	0.26	0.90	0.16	2.0	4.0
North Dakota			0.54	0.62			
Oklahoma	0.4						
Texas						1.3	
Utah	0.6	3.9	0.09	0.23	0.08		4.4
Wyoming	10.7	9.7	0.43	1.36	0.47	8.5	7.8

Source: A.M. Church, "Market and Non-Market Evaluation of the Mining Firm," in *Non-Renewable Resource Taxation in the Western States*, Lincoln Institute of Land Policy Monograph 77-2 (Cambridge, Mass., 1977).

The mine-on-wheels concept evolved from detailed studies of typical or model mines undertaken by the U.S. Bureau of Mines (1972). Although the resulting data are rich in detail, they produce misleading and incorrect results. The first problem is that the *price* of the resource is assumed either to be fixed as net of taxes or to include federal and state taxes and is invariant by location. The error rendered in the case of copper and other commodities which are valuable per unit weight is not serious because transportation costs are relatively low. However, transportation costs for coal, as for other inexpensive and bulky commodities, are often several times the resource's f.o.b. mine-mouth price and thus are a major determinant of market price at the point of consumption.

A second methodological problem rises because the model mine holds technology and the quality of resource (ore grade, seam thickness, overburden ratio, impurities, and so on) constant because a single mine site and size are hypothesized. The theory of nonrenewable resources indicates that size of the stock, expected price, costs of extraction and processing, and interest rates are the major determinants of the rate of extraction (Hotelling 1931). Furthermore, since the lowest-cost resource is mined first, current extraction cost is a cumulative function of past extraction and is dependent on the quality and extent of resource stocks. Investment in (unmined) resource stocks, which must compete for capital, creates a rising rental price for the resource (that is, exhaustion implies a positive "user cost"). Since resource quality, transportation costs, and the prices of inputs vary by location, and factor substitution is undoubtedly possible at least to some degree, one would also expect technology, prices, and optimal mine size and extraction rates to vary significantly among various locations. Underground- and surface-mining techniques for copper and coal, which are detailed later, and variation in site production capacities support this contention.

A related problem is that the price of inputs purchased also varies from location to location. The existence of an efficient national capital market means that the cost of capital is uniform, but distortions engendered by different risks and state taxes mean that capital costs also vary from site to site. The prices of materials and supplies are sensitive to location, as are the prices of large pieces of mining equipment when remote sites are involved. However, the largest price variations are in labor costs. Wage rates are sensitive to local labor markets and vary across the United States in all industries. The variation, however, is less than if labor markets were purely competitive. National or regional bargaining by a number of unions (United Mine Workers, Operating Engineers, and the Chemical and Petroleum Workers) dampens variations. On the other hand, union work rules, informal practices, and implicit agreements coupled with state health laws and their enforcement and geologic characteristics (ore grade) cause labor productivity to vary widely from state to state. In fact, variation in labor productivity in some extractive industries owing to these factors as well as the technologies employed in extraction, processing, and environmental quality control is far greater than the variation in nominal wage rates. It is the output per dollar of labor input which is critical to resource firms, but noncompetitive elements cause productivity and wage rates not to offset each other. The result is that the productivity price of labor varies to a larger degree than nominal wage rates.

Another compounding factor in wage costs is that the sites of mines are frequently remote and inhospitable. Offshore oil sites, the Alaska North Slope, and Western boomtowns lack private and public goods and services. Theory and empirical evidence indicate that workers must be compensated for these undesirable work and living environments, which in turn creates greater variation in wage rates (Cummings and Mehr 1977).

Another problem with the mine-on-wheels methodology is that tax-shifting assumptions are implicitly made. Taxes either are shifted forward to consumers when an effective tax on output is estimated (table 5–37) or are shifted entirely onto capital when an effective tax on profitability is estimated (Laing 1976; Gentry 1977). While little empirical work on shifting of state taxes has been completed, the shifting assumptions made in these studies should be made explicit by researchers.

It is reasonable that consistent and reasonable estimates of effective tax rates require all assumptions be stated and that the model mine be appropriate to each particular location. This means that overburden must be removed or shafts sunk and the resource extracted and beneficiated at each location to give reasonable cost estimates. Variable technology and resource quality is considered in our approach when location-sensitive models for copper and coal production are developed. For coal, a complex model of supply is estimated. For copper, site-specific data on past and expected future mining costs are utilized to estimate a supply function by state. Foreign supply and demand are estimated econometrically. This task is a necessary prerequisite to comprehensive and current estimates of the effective tax rates which would be enlightening to policymakers and those wishing to forecast economic effects and ultimate tax incidence.

The Coal Model and Effective Tax Rates

The coal model, developed by Martin B. Zimmerman at the Massachusetts Institute of Technology (reported in detail in chapter 6), is segregated into several producing regions. Two technologies are assumed: underground and surface mining. The optimal-size mine in each region is a function of both local geological conditions and the economies of scale inherent in the two technologies. Separate econometric equations describing capital [lasting five, ten, or twenty years (primarily mine-development expenses)] and costs of labor, supplies, and materials are estimated to be a function of the rate of extraction and the seam thickness for underground mining or the overburden ratio (the ratio of overburden to seam thickness) for surface mining. Input quantities are transformed into costs by use of model-mine data; overhead costs are added to these estimates along with certain fixed costs such as loading, hauling, shaft, and ventilation costs. The annual cost of capital is calculated by accounting for the effect of federal tax policy, including the maximum percentage-depletion rate (50-percent-of-profits limitation is assumed), the investment tax credit, and straight-line depreciation rates utilizing a 10 percent discount rate. Estimated average-cost curves are U-shaped, and optimal mine size is assumed to correspond to the minimum point on this curve. Since user costs are low for coal because of vast domestic reserves, utilizing this point for output determination (rather than the theoretically correct point for maximization of the present value of profits which occurs at an output greater than minimum average cost) introduces no serious distortions.

Introducing geologic variables describing the quality of the resource allows depletion (resulting from cumulative output) to affect costs over time. The quality and extent of coal reserves, the level of state taxation, and the cost of railway and barge transportation are location-sensitive factors affecting costs.

Effective tax rates are estimated by applying nominal tax rates to the technologies forecast for a greenfield-investment coal mine. The technological relationships among inputs (at the minimum average cost at mine mouth) for surface and deep mining (expressed as a percentage of total costs) are presented in table 5–38 in 1977 dollars. The mining technologies are identical for the Midwest and northern and southern Appalachian regions for surface and deep mining; and only surface mining is competitive in the northern Plains and Southwest regions. The twenty-year prime capital refers to capital lasting twenty years and qualifying for the federal investment tax credit. The remaining twenty-year capital is primarily made up of mine-development costs. Equilibrium mine-mouth prices predicted by the coal model are shown in the last column and are expressed as net after state taxes. They describe new long-term contract prices. The magnitudes are not uncharacteristic of new contract prices (in 1977), and the estimated technologies confirm expectations that surface mining is relatively capital-intensive and deep mining is labor-intensive.

Table 5–38
Average Cost of Coal at Mine Mouth
(1977 dollars)

Region	Capital (%)					Labor (%)	Materials and Supplies (%)	Price per Ton[b] according to Sulfur Content
	5 Years	10 Years	20 Years	20-Year Prime[a]	Total			
Southern Appalachia								Low: 34.53 Medium: 27.94 High: 21.83
Northern Appalachia								Low: 34.53 Medium: 27.79 High: 21.76
Midwest (Illinois, Indiana, Western Kentucky)								
Surface	2	3	3	44	51	18	31	High 22.71
Deep	1	16	7	NA	24	50	26	
Northern Plains (Montana, Wyoming)								
Surface	7	2	2	27	39	16	45	Low 5.70
Mountain (Utah, Colorado)								
Surface	3	3	3	41	50	21	29	Low 17.08
Deep	1	14	4	NA	19	50	31	
Southwest (Arizona, New Mexico)	2	2	2	42	48	15	36	Low 13.15

Source: Martin B. Zimmerman, Massachusetts Institute of Technology. Used with permission.
NA = not applicable.
[a]Eligible for investment tax credit.
[b]At mine mouth net of output taxes.

Copper Technology

Copper extraction and beneficiation technology is complex because of the interdependence of the mining, milling, concentrating, and smelting processes. The effect of state taxation is complicated by the fact that smelting and refining may take place in different states either by a fully integrated firm or through toll arrangements. Site-specific data on costs for these various stages in the production process were made available from public and industry sources. Joel P. Clark and Patricia T. Foley of the Massachusetts Institute of Technology in 1979 utilized these cost and engineering data to describe average- and marginal-cost functions by site. This model is described in detail in chapter 6.

Methodology for Computing Effective Tax Rates

The models of technology of coal and copper production are utilized to convert the array of various tax bases and rates (table 5-10 through 5-36) to an overall effective tax rate based on mine-mouth output. The base for each nominal tax is applied to the most appropriate component found in the models of extracting and processing to yield total taxes per ton of coal or per pound of cathode copper. The unit taxes were then divided by either an estimated or an actual (1977) price per unit of output, to yield tax rates as a percentage of price.

Each tax is classified as to whether the tax falls on output, physical capital, profit, or land (see tables 5-10 through 5-36). The following list indicates the types of taxes commonly classified under each base:

Output base: Severance, sales, and property taxes (when the assessed value is a funciton of the value of gross revenues)

Physical-capital base: Property tax when assessed on new replacement cost or replacement cost less depreciation and sales and excise taxes on machinery

Profit base: Income, franchise, excise, license, privilege, and net-proceeds taxes.

Land base: Property tax when assessed on the present value of future profits.

The tax bases for each state are derived from the production costs of a hypothetical coal mine or copper mine/mill located in each state. It should again be noted that these *costs* are location-specific. In short, adjustments to the production costs for standard-sized mines are made to reflect the physical mining conditions in the individual state.

The coal model described previously yields per-ton cost equations for each major coal-producing region in the United States in the following form:

$$C_5 + C_{10} + C_{20} + C'_{20} + L + M = 1$$

where C_5 = annualized cost of physical capital with 5-year life

C_{10} = annualized cost of physical capital with 10-year life

C_{20} = annualized cost of physical capital with 20-year life

C'_{20} = annualized cost of capital not subject to investment tax credit

L = annual cost of labor

M = annual cost of materials and supplies

Each cost allocation is expressed as a percentage of the minimum average cost of mining one tone of coal. Both surface- and deep-mine cost equations, if applicable, are developed for each of the three sulfur-content categories for each producing region.

In the case of copper-producing states, it is necessary to obtain production cost for mine/mill operations from two sources. Capital costs are derived from a 1970 Bureau of Mines model (Bennett et al. 1973). Operating costs are based on empirical observations derived from the short-run marginal-cost equations estimated by the Massachusetts Institute of Technology (see chapter 6). The Bureau of mines model estimates capital and operating costs for various sizes of mines and mill/concentrators. The appropriate size mine/mill was chosen for each state based on the average "normal" annual production for the mines in that state and the stripping ratio (ore divided by material mined). Operating costs in the Bureau of Mines model are not used here since they reflect costs for a 1970 mine operating at capacity. Real-world numbers present a more accurate picture of capacity utilization and current technology. The capital costs for the mine and mill/concentrator are inflated to 1977 dollars by using the Marshall and Stevens price index.

Mine-site costs are converted to a per-pound basis by using the average normal annual production in the relevant state. (Note that costs for molybdenum and other by-product recovery are not included in the mill/concentrator costs.) Also, it is assumed that all concentrating is accomplished by the flotation process. These costs are broken down into component parts (labor, fuel, repair supplies, and so on) by using average copper-mining percentages for the mine and percentages derived from the appropriate Bureau of Mines model for the mill/concentrator. The per-pound capital and operating costs for each producing state are summarized in table 5–39.

Taxes on output fall on either the value or the quantity of output. In either case, the tax base is evident and taxes can be applied directly to the appropriate base. Taxes on physical capital are generally applied to the original cost, the replacement cost, or the depreciated cost of the capital. For this book, original cost and replacement cost are assumed to be the same (that is, a world without inflation). Depreciation is calculated on a straight-line basis with salvage value assumed to be zero. Thus, the starting point for all capital-tax bases is the original cost of capital per ton or per pound of output.

In order to calculate the original cost of capital for coal mining, annual capital costs are determined from the coal-mine model and are then converted to original costs. The annual cost of capital is found by taking the annual cost percentages C_i times the price per ton of output. For example, the annual cost of capital with a 5-year life is

$$(C_5)(\text{price per ton})$$

The original cost of physical capital used in coal mining is then

$$\left(\frac{\text{Annual cost of capital with } i\text{-year life}}{a_i}\right)(1 - 0.28)$$

Table 5–39
Copper Mine/Mill Costs
(1977 dollars)

	Arizona[a]	*Michigan*[b]	*New Mexico*[c]
Total Capital Costs[d]			
Mine equipment	1.8		7.4
Plant and buildings	1.7		2.0
Concentrator	47.3	72.6	41.1
Surface-mine plant and equipment		3.9	
Underground-mine plant and equipment		14.9	
Total Operating Costs			
Mine	11.7	38.1	18.7
Concentrator	14.6		17.5

[a]In cents per pound.

[b]Underground mine.

[c]Surface mines.

[d]Note that these are total figures and not annual figures.

where a_i = annual return necessary to realize after-tax return of r percent

   i = 5-, 10-, or 20-year life

The adjustment $(1 - 0.28)$ accounts for the initial engineering, overhead, and construction costs built into the annual cost-of-capital percentages. These costs are estimated in order to approach the cost of physical capital. The value of a_i (the capitalization rate) is determined from

$$a_i = \frac{[1 - (U - UV)(1 - F/T)[2/(rT)]}{F(1 - U + UV)} (1 - 0.1)$$

where U = corporate income tax rate

   V = depletion allowance expressed as percentage of gross profit

   r = after-tax rate of return desired

   T = life of capital

   $F = (1 - e^{rT})/r = \int_0^\infty (T_e - rT)\,(dT)$ (discount factor)

The adjustment $(1 - 0.1)$ is an allowance for the investment-tax-credit provisions of the federal tax code. However, this factor is not applied to C_{20} since this represents capital *not* subject to the investment-tax credit. For this book, it is assumed that $r = 0.10$ (that is, all equity is financed at a cost of 10 percent), $U = 0.48$, and $V = 0.50$, with the resulting values for a_i being

   $a_5 = 0.2394$ $a_{10} = 0.1543$ $a_{20} = 0.1183$ $a'_{20} = 0.1314$

For example, the original cost of capital with a 5-year life (K_5) would be found from

$$K_5 = \left(\frac{C_5 P}{a_5}\right)(1 - 0.28)$$

where P = price per ton.

   The tax base B for taxes on the original or replacement cost of capital used in coal mines is then

$$B = (K_5 + K_{10} + K_{20})(R)$$

where R = the assessment ratio. for taxes on the current (depreciated) value of capital, we assume that the coal mine has a 20-year life and is halfway through that life (10 years old). Thus, the tax base is

$$B = (K_5 + K_{10} + \tfrac{1}{2}K'_{20})(R)$$

Capital not subject to the investment-tax credit (K_{20}) includes such items as land for which the tax base is calculated differently.

In the case of copper mine/mill operations, the Bureau of Mines model yields original costs directly. These costs are inflated to 1977 dollars and put on a per-pound basis, as described earlier. The tax base for taxes on the original or replacement cost of capital is then

$$B = (E + P + C)(R)$$

where $E =$ original mining-equipment cost per pound of output

$P =$ original mine plant and building cost per pound of output

$C =$ original concentrator cost per pound of output

The capital costs per pound are for surface-mine plant and equipment (except in the cases of Michigan and Montana, which have an underground mine) and the mill/concentrator. Mine plant and buildings and the mill/concentrator are assumed to have a 15-year life, and mine equipment is assumed to have a 10-year life. For the underground mine, all physical capital is assumed to have a 15-year life. These time spans are chosen since they were used for depreciation purposes in the Bureau of Mines copper model. If we assume that the mine and mill/concentrator have a 15-year lives and are 10 years old, then the tax base for taxes on the current (depreciated) value of capital is

$$B = [F + \tfrac{5}{15}(P + C)](R)$$

For the underground mine in Michigan, the base is

$$B = \tfrac{5}{15}(\text{all capital costs})(R)$$

Finally, each of the above bases is multiplied by an assessment ratio R to arrive at the final state-tax base BR.

Taxes on profits are in the form of an income tax. Thus, the determination of the tax base requires first the determination of a base-income figure and then adjustment of this figure to account for additions and deductions required by the various states.

For both coal and copper, annual net income is assumed to be 10 percent (10 percent rate of return) of the value of initial capital invested in the mine or mine/mill. Thus, net income I in coal-mining states is

$$I = (0.1)(K_5 + K_{10} + K_{20} + K'_{20})$$

In copper-mining states, gross income is

$$I = (0.1)(E + P + C)$$

In the case of the underground copper mine in Michigan, gross income would be 10 percent of all underground-mine and mill/concentrator capital costs.

The net-income estimate is adjusted to account for various federal deductions and credits including the depletion allowance, by using estimates computed from corporate tax returns (U.S. Department of the Treasury 1978). The resulting federal-income-tax base B_f for coal-mining states is

$$B_f = 0.59I$$

and for copper-mining states is

$$B_f = 0.61I$$

Four variations of the federal income tax base are utilizd by states in determining their tax base. The first variation is simply that the state-tax base is the same as the federal-tax base, so that

$$B = B_f$$

The second variation is that the state allowed the deduction of federal income taxes from the state-tax base, so that

$$B = B_f - 0.28I \qquad \text{for coal mines}$$
$$B_f - 0.294I \qquad \text{for copper mines}$$

The 28 and 29.4 percent adjustments are the effective, average federal corporate income tax rates for coal- and copper-mining corporations, respectively. The third variation is that the state does not allow the deduction of state income taxes (deductible for the federal base) in calculating the state bases. Therefore, the state-tax base is

$$B = (1 + t)(B_f)$$

where t = nominal state income tax rate.

The final variation is that the state does not allow the deduction of the federal depletion allowance in calculating the state-tax base. The state-tax base in this case is

$$B = B_f t(dP)$$

where d = effective coal-depletion allowance and P = price of coal. There are no copper-mining states using this variation.

Taxes on land are based on the value of minerals remaining in the ground. However, it is difficult or impossible to make accurate estimates of the quantity of minerals in place, the costs of recovery and processing, or future market prices. As a result, many states base property-tax assessments on proxies for the present value of the income stream expected to be generated by production out of a mineral deposit. The proxies are usually a multiple of gross sales or of net profits, computed by a standardized formula or in accordance with a firm's accounting practice. In the former case, the tax base is actually output rather than the value of mineral deposits. The major exception to this practice is in Arizona, where great administrative and judicial energy is spent determining and arguing about estimates of the present value of net-income streams from each site. We assume an annual return on capital of 10 percent and capitalize it over 10 years. This is multiplied by Arizona's 60 percent assessment ratio to determine the property-tax base.

The tax-base calculation methods discussed above yield tax bases (B) in terms of dollars per ton of coal produced or per pound of copper produced. As a result, determination of the effective tax rate is a simple matter of multiplying the nominal tax rate by the tax base and then dividing the result by the price per ton or per pound:

$$T = tB$$

$$T' = T/P$$

where t = nominal tax rate

T = total taxes per ton or per pound of output

P = price per ton or per pound of output

T' = effective tax rate

For coal, P is the price per ton f.o.b. mine mouth (see table 5–38). Table 5–40 summarizes the total effective tax rates for both underground and surface mining in the coal-producing states.

Table 5–40

Effective State and Local Tax Rates on Coal in 1977

(percent)

State	Total Taxes		Property Tax	
	Surface	*Underground*	*Surface*	*Underground*
Alabama	5.74	3.49	1.06	0.53
Arizona	15.92	—	9.89	—
Colorado (1978)	8.90	5.48	3.32	1.94
Illinois	13.19	8.84	6.10	3.00
Indiana	2.57	1.14	1.43	0.70
Kentucky	11.79	8.76	3.22	1.59
Montana	45.99	—	7.02	—
New Mexico	6.27	—	1.72	—
North Dakota	16.63	—	0.18	—
Ohio	5.71	3.19	4.77	2.55
Pennsylvania	3.12	2.32	0.36	0.63
Tennessee	10.04	6.31	4.23	1.81
Utah	2.56	—	1.81	—
Virginia	2.73	2.44	2.10	2.00
West Virginia	6.23	5.02	2.00	1.03
Wyoming	19.86	—	2.32	—

Note: Tax rates are with respect to gross revenue.

Table 5–41

Effective State Taxes on Copper in 1977

(percent)

State	Tax	Published Price	Operating Costs
Arizona	Education and special excise, transaction privilege (output base)	2.50	2.50
	Corporate income (capital base)	0.48	0.41
	Property tax (capital base)	4.41	3.66
		7.39	6.57
Michigan	Single business tax (capital base)	0.32	0.26
	Property tax (capital base)	4.36	3.62
		4.68	3.88
New Mexico	Severance and natural-resource excise (output base)	1.25	1.25
	Corporate income tax (capital base)	0.37	0.30
	Property tax (capital base)	0.71	0.68
	Gross-receipts tax (capital base and on O+M)	1.64	1.36
		3.97	3.59
Utah	Occupation tax and sales-and-use tax (output base)	3.12	3.12
	Corporate income tax (capital base)	0.15	0.12
	Property tax (capital base)	1.86	1.54
		5.13	4.78

Note: Published prices are in *Engineering and Mining Journal*, the U.S. domestic refinery price with adjustment. Total operating costs do not include depreciation; a 10 percent rate of return is assumed.

For copper, P is based on the U.S. domestic refinery price published in the *Engineering and Mining Journal* or total operating costs. Since we are considering taxes on only the mine/mill operation, the costs of smelting and refining are deducted from the refinery price. By using proprietary empirical data, the average smelting costs are $0.25 per pound. Subtraction of this cost from the 1977 U.S. domestic refinery price of $0.65808 per pound yields a mine/mill price of $0.40808 per pound. Nominal tax rates are found as described above. The final, overall effective tax rates are summarized in table 5–41.

Readers are cautioned that since hypothetical mine models were used to calculate portions of the effective tax rates and prices in the case of coal, the rates may not be consistent with those derived from other estimating techniques and should not be used out of context. However, a consistent methodology was used for all states in calculating the tax rates, and thus the relative effective tax rates are accurate.

In chapter 6, the coal and copper industry models are described in detail, and various tax regimes are simulated in order to evaluate efficiency losses and who bears the burden of the tax.

6 Simulating Tax Policy in the Copper and Coal Industries

In the previous five chapters the economics of tax policy and how the effects of tax policy may be estimated and evaluated are developed. However, data restrictions preclude accurate quantifications of how taxes affect extractive industries. Taxes are only one among many variables which affect the resource industries. Consequently, it is difficult to separate tax-induced effects.

In this chapter, combined economics and engineering models of the domestic copper and coal industries are employed to simulate the outcomes of existing (1977) tax policy and alternative tax regimes. The predictions of these simulation exercises are evaluated in order to determine who bears the tax burden and the magnitude of the resource dislocation imposed on society. Recall that the effects of tax policy are evaluated in two ways: how the final incidence of the tax affects the distribution of income (tax burden) and how the tax-induced reallocation of resources affects economic efficiency (by using the Pareto-optimality criterion). The only tax which has no effect on resource utilization is on economic rent. However, as pointed out previously, it is virtually impossible to specify and administer a tax which falls solely on this real, but elusive, quantity. Nevertheless, rents are a seductive target for governments and other parties which seek to increase their incomes while minimally disrupting the economic system. Individuals and groups in the private sector seek to usurp rents by means of bargaining power and through changes in property rights (the rules of the game as expressed in legislative statutes and common law). Governments both create and socialize them by means of taxes and expenditure and regulatory policies. The coercive power of government makes its policies ideal for creating, redistributing, and destroying economic rents. We concentrate on how state taxes may achieve these ends in the copper and coal industries.

The copper industry has been plagued by an unstable national and international marketplace in the 1970s and the domestic requirements for large-scale, pollution-related investment. Economic rents accruing to the industry may be essential to enable it to achieve mandated pollution standards and compete in world markets. The coal industry is also faced with numerous problems. Among these are declining productivity, increasing regulation (and regulation of its consuming industries), and labor problems. The result is rapidly expanding output in the Western states and stagnant conditions in the Midwestern and Appalachian producing areas. Economic rents are growing for the low-cost, low-sulfur producers in the Powder River

Basin of Montana and Wyoming. The result is a scramble for others than the coal producers to appropriate those rents. To date, the more successful usurpers have been the railroads via increased rates for coal unit trains (government-regulated rates), states via increased taxes (as reported in chapter 5), and to a lesser extent consumers (via lawsuits contesting Montana's high-severance-tax structure and proposed federal legislation to limit coal severance taxes to 12.5 percent). This exercise in tax-policy simulation is constructed in order to add information for those considering these questions. The copper and coal models are summarized next, and then the tax simulations are described and evaluated.

The Copper Model[a]

U.S. Copper Supply: An Economic/Engineering Analysis of Cost-Supply Relationships

The long-term economic viability of the U.S. copper industry is of concern to industry analysts, government policymakers, and others interested in the financial health of this basic industry. Much of this concern results from the dismal financial performance of the industry from late 1974 through mid-1979. Such a long period of extremely poor financial returns had not existed in the U.S. copper industry in the previous four decades. Reasons postulated for the poor performance include an unfavorable supply/demand balance owing to "excess" production in developing countries, often not entirely motivated by profit considerations, and increased domestic costs owing to higher state taxes and more stringent state and federal regulations (Foley 1979; Foley and Clark 1980).

The purpose of this model is to develop short- and long-run supply schedules for the primary U.S. copper industry and combine these with demand estimates so that alternative tax policies may be simulated. The supply portion of the model is derived from site- and input-specific cost data and can be used to evaluate the effects of changes in policy, market, and other variables on U.S. copper supply.

Supply and Cost. In chapter 1, the supply function for coal producers was hypothesized as being linear and of infinite elasticity. This assumption is overly restrictive, for it allows for neither depletion nor site-to-site differences

---

[a]This description of the copper model is adapted from research undertaken by Patricia T. Foley and Joel P. Clark of the Department of Materials Science and Engineering, Massachusetts Institute of Technology. I wish to express my appreciation to them for having made this material available.

within each state. Since copper is produced in a handful of states, site-specific considerations within each state should be accounted for. Basic economics holds that the marginal-cost (MC) function for each site and for each firm "acts like" the supply curve because the profit-maximizing firm produces at the point where marginal cost is equal to the price current in the marketplace. In the short run, when plant and equipment are fixed, the firm will shut down a site if price is less than average variable cost (AVC); in the long run, shutdown occurs if all average (unit) costs are less than price.

The supply curve for the entire industry (or state) is the horizontal summation of the relevant portion of each firm's site-specific marginal-cost curves. Thus the supply curve is a function of costs of production. However, in the case of nonrenewable resources, the profit-maximizing firm follows a somewhat different path by operating at a point between where price intersects marginal cost and the lowest average cost, because of depletion and eventual exhaustion. The copper model is static so that depletion cannot be explicitly accounted for. However, we analyze individual copper mines so that the effect of lower ore grades and higher costs in newer mines and those being evaluated for greenfield investment captures in large part the effects of depletion on costs.

Because it is difficult to measure how marginal costs are affected by the level of output at each site, we assume that marginal cost is a constant up to the capacity of each mine/mill/smelter complex. The effect of this assumption is shown in figure 6–1 for three sites of three firms (1, 2, 3). Marginal cost is constant up to capacity output for each site and thereafter is infinite. The sites are arrayed in order of their costs, and the result is the supply curve. This function is estimated for each site and aggregated to derive supply curves for each state and for the United States. The effect of tax

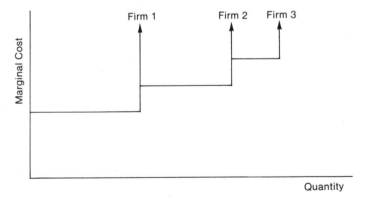

Figure 6–1. Marginal-Cost Curves Based on the Assumption of Constant Marginal Costs to Maximum Capacity and Infinite Marginal Costs Thereafter

structure is incorporated into the cost structure for each site in each state. In the short run, taxes which are fixed and invariant with respect to output do not affect the marginal-cost schedule; however, taxes affecting operating expenses shift it upward by the amount of the tax (the effective tax rates estimated in chapter 5 are applied in the tax scenarios). In the long run, all inputs are variable. Another way of stating this is that capital and labor are perfectly mobile and mine sites are operating only where positive profits can be earned. The area between marginal cost (in figure 6–1), including taxes and the market price, measures economic profits for each site. Sometimes this area is called *producers' surplus*, or *quasi-rents*.

Engineers, accountants, and economicsts each view costs in their own way. To the engineer, costs are the actual dollar values of materials and labor necessary to construct and operate a mine or plant; to the accountant, costs are the bookkeeping numbers used for tax calculations and reports to shareholders; and to the economist, costs are the payments to resources and factors of production necessary to keep them in their present employment.

In industrial practice, costs are separated into two categories. Capital or investment costs imply expenditures that lead to assets of lasting value. Operating or production costs are costs that are continually or periodically being spent to keep the operations going.

Capital costs usually are concentrated at the beginning of a venture. For accounting purposes, many capital costs are depreciated throughout the economic life—not necessarily the actual engineering life—of a project. Depreciation costs can be deducted when net taxable income is calculated, thereby reducing the corporate-tax bill for the firm. Because of this, depreciation must be considered in calculating the income-tax component of operating costs. Depreciation, however, is not a cash expenditure and should not be included in cash-flow analysis.

Capital-cost estimates for future mines are needed to calculate the opportunity costs included in the longrun supply function. The capital costs of existing mines have already been incurred and do not affect unit production costs. Therefore, except for depreciation effects, these costs can be ignored in the development of the short-run supply functions.

Operating costs are classified as fixed, variable, and semivariable. Fixed costs are the inevitable costs that must be paid regardless of the level of output or the amount of resources used. Examples of fixed costs are property taxes (which in Arizona must be paid for three years after a mine ceases production) and the salaries of top management and administrative personnel who are paid on a yearly basis, independent of seasonal fluctuations in the volume of production.

The distinction among fixed, variable, and semivariable costs depends on the time horizon. In the long run, when new plants can be built and mines developed, all costs are variable. In the short run, variable costs are those that change with the level of output. For the copper industry, they include such expenses as wage labor, power, operating supplies, and fuels. Semi-variable costs are costs that can be changed without building new facilities, but cannot be adjusted easily when output changes. Contract provisions for hiring and firing untion labor and agreements to purchase a specific amount of electricity from local utilities tend to make costs semivariable. While the distinction between variable and semi-variable costs is important in very detailed engineering/economics studies, in this analysis semivariable costs are considered variable in the short run.

In addition, operating costs can be classified as direct and indirect. Direct costs are those associatd with the manufacture of a specific product. Direct costs include materials, utilities, direct and maintenance labor, and payroll overhead. Indirect costs are those expenditures which either have a supportive function or are associated with one or more departments or products of the firm. They include technical and clerical labor, administrative costs, facilities' maintenance and supplies, and research. The borderline between direct and indirect variable and fixed costs is often arbitrary.

Because fixed costs are constant in the short run and because they are incurred even if the firm produces no output, they are irrelevant in the production decision for an existing plant. What is important in the short run is that the point where price equals marginal cost be above the level of average variable cost.

In the long run, total costs must include fixed costs and opportunity costs. If the market price is expected to be below the marginal cost at an existing mine for a significant time, that facility will close down. In addition, new deposits with costs expected to be higher than long-run price will not be developed.

*Cost Estimation: Analysis of the Process and
Statistical Approaches*

Just as the concept of cost has a different meaning to engineers and to economists, so does the notation of cost estimation. Generally, engineering cost estimation involves a detailed assessement of the production process, whereas the economic or statistical approach focuses directly on sample observations of costs and outputs to estimate the cost function.

In engineering cost estimation, flow diagrams are used to describe complex production processes in terms of smaller units of production. (For example, in open-pit copper mining, the units of production might be drilling, blasting, loading, and haulage.) Inputs are estimated for the individual units of production. Often, these inputs are specified in terms of their technical characteristics, such as the number of British thermal units of fuel.

Economists have attempted to construct production and cost functions from this technical information. In the economics literature, this is referred to as the *process-analysis approach* (Walters 1963; Griffin 1972). Its advantages are the known range of applicability of the data and the relative ease with which technological changes can be incorporated. Process and plant produciton functions are useful for a wide variety of purposes. They are useful, for example, in constructing estimates of plant (or mine) cost curves and in forecasting raw-material and labor requirements.

The use of process analysis has been aided by the development of linear-programming (LP) models. In LP models, the technological opportunities of the firm are represented by a finite number of activities, which attempt to describe in engineering or physical terms the possible options open to the producer in selecting the optimal production combinations. Technical limitations are introduced in the form of contraints, such as materials and energy balances, limitations of raw-material inputs, quality specifications on production, and so forth. Once the production function and constraints are described, the cost function can be derived by solving the linear-programming problem. This is done by assuming optimization behavior, that is, by minimizing the cost of various output levels, given the input prices.

When linear programming is used for process analysis, its disadvantages must be kept in mind. One of the disadvantages of an LP description of the production function is that the matrix of technical coefficients is fixed for all output levels. (This problem may sometimes be overcome by approximating nonlinear functions by linear segmens.) The second and (for our purposes) most critical disadvantage is that the LP results describe what a firm "ought" to do to minimize costs, not what it actually does.

Rather than gathering engineering data to measure the production function and then using the process analysis to derive the cost function, many economists have attempted to estimate cost and supply functions directly by using accounting (and occasionally engineering) cost data (Walters 1963; Johnston 1970). In practice, measuring cost curves is more convenient than estimating production functions since the available accounting data are normally reported in monetary terms. The major motivation behind this type of cost analysis has been to statistically test economics theories regarding the shape of short-run cost functions. In addition, almost all the existing copper supply-demand models have statistically estimated supply functions (Hartman 1977; Charles River Associates 1970; and Fisher, Cootner, and Baily 1972). [The execptions are systems dynamic (Pugh-Robert Associates 1976) and linear-programming models (Hibbard et al. 1977).]

The Economics/Engineering Approach Determining the
Shape of the Supply Curve

The supply functions presented here are intended to represent the production processes of the copper industry. It is important, therefore, that the approach taken in deriving these supply functions be compatible with engineering aspects of mining and production as well as with economics theory. The approach used here combines techniques from engineering process analysis and statistical cost estimation to build a supply curve compatible with economics theory and engineering reality.

Economics theory suggests U-shaped marginal- and average-cost curves, but empirical evidence presented in the economics literature has supported the alternative hypothesis of constant-marginal-cost curves for firms in some industries (Walters 1963; Johnston 1970). Attempts were made to test the theory of U-shaped cost curves in copper mining by using statistical regression techniques. Time-series data on mining cost, average daily production, and yearly production were available for three open-pit mines. Six observations were available for two of the mines and four for the other one. The following regression equations were estimated for each of the mines:

$$\mathrm{MNCST} = \alpha_0 + \alpha_1 \mathrm{DPROD} + \alpha_2 \mathrm{DPROD}^2 + \alpha_3 \mathrm{DPROD}^3$$

$$\mathrm{MNCST} = \beta_0 + \beta_1 \mathrm{YPROD} + \beta_2 \mathrm{YPROD}^2 + \beta_3 \mathrm{YPROD}^3$$

where MNCST = mining cost in constant 1972 dollars

     DPROD = average daily production in each year

     YPROD = yearly production

The results are inconclusive. Most of the estimated equations had low correlation coefficients and t statistics that were insignificant at the 90 percent confidence level. There were also indications of multicollinearity in the data.

The poor results may be partially attributable to the inapplicability of the classical cost-output hypothesis to natural-resource economics. Mining costs are negatively correlated with the ore grade mined, and the grade of ore mined decreases over time. In the United States, for example, the average yield of copper from ore as reported by the Bureau of Mines decreased from 0.55 percent in 1971 to 0.47 percent in 1975. Therefore, when the cost-output relationship is estimated over a six-year period, changes in ore grade have important effects, and these must be considered. An easier way to test the cost-output hypothesis, if the data were available, would be to conduct the analysis during a period (such as a week, month, or quarter) when the ore grade did not change significantly.

Although the attempt to verify or disprove the economics hypothesis of U-shaped cost curves was inconclusive, the alternative hypothesis of constant marginal costs over a range of output was adopted for this book. There are a number of reasons for this assumption. One is that constant marginal costs have been observed in other industries, and there is reason to believe that this is also true for copper production. In the short run, copper mines and mills vary output by increasing or decreasing the number of shifts worked per week, thus keeping marginal costs fairly constant. Inventory control of concentrates is used to keep daily smelter production at a reasonably steady level. When a serious oversupply or undersupply of concentrate exists in inventory, the mine or smelter is shut down for a short period. A second reason is that operating personnel within the copper industry seem to believe it. (During interviews people at different copper companies were asked about the cost-output relationship. No one was aware of internal studies concerning short-run cost-output functions. However, in all cases it was thought that the assumption of constant marginal cost was valid.)

In a competitive market, the horizontal summation of marginal-cost curves of production units within the industry forms the supply function for that industry. In the case of primary copper production—which is now considered to be competitive in the world market (Mikesell 1979) and where firms may have one or more production sites, each with independent production costs—the supply schedule for the industry is a summation of the marginal-cost curves of each mining, milling, smelting, and refining complex.

Determining the Site- and Input-Specific Costs. Site-specific open-pit mining and sulfide milling costs were derived by using regression equations, and input-specific costs were derived by using detailed cost data provided by a number of companies. The data were grouped by type of operation—open-pit mine, underground mine, sulfide mill, vat leaching, heap leaching, waste-dump leaching, precipitation plant, or solvent extraction-electrowinning plant—and by input—wage and salary labor, mine, mill or plant supplies, power, property tax, administrative costs, and townsite costs. Next, input-cost factors (the percentages of total mining and milling costs spent on a specific input) were calculated for the sites and operations for which data were available. These factors were used to calculate input-specific costs for all the operations at the mine/mill sites listed in table 6–1. The input-cost factors that were applied at each site were derived from detailed data at similar facilities. For example, input-cost factors calculated from the detailed

Table 6–1
Mines Included by Type and State

Open-Pit Sulfide Mines
Arizona
   Anamax   Twin Buttes
   ASARCO   Mission
                                   Silver Bell
                                   Sacaton
                                   San Xavier North
   Cities Service   Pinto Valley
   Cyprus   Bagdad
                                   Pima (closed in 1977)
   Duval   Mineral Park
                                   Sierrita
                                   Esperanza (closed in 1977)
   Inspiration   Inspiration Area (sulfide ore)
                                   Inspiration Area (dual-process ore)
   Kennecott   Ray
   Phelps Dodge   Metcalf
                                   Morenci
                                   Ajo
Montana
   Anaconda   Butte
New Mexico
   Kennecott   Chino
   Phelps Dodge   Tyrone
   UV Industries   Continental
Utah
   Kennecott   Utah

Underground Sulfide Mines
Arizona
   Magma   San Manuel
                                   Superior
   Phelps Dodge   Safford
Michigan
   Louisiana Land and Exploration   White Pine
Montana
   ASARCO   Troy
New Mexico
   UV Industries   Continental
Utah
   Anaconda   Carr Fork

Heap Leaching
Arizona
   Anamax   Twin Buttes
   Cities Service   Miami
   Cyprus   Bagdad
Nevada
   Duval   Battle Mountain

Vat Leaching
Arizona
   Inspiration   Inspiration Area (dual-process ore)
   Kennecott   Ray

Table 6–1 *(continued)*

Waste-Dump Leaching	
Arizona	
ASARCO	Silver Bell
Duval	Mineral Park
	Esperanza
Inspiration	Inspiration Area
Kennecott	Ray
Phelps Dodge	Metcalf
	Morenci
New Mexico	
Kennecott	Chino
Phelps Dodge	Tyrone
Utah	
Kennecott	Utah

data of a small, open-pit sulfide mine were applied to other small, open-pit sulfide mines, and so on.

Total input costs per year at each mine or mill then were divided into fixed and variable costs. In the short run, salary labor, property taxes, administrative costs, and townsite costs were assumed to be fixed. Wage labor, mine, mill and plant supplies, and power costs were considered variable.

Production costs for refined copper were obtained by tracing the flow of copper-bearing material from the mine to the smelter to the refinery. Fixed and variable smelting, freight, and refining charges were added to the mine/mill cost at each site, yielding a cost of refined copper production at each mine. Smelting-cost data used in this book are site- but not input-specific. (The term *smelting cost* refers to the cost of smelting plus freight and refinery charges.) For mines that have their concentrates custom- or toll-processed, smelting charges are assumed to be a variable cost. For vertically integrated producers, smelting costs have both fixed and variable components.

Mine/Mill and Leaching Costs. For some mine/mill or leaching operations, the actual production costs were known; for others they had to be estimated. Econometric techniques were used to estimate the production costs for open-pit sulfide mine/mill operations.

To a large extent, mining and milling costs depend on the physical parameters of production such as ore grade, stripping ratio, percentage recovery of copper, and capacity utilization. Data for these parameters at each mine are generally available in corporate annual reports or SEC form 10-K. If the functional relationship between cost and these physical parameters were known, these available data could be used to estimate site-specific production costs.

An accurate and internally consistent set of data containing mining and milling costs along with the physical parameters of production at four mines over six years (1972–1977) was made available by one of the major copper producers. These data were used to estimate econometric-cost equations, which in turn were used to estimate production costs at other mines and mills. Data were available on mining costs, milling costs, ore grade, stripping ratio, percentage recovery of copper from ore, and yearly and daily capacity utilization at the mill. The time-series and cross-sectional data were pooled, and the resulting seventeen data points (not all the mines were in operation over the entire six years) were sufficient for a multiple-regression analysis.

Pooling time-series and cross-sectional data often adds a new dimension of difficulty to the problem of regression analysis. The structure of the disturbance term in the equation to be estimated may become quite complex, since the disturbance term is assumed to result in part from the effects of omitted explanatory variables. The difficulty arises because with pooled data, the disturbance term is likely to consist of time-series-related disturbances, cross-sectional disturbances, and a combination of both. The estimation technique used involved the addition of dummy variables to the equation to be estimated to allow for changing cross-sectional and time-series intercepts. The covariance model was used to improve the estimation of mining- and milling-cost equations.

In our analysis, mining costs were postulated to depend on the ore grade mined, the stripping ratio, and the percentage recovery of copper. This may be expressed quantitatively as

$$MNCST = f(ORGRDMN, STPRT, PCTRCV)$$

where $MNCST$ = mining cost in 1972 dollars

$ORGRDMN$ = ore grade mined

$STPRT$ = stripping ratio

$PCTRCV$ = percentage recovery of copper

The following equation was used to test this functional hypothesis:

$$MNCST = \beta_0 + \beta_1 ORGRDMN + \beta_2 STPRT + \beta_3 PCTRCV$$

Several alternative forms of this equation were tested and rejected. The form utilized in this simulation is (corrected for second-order autocorrelation)

$$MNCST = 45.0 - 18.97 ORGRDMN + 1.98 STPRT - 0.36 PCTRCV$$

$$(4.3) \qquad\qquad (6.3) \qquad\qquad (4.1) \quad \bar{R}^2 = .92$$

These equations were tested by estimating the mining cost from the physical parameters of mines for which mining costs were known. The results were satisfactory when the values for ore grade, stripping ratio, and percentage recovery were within or close to the range of those used to derive the estimating equation: ORGRDMN, 0.58 to 0.74; STPRT, 0.52 to 8.4; PCTRCV, 76.94 to 88.49.

It was more difficult to formulate a hypothesis for milling costs because most of the parameters which determine milling costs (such as the hardness and grindability of ore) are not easily quantified and are not published in nonproprietary sources. Milling costs were postulated to depend on ore grade, percentage recovery of copper, and capacity utilization at the mill. Capacity and capacity utilization were measured on an empirical rather than a design basis. Both daily and yearly capacity utilization were tested in the equation.

$$MLCST = f(ORGRDML, DCAPUE, PCTRCV, YCAPUE)$$

where MLCST = milling cost in 1972 dollars

ORGRDML = ore grade milled

PCTRCV = percentage recovery of copper

DCAPUE = daily capacity utilization, defined as the average daily production in a given year divided by maximum average daily production from 1972 to 1977.

YCAPUE = yearly capacity utilization measured on an empirical basis, defined as yearly production for a given year divided by the maximum yearly production from 1972 to 1977

The following equation was used to test this hypothesis:

$$MLCST = \alpha_0 + \alpha_1 ORGRDML + \alpha_2 DCAPUE + \alpha_3 PCTRCV + \alpha_4 YCAPUE$$

Numerous alternative formulations were estimated; problems were encountered, however, in applying the results of the covariance model to the actual data. Attempts were made to discern which factors made mill 1 different from mill 2, and mill 2 different from mill 3. The differences could not be attributed to size or to the presence of plants for by-product recovery. Without knowing the factors responsible for the variability in milling costs among the three mills, it was impossible to use the covariance model to estimate costs for other mills. Therefore, this equation, which exhibited good test statistics and coefficients similar to those used in the covariance model, was used to estimate milling costs:

$$MLCST = 24.3 - 15.97ORGRDML - 5.20YCAPUE$$

$$(3.8) \qquad\qquad (4.6) \qquad\qquad \bar{R}^2 = .76$$

Costs for underground mining and for leaching operations were derived by modifying actual estimates at similar operations to take into account site-specific differences. This method is not as satisfying as the regression technique, but only about one-fourth of U.S. copper is supplied by underground mines or by leaching operations.

After total costs were estimated for mining, milling, and leaching operations, they were broken down into fixed and variable costs. This was done by using input-specific factors.

The input-cost factor is the percentage of total cost (or, in some cases, direct cost) spent on a specific input. For example, if the total cost at a waste-dump leach operation were $2.7 million and the scrap-iron costs were $900,000, then the input-cost factor for scrap iron would be 0.33. Input-cost factors were calculated for operations for which input-specific cost data were available and then used with total (or direct) cost estimates at similar operations to estimate input-specific costs for these operations. In some cases, the input-specific costs were adjusted before they were applied at other operations. For example, at one open-pit mine/mill operation, the input-cost factors for both mine and mill supplies might be 0.20; these factors may be adjusted to 0.18 for mine supplies and 0.22 for mill supplies before they are applied to an open-pit mine/mill with a low stripping ratio and a low-percentage recovery of copper.

After input-specific costs are estimated at each mine/mill/leach operation, they are summed to give fixed and variable costs. Fixed costs include salary labor, property taxes, administrative costs, insurance costs, townsite costs, and certain royalties. Variable costs include wage labor, mine supplies, mill supplies, leaching and precipitation plant supplies, power costs, and sales and severance taxes. Table 6–2 gives a hypothetical example of the cost data developed for each of the forty-five mines.

Smelting Costs. Smelting, freight, and refining charges were estimated and assigned to the appropriate mine/mill/leach operations in order to determine the total production cost of refined copper for each operation. Copper flows were traced from mines to smelters, and site- but not input-specific smelting costs were estimated for fourteen smelters. (The term *smelting cost* refers to the cost of smelting plus freight and refining.)

Table 6–2
Hypothetical Cost Data of Open-Pit Sulfide Mine/Mill

Data	Total Cost (May 1978 Dollars)	Fixed Cost (May 1978 Dollars)	Variable Cost (¢/lb)	Percentage of Direct Cost	Percentage of Total Cost
Copper produced: 41,700,000 lb					
Costs					
Labor (including fringe)					
Wage	5,921,000		14.2	37.50	34.54
Salary	1,208,000	1,208,000		7.65	7.05
Mine supplies					
(total)	3,150,000		7.56	19.95	18.38
Mill supplies					
(total)	2,978,000		7.14	18.86	17.37
Power	2,533,000		6.07	16.04	14.78
Property tax	700,000	700,000			4.08
Administrative					
costs	450,000	450,000			2.63
Townsite costs	200,000	200,000			1.17
Total	17,140,000	2,558,000	34.97	100.00	100.00

Total smelting costs vary depending on whether the smelting is done on a captive or toll basis. Captive smelting costs are transfer prices between an integrated producer's mine and smelting operations. Total captive smelting costs were computed on a per-pound basis, excluding interest and depreciation. The industry average was $0.2609 per pound of copper produced. Data were not available to differentiate between copper from concentrates and copper from precipitates. For the integrated producer, smelting costs have both fixed and variable components in the short run. Based on detailed input-specific costs for hypothetical "model" smelters (Bennett et al. 1973; Mean and Bonem 1976), it was assumed that one-third of the captive smelting cost was fixed and two-thirds was variable.

Mining companins that are not vertically integrated have contractual arragements with toll smelters to treat their concentrates. The contract enables smelters to depend on certain mines for a constant source of supply and also serves as a vehicle for the smelters to obtain minimum and maximum quantities of concentrates. Charges for toll smelting depend on the terms of the contract between individual mines and smelters. In addition to long-term contracts, some smelters will obtain concentrate on a spot or an occasional basis. Based on discussions with producers, we approximated the charge for toll smelting as the cost of captive smelting including interest and depreciation plus 10 percent. Because of the contractual arragements, toll smelting costs are assumed to be completely variable in the short run.

Adding the appropriate fixed and variable smelting costs to the mine/mill/leach costs gives the total cost of producing refined copper. However, by-product credits from molybdenum, silver, and gold also must be considered.

By-Product Credits. Neither economists nor financial analysts have found a way to account satisfactorily for by-product credits. In economic studies, the joint-production question is avoided by assuming an output aggregate (Griffin 1977); in accounting reports, by-product credits are allocated consistently, but arbitrarily.

In terms of the copper supply curve, by-products must be considered because they make production more profitable, even though they do not lower the actual cost of producing copper. By-products are also considered in mine planning decisions. Their presence in the ore and their price in the marketplace affect the cutoff grade of the copper ore that is mined. By-products affect production decisions as well as the shutdown point of the mine.

As discussed earlier, the shutdown point for the firm occurs when price falls below average variable cost for a "significant" time. By-products change the shutdown point because they increase the revenue associated with producing a pound of copper. In the construction of the cost function, the price of copper is not a consideration. Therefore, by-product credits must be accounted for by a shift in the cost curve. For a given price of copper, the effects of by-product credits on the shutdown point can be taken into account by subtracting by-product credits from production costs.

In sum, the total cost of producing refined copper is equal to the fixed and variable costs at the mine/mill or leaching facility plus the fixed (if any) and variable costs of smelting minus the variable by-product credits. All variable costs are assumed to vary linearly with production; marginal costs are assumed to be equal to average variable costs and to be constant up to maximum capacity and then infinite.

Maximum Capacity. To construct the marginal-cost curves, a measure of maximum capacity was needed. The maximum capacity of an operation determines the point of discontinuity at which the horizontal constant-marginal-cost curve becomes infinite. Maximum capacity of any operation represents the point at which no more copper can be produced at any price. It is measured in pounds per year of copper.

Mine capacity depends on mill size, amount of mining labor and equipment available, and physical limitations at the mine, such as the hardness of the ore, required pit-wall slopes, and so on. Mill capacity depends on labor and supply availability and physical limitations such as the grindability of the ore and electric power availability. Generally, in annual reports and SEC Form 10-K, either mine or mill size is reported in terms of the tons of ore mined or treated in that year on a daily or an annual basis. When this number is multiplied by the ore grade and percentage recovery of copper, one obtains the amount of copper produced per year. For our purposes, maximum capacity for open-pit and underground mines and for vat and heap leaching operations is defined as follows:

> Maximum capacity = largest amount of ore mined or milled from 1972 to 1977) × (ore grade mined in 1977 or average ore grade of reserves, whichever is larger) × (largest percentage recovery from 1972 to 1977)

If the capacity at an operation did not change in the past six years, then the largest amount of ore mined or milled in this period represents the maximum short-run throughput of ore. If capacity did change in the last six years (for example, at Cyprus Bagdad), the changes were considered in calculating maximum capacity. The highest percentage recovery in the last six years was used on the assumption that if mines were operating at maximum capacity, then the price was above their average variable cost. If this were true, producers would attempt to recover as much copper as possible.

This line of reasoning was not followed in selecting the appropriate ore grade to use in the calculation. For ore grade, the 1977 value or the average grade of reserves—whichever was published or, if both were published, whichever was larger—was used. Ores with a higher-than-average grade may be present, but in the short run they may not be accessible for mining. Moreover, when demand and price are high, companies will often mine low-grade, high-cost ores. This implies that when mines are operating at maximum capacity, they will not be mining higher than average-grade ores.

For waste-dump leaching, the amount of copper produced depends on the availability of sulfuric acid. Production of copper in any year is not closely related to the amount of ore mined in that year. Therefore, for waste-dump leaching, maximum capacity was assumed to be the largest level of production in the last six years.

Total-Operating-Cost Equations. Production costs for refined copper were obtained by tracing the flow of copper-bearing material from mine to smelter to refinery. Fixed and variable smelting, freight, and refining charges were added to the mine/mill cost at each site, yielding a cost of refined-copper production at each mine. Smelting-cost data used in this analysis are site- but not input-specific. For mines that have their concentrate custom- or toll-processed, smelting charges are assumed to be a variable cost. For vertically integrated producers, smelting costs have both fixed and variable components.

In the short run, constant fixed costs are incurred regardless of the amount of copper produced. Variable costs are incurred only when copper is produced, and they are assumed to vary linearly with output up to the capacity constraint. Once maximum capacity is reached, it is assumed that production is not possible at any cost. These assumptions yield a total-cost (TC) equation of the following form:

$$TC = E + Vq \qquad \text{for } 0 \le q \le qc$$
$$\infty \qquad \text{for } q > qc$$

where TC = total cost

F = fixed cost

V = variable cost

q = quantity

qc = maximum capacity defined for each operation

Marginal costs (MC) are the first derivative of total costs with respect to quantity. For linear variable costs, marginal costs are constant up to capacity and then infinite.

The short-run total and marginal costs for the hypothetical mine shown in table 6–3 are

Mine/mill	$5,722,000 + 0.3241q$
Smelter	$8,674,000 + 0.1733q$
By-products	$-0.0788q$
	$14,396,000 + 0.4186q$

$$MC = \begin{cases} \$0.4186/\text{lb} & 0 \le q \le 19{,}995{,}000 \text{ lb} \\ \infty & q > 199{,}995{,}000 \text{ lb} \end{cases}$$

The long-run situation is analagous, except that in this case all costs are variable and the cost of capital is included. It is assumed that in the long run, new capacity is introduced by developing new mine sites rather than expanding existing capacity.

Capital Costs. Estimates of capital costs were needed to determine the cost of capital for staying in or entering the copper industry. As explained above,

Table 6–3
U.S. Copper Deposits Likely to Be Developed in the Near Future

Project	State	Type of Mine	Grade of Ore and Associated Minerals	Amount of Ore (Million Tons)	Copper Production Capacity (Thousand Short Tons per Year)	Addition of Copper to Existing 1977 Capacity (Thousand Short Tons per Year)	Year of Likely Production	Likelihood of Production in 1980s	Capital-Cost Estimate Millions (Total in 1977 Dollars)	Capital-Cost of Copper (Cents per Pound)[v]
Amax										
Eisenhower (see below)										
Mount Tolman	WA	OP	0.13% Cu 0.13% MoS$_2$	300	25	25	1984	Medium-high		
Puerto Rico			0.82% Cu	104				Low		
Anaconda										
Carr Fork	UT	UG	1.84% Cu	61	56	56	1979	Definite	216	29.74
Yerington[a]	NV	OP			25	25				
Victoria[b]	NV	OP	2.1% Cu 0.25 oz Ag/short ton of ore	1.5	6	6		High		
Anamax										
Helvitia	AZ	OP	0.64% Cu	320	90	90		Low		
ASARCO										
Troy	MT	UG	0.74% Cu	64	20	20	1981	High	80	33.8
Sacaton	AZ	UG	1.25% Cu	15				Medium		
Casa Grande										
Casa Grande	AZ	UG	1% Cu	350	100	100	Mid-1980s	Medium		
Cities Service										
Pinto Valley	AZ	WDL-SX-EW			6	6		Medium		
Pinto Valley[c]	AZ	OP				12		Medium		
Miami East	AZ	UG	1.95% Cu	50	13	13		Medium		

Company / Project	State	Type	Grade				Start	Certainty	
Continental Oil Florence[d]	AZ	OP	0.4% Cu	800	100	100	1980	Medium	
Continental Materials– Union Miniere Oracle Ridge[e]	AZ	UG	2.2% Cu 0.64 oz/ton Ag	11	14	14		Uncertain	40
Cyprus Pima Pima[f]	AZ	OP	0.497% Cu + MoAg	147	60	60	Mid-1980s	High	
Duval Esperanza[g]	AZ	OP	0.28% Cu 0.029% Mo	29	17	17	1979	Definite	
Battle Mountain[h]	NV	HL-SX-EW	0.23% Ag 0.089% Au same Cu	6.6	7	2.3	1979	Definite	18
Eisenhower Eisenhower[i]	AZ	OP	0.63% Cu	157	26	13	1979	Definite	
Emmon Cranden	WI	UG	5% Zn some Au, Ag, Pb 1% Cu	70	30	30	1986	Medium	350
Pinos Altos	NM	UG	2% Cu + Ag + Au 3% Zn	7.7	10	10	Late 1980s		94.3
Inspiration Christmas[j]	AZ	OP	0.77% Cu	19	8	8	1979	Definite	
	AZ	UG	1.78% Cu	20					
Ox Hide[k]	AZ	HL	0.31% Cu	30	5	5			
Sanchez	AZ		0.36% Cu	79			1979	Definite	

Table 6–3 (continued)

Kennecott									
Nevada	NV	WDL	0.67% Cu	85	20	20	1980	High	300
Chino[l]	NM	OP	0.74% Cu	415	30	40(?)	1980	High	
Ray[m]	AZ	VL-SX-EW	0.79% Cu	645		10			
Utah[n]	UT	OP	0.71% Cu	1,505					350
Flambeau[o]	WI	OP	0.45% Cu	4	13	13		Low	
Newmont/Magma									
Kalamazoo[p]	AZ	UP	0.72% Cu	565	32	15	1981	High	
Vekol[q]	AZ	OP	0.56% Cu	102		32		High	
Rancher's									
Bluebird[r]	AZ	SX-EW			8	8			
Old Reliable[s]	AZ				2	2			
Big Mike[t]	AZ				1				
Noranda									
Lakeshore[u]	AZ	UG	1% Cu	500	65	65	1979	High	
Phelps Dodge									
Ajo[v]	AZ	OP			20				
Metcalf[w]	AZ	OP			35				
Safford	AZ	UG	0.7% Cu, 0.55% Cu, 0.02% Mo	700	60	60	Mid-1980s	Medium	225
Copper Basin	AZ	OP		175				Low	58.30
Lousiana Land Exploration and Superior Oil									
Co.[x]	ME		Cu, Zn, Ag, Au in massive sulfide deposit	37			1985	Medium	

Sources: "U.S. Copper Production Could Be Increased by 53 Percent in 1980s, Survey Shows," *Arizona Pay Dirt*, April 1979, p. 4.

Arizona Pay Dirt, 1977–1979.

Engineering and Mining Journal, 1975–1977.

Survey of Planned Increases in World Copper Mine, Smelter and Refinery Capacities, 1974–1980 (London: International Wrought Copper Council, 1975).

Various company annual reports and SEC Forms 10-K.

EW = electrowinning   UG = underground
HL = heap leaping   VL = vat leaching
OP = open pit   WDL = waste-dump leaching
SX = solvent extraction

a Could resume operation for about 3 years before reserves depleted.

b Purchased from Anaconda by Day Mines for $500,000 in 1978. Day Mines also acquired the surface plant and mining and milling equipment from Nelson Machinery Co. for $1.4 million plus 20 percent of profit after recoupment of cost. Nelson Machinery will also receive $0.25 per stoke of ore processed in the mill up to a maximum of 1.5 million stokes.

c Incremental expansion.

d 35,000 tons/year electrowon from oxide ore; 65,000 tons/year concentrate in sulfide ore.

e Problems came up in development of ore body; production is uncertain.

f Reactivation production will begin at 8,500 tons and will build up to 66,000 tons gradually over 2 or 3 years.

g Reactivation.

h Converted from precipitation of solvent extraction (SX)-electrowinning (EW) plant; production increased 150 percent.

i This is a joint venture of Anamax and ASARCO. The ASARCO portion will not add to immediate production but will increase the life of the Mission mine.

j Reactivation.

k Reactivation.

l Kennecott plans to spend $300 million for modernization and expansion, anticipating that this will lead to substantially lower unit production costs.

m Incremental expansion.

n Plans to modernize concentrator and replace antiquated mobile equipment will reduce costs and may increase production slightly.

o Numbers based on one proposed mining scheme in 1974.

p Really only an incremental expansion of San Manuel.

q Agreement has been signed with Papago Indian tribe and Vekol Mining Company.

r Reactivation.

s Reactivation.

t Reactivation.

u Reactivation change in ownership and perhaps a change in mining plans.

v Possible incremental expansion.

w Possible incremental expansion.

x Mining at rate of 6,000 stokes/day.

y Our estimates.

in the long run, a firm can vary its production capacity and will make its long-run production decisions based on the costs of adding new capacity or developing new deposits as existing mines are depleted. In the copper industry, new capacity usually means development of deposits owned or controlled by firms already in the industry. The costs of developing and operating new deposits depend on many factors, such as grade and tonnage of copper ore and by-products present, hardness and complexity of the host minerals, infrastructure requirements, and availability of power and water.

Table 6–3 lists U.S. copper deposits likely to be developed in the next decade. This table gives the location and type of mine, the grade and tonnage of ore, the production capacity for new mines in addition to the capacity for existing mines, the likely year and change of production, and capital-cost estimates both in millions of 1977 dollars and in cents per annual pound of copper.

The capital-cost estimates in millions of dollars were either taken from the literature or calculated based on comparison with similar deposits and similar-type mines. Even though the physical parameters of production were known, in some cases it was not possible to estimate capital costs because other factors such as geologic conditions and accessibility of resources were not known.

The cost estimates in cents per annual pound of copper were calculated by using the following formula (Zimmerman 1979a):

$$I = \int_0^T (C - \text{taxes})e^{-rt}\, dt$$

where I = initial investment

C = annual return necessary to realize after-tax return of r percent

T = life of capital good

Let u = corporate income tax and v = depletion allowance in percentage of gross profit, so that total depletion = $v(C - \text{depreciation})$. Then

$$\text{Taxes} = u(C - \text{depreciation} - \text{depletion})$$

and

$$I = \int_0^T [C - u(C - \text{depreciation} - \text{depletion})]e^{-rt}\, dt$$

$$= \int_0^T ce^{-rt}\, dt - \int_0^T uce^{-rt}\, dt + \int_0^T u(\text{depreciation})e^{-rt}\, dt$$

$$+ \int u(\text{depletion})e^{-rt}\, dt$$

$$= CF - ucF + u \int_0^T (\text{depreciation})e^{-rt}\, dt$$

$$+ u \int_0^T v(C - \text{depreciation})e^{-rt}\, dt$$

Where

$$F = \int_0^T e^{-rt}\, dt$$

we have

$$I = CF - ucF + uvcF + (u - uv)(\text{present value of depreciation}) \quad (6.3)$$

The last term on the right-hand side is

$$(u - uv)I \left(\frac{2}{rt}\right)\left[1 - \frac{1}{rt}(1 - e^{-rt})\right]$$

Substituting the above expression into equation (6.3) yields

$$\frac{C}{I} = \text{cost of capital} = \frac{1 - (u - uv)(1 - F/t)[2/rt)]}{F(1 - u + uv)}$$

The value of I is the capital cost in millions of dollars, taken from table 6–2. A value of 48 percent was used for u; 50 percent, for v. This assumes that the provision that limits percentage depletion to 50 percent of the taxable income from the property is the constraint. The after-tax rate of return r was assumed to be 15 percent, and the life of the capital good t was calculated based on reserves, production rate, and percentage recovery of ore. In addition, it was assumed that an investment tax credit of 10 percent was applicable to most projects.

Short-Run and Long-Run Supply Curves

The short- and long-run supply curves for the U.S. copper industry are shown in figures 6–2 and 6–3, respectively. In figures 6–4 to 6–7, the supply functions are separated into copper operations within and outside Arizona to

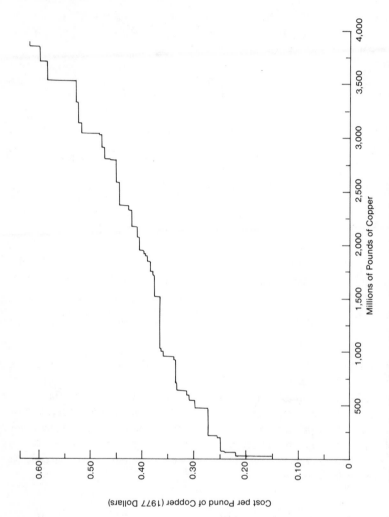

Figure 6–2. Short-Run Supply Curve of the U.S. Primary Copper Industry

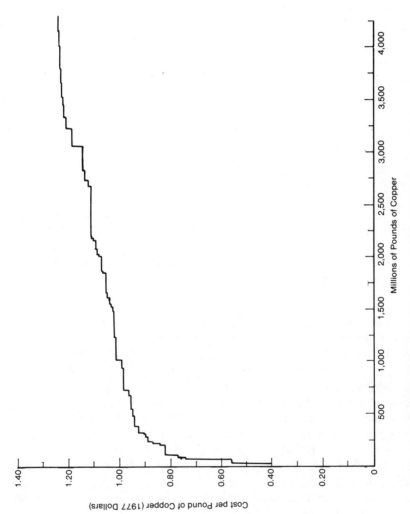

Figure 6-3. Long-Run Supply Curve of the U.S. Primary Copper Industry

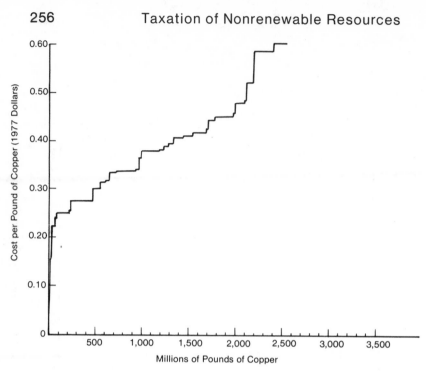

Figure 6–4. Short-Run Supply Curve of the Primary Copper Industry in Arizona

permit their use for tax-policy assessment. All costs are given in 1977 dollars. Care should be taken in interpreting these functions since each plateau does not necessarily refer to an entire site. For instance, there is not a site in this country that produces its entire output at $0.15 per pound in terms of short-run variable cost. This part of the curve refers to the dump-leaching component of one particular operation, and it was separated from the other flows at that site because of the distinctive nature of this process.

Estimations of the price elasticity of supply based on these curves show the following results for the United States was a whole. In the short run, at levels of production in the range 2.8×10^9 to 3.8×10^9 pounds of copper annually, the price elasticity of supply is essentially unitary ($E_S = 1.2$). At lower levels of production (1.5×10^9 to 3.0×10^9 pounds), the short-run price elasticity of supply is higher, roughly 2.5.

In the long run, the price elasticity of supply is very high. In the range from 1.0×10^9 to 3.3×10^9 pounds, the price elasticity is roughly 6. At levels of production greater than approximately 3.3×10^9 pounds, the price elasticity increases to even larger values at prices greater than $1.20 per pound in 1977 dollars (corresponding to roughly $1.55 per pound in first quarter 1980 dollars). This says that there are a number of copper deposits,

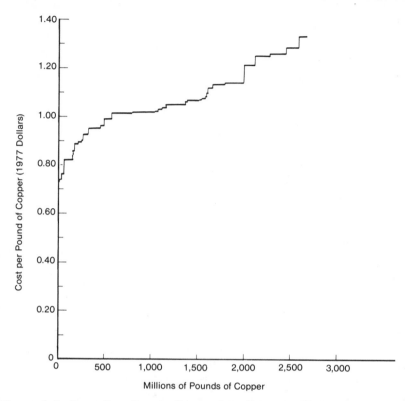

Figure 6–5. Long-Run Supply Curve of the Primary Copper Industry in Arizona

potentially producing many times current production tonnages, which can be brought into operation at these higher prices.

The values of the price elasticity of supply calculated from the actual short- and long-run supply functions are larger than those estimated from the econometric models of the copper industry. For instance, the classic study of Fisher, Cootner, and Baily (1972) produced the following estimates of the price elasticity of supply of U.S. refined copper (based on an estimation period of 1949–1958 and 1962–1966):

$$E_S = \begin{cases} 0.453 & \text{short run} \\ 1.670 & \text{long run} \end{cases}$$

Other estimates based on econometric models are within the same range as those of Fisher, Cootner, and Baily.

There are a number of reasons why the estimates of the price elasticity of supply calculated from the engineering and econometric studies are different.

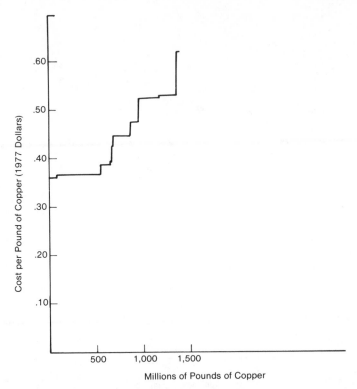

Figure 6–6. Short-Run Supply Curve of the U.S. Primary Copper Industry
outside Arizona

Most of the differences may be explained by the nature and fundamental
assumptions of the two approaches. First, the sources of data are different.
Econometric models are, by necessity, based on observations of past data,
and these data are available for only a limited range of output and prices.
Moreover, since econometric models predict best at mean values, the
elasticity estimates reported above are calculated at the mean. The engineering approach, which is based on the costs of production estimated from
actual characteristics of existing and new mine sites, does not depend on
historical prices. Therefore, the engineering approach is expected to be much
better for analyzing and forecasting otuside the historical range of data.

Second, the fundamental assumptions of the two approaches are
different. The econometric models estimate supply responses, assuming that
future behavior will be the same as historic behavior. The engineering
approach, as presently conceived, does not consider past behavior regarding
price/output decisions explicitly. Rather, it assumes that if the marginal costs
of production are less than the price, then the uneconomic firms will not

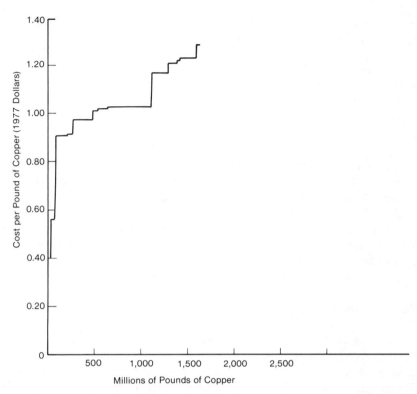

Figure 6–7. Long-Run Supply Curve of the U.S. Primary Copper Industry outside Arizona

produce. This is clearly not the case, since in the past a number of firms have continued to produce for significant periods (one to two years) when their marginal short-run costs exceeded the price of copper, presumably in the expectation of higher future prices. This is especially true of underground mines with high shutdown costs. However, it should be pointed out that when copper prices were in the range of $0.50 to $0.60 per pound, the four high-cost, open-pit mines were closed and the two high-cost, underground mines cut back production.

The horizontal segments of the separate site-specific cost functions may also be too long in some cases because of the assumptions about maximum capacity. This will tend to extend the supply curve to the right. Thus, the engineering approach is likely to overestimate short-run price elasticities, especially in the range of historical data. The engineering approach could be modified to model this situation more accurately by calculating a shutdown point for each mine/mill/smelting complex. However, this is an exceedingly time-consuming and expensive task, for shutdown costs vary from mine to mine and state to state.

In summary, it is believed that the econometric and engineering approaches have different strengths and weaknesses and are thus useful complements. The econometric approach is probably more useful for estimating short-run elasticities, especially within the historical-data sample; the engineering approach is more valid for analyzing rational producer behavior and for estimating long-run elasticities, especially outside the historic-data sample.

The estimated short- and long-run supply curves are combined with estimates of demand to simulate various tax regimes. Short- and long-run demand functions are concensus estimates derived by other researchers (Charles River Associates 1970; Fisher, Cootner, and Baily 1972; Hartman 1977). The estimated tax rates (chapter 5) were added to the cost estimates to reflect the effect of taxes. The results of tax simulations are shown in the final section of this chapter.

The Coal Model[a]

In modeling coal, one is confronted with a more complex problem than for copper. Transportation costs are significant (as much as several times the mine-mouth price) and quality differences remain after processing (sulfur and Btu content), whereas these complexities are either nonexistent or minor in the case of copper. Furthermore, the upheaval in recent years in fuel markets has altered competition among fuels. Rising oil prices have eliminated oil from consideration as a fuel for new electric utility plants. In the past, natural-gas price controls produced a similar result for gas power plants. In the future, under deregulation, gas prices are expected to be high enough to again eliminate gas from consideration. Finally, public acceptance of nuclear power has diminished to the point where coal is the only practical alternative source of energy for new base-load generation facilities (Zimmerman and Ellis 1980). This change in market structure creates opportunitiers for those with market power. In essence, the demand for coal has become less elastic, and those in a position to exert market power can take advantage of the situation to raise prices and create or usurp economic rents (Alt 1978; Zimmerman 1979a, 1979b, 1980a, 1980b).

The coal industry itself is competitive. Coal producers will, in the long run, sell at long-run marginal cost where marginal cost reflects the cost of production as well as the competitive rent accruing to owners of depletable

---

[a]This description of the coal model is adapted from research undertaken by Martin B. Zimmerman and Christopher Alt of the Sloan School, Massachusetts Institute of Technology. I wish to express my appreciation to them for having made this material available.

resources. Other elements of the industry, however, are more favorably situated. To some extent, labor supply is controlled by the United Mine Workers. However, competition from nonunion miners, largely in the Western states, will limit the union's ability to appropriate rents. Those with the most favorable market positions are the railroads and state legislatures.

Not surprisingly, these developments are focused on coal coming from the Powder River Basin of Montana and Wyoming. This coal is low in sulfur content. The coal lies in thick beds close to the surface. Mining costs are low, and potential rents are consequently high. How far can this process go? How high can the state legislatures raise taxes, and how far can railroads raise rates?

The Model

The model and its constituent submodels are described in detail elsewhere (Alt 1978; Zimmerman 1979a, 1980a). It is beyond the scope of this section to repeat that documentation. Here we provide an overview of the estimation and construction of the model.

Figure 6–8 provides a schematic view of the coal-policy model. Given an initial set of prices, the initial demand for coal by region is determined. These demands are passed to a linear program as constraints. The supply model establishes the cost of production for coal as a function of output by sulfur content and by region. The transport model determines the cost to transport coal from mine mouth to demand center. The linear program then minimizes the cost of mining and transporting the coal necessary to meet the regional-demand constraints and subject to limitations on sulfur emissions.

The solution of the linear program yields output by region and by sulfur content, flows of coal from supply region to demand region, and the delivered price of coal to each region. The delivered prices are passed back to the demand model to estimate a new set of regional demands. The coal outputs are fed back to the supply model to establish a new set of supply prices, and the process is repeated. In theory, we could iterate until convergence to a set of prices consistent with instantaneous equilibrium in supply and demand. In practice, the models run recursively. The demand in year t is a function of price in year $t - 1$. This reduces considerably the computational complexity of the model without introducing any serious distortion. This is so because the demand response to price change is slow. In any given year, the demand is quite inelastic with respect to prices; however, the longer the period considered, the more demand becomes elastic, because alternative sources of energy are developed, conservation is more easily achieved, and capital becomes mobile with respect to location.

The linear program is solved each year, so the model is dynamic (years from 1980 to 2000 are considered). However, the output in any year affects

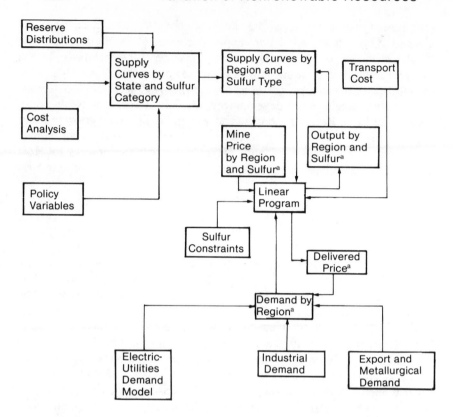

^aModel outputs.

Figure 6–8. Coal-Policy Model Structure

coal prices throughout the simulation period. The essence of the supply model is the *cumulative* cost function. This function estimates how costs increase owing to depletion as output cumulates over time. In each year, the cost of producing any given amount of coal is updated to reflect the impact of cumulative output through that year.

Cost and Supply of Coal

The structure of the coal industry is continually changing. Depletion of lower-cost deposits raises costs. Changing regulations affect what coal can be produced and what coal can be burned. This changing structure presents problems for the modeler. The supply model attempts to deal with these problems by deviating substantialy from more traditional econometric modeling efforts. First, there is significant disaggregation according to segments likely to be affected differently by policy actions. Second, the model attempts to capture the effects of depletion by integrating geological information into the estimation methodology.

There are three steps to the estimation of the coal-supply model. In the first step, the relationship between geology and the productivity of a mining unit is estimated. The unit is an appropriately defined complement of miners and machines. The relationship is estimated by using observations on actual mines. Once the productivity of a unit is known, given the geology of a deposit, we can calculate the number of units required for any given level of production. An equation of the following type is estimated:

$$\frac{Q}{u} = f(Q,G) + \varepsilon$$

where Q = output of mine

u = number of units

G = observable geological characteristics

ε = disturbance term reflecting unobservable variables

A unit in an underground mine consists of a continuous mining machine and a complement of miners plus shuttle cars to haul the coal away. In strip mining, units measure the size of the earth-moving machinery.

The next step is to assume that expenditures on labor, materials and capital are a function of the number of units. The estimation of the relationship uses engineering models of mines. These are estimates prepared by the Bureau of Mines for hypothetical mines producing a given amount of coal in a coal seam of given characteristics. The most questionable part of these hypothetical mines is the productivity assumptions underlying the estimates. We use our own productivity estimates derived from the econometric estimates described in the text. The only information we use from the engineering estimates is the relation between the number of producing units and expenditures (U.S. Bureau of Mines 1972, 1974a). The expenditure functions are combined with the productivity equation to yield cost as a function of geology. From the productivity equation we know the number of units necessary to produce a given amount of coal in a seam of given characteristics. We know how much it costs to produce a given amount of coal. We have, in this way, an equation that yields cost as a function of output and geology.

In the long run, the marginal cost of coal is the marginal cost of opening a new mine. This is the cost incurred at the rate of output that minimizes average cost. We can solve the cost equation for the rate of output that minimizes the average cost of production, given the characteristsics of the seam. The result is an equation yielding long-run marginal cost as a function of geological characteristics of the coal seam. The resulting functions are of the following form:

Deep mining:
$$MC_D = \frac{K_D}{(th)^\gamma(\varepsilon)}$$

Strip mining:
$$MC_S = (K_S)(R)^{1/\alpha}(\eta)$$

where MC = marginal cost

    γ,α,K = constants resulting from estimation

    th = thickness of seam

    R = ratio of overburden in feet to feet of coal seam

    ε,η = disturbances reflecting unobservable geological characteristics for deep mining and strip mining, respectively

The distribution of coal in the ground according to the cost of mining is thus a function of the distribution of ε and th for deep mining and η and R for strip mining. The disturbance terms are assumed log-normal. The parameters of the distribution for these terms are estimated in the process of estimating the cost functions. The final step in the process is to determine the distribution of coal in the ground according to seam thickness and overburden ratio.

The data are, however, scarce. We use detailed data on a few key areas to establish the shape of the distribution. The log-normal distribution appears an appropriate approximation for both the seam-thickness and the over-burden-ratio distributions. The variance for the distribution of tons according to seam thickness is estimated from data for east Kentucky. It is assumed this variance applies everywhere. The mean of the distribution is estimated separately for each state. The total amount of coal also varies by state and by sulfur category. The total amount of coal is taken from the U.S. Bureau of Mines (1971, 1974a), which gives the amount of coal in the reserve base lying in seams less than and greater than 40 inches. Given the assumption of log-normality, the estimated variance, and the percentage of coal in seams greater than 42 inches thick, the mean can be calculated by

$$U_{42} = \frac{\log 42 - \log th}{\sigma_{\log th}}$$

where U_{42} = point on standard normal distribution corresponding to log 42

log th = mean of logs of seam thickness

$\sigma_{\log th}$ = variance of distribution of coal according to log of seam thickness

The distribution of tons according to overburden ratio is also approximated by a log-normal distribution. The variance of the distribution is estimated from data for Illinois and Montana. The mean is also separately estimated for each state.

The long-run cost function is combined with the distribution of coal according to physical characteristics to yield a distribution of coal in the ground according to the cost of mining. There is an estimated distribution for each state, sulfur category, and type of mine.

Each distribution is truncated by the economics of the mining industries. The cheapest deposits are mined first. Therefore, coal remaining in the ground must be at least as expensive as the current cost of mining. The truncation point is set equal to the cost of coal today. This is done separately for each state and sulfur category. Figure 6–9 shows a typical distribution of coal in the ground according to the cost of mining.

Transport Costs

The transport cost from mine to consumption point is estimated from actual transport rates. The model considers factors that determine cost (miles shipped, annual volume, loading and unloading time) as well as market power

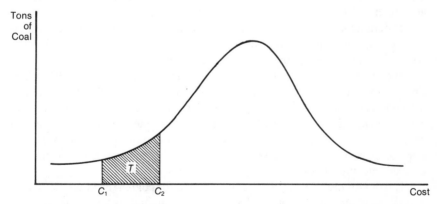

Figure 6–9. Distribution of Coal according to the Cost of Mining

of the railroads (Zimmerman 1979*b*). The rates reflect the competitive position as of 1975 and consequently already include a certain amount of rent. We escalate the rates by the American Association of Railroads cost index to get the rates into 1977 dollars.

Demand Model

The supply model and the transport-rate equation yield a delivered cost of coal. The missing elements are demand and a market-clearing mechanism. To capture demand, we have adapted the well-known model of Baughman, Joskow, and Kamat (1979) for our purposes. That model and its adaptation are described in this section. The following section describes the market link.

The demand for coal is derived from the demand for electricity and from the demand for industrial products. The model we have adapted to generate coal demand deals with its derived nature. The model used is the regional-electricity model (REM) developed at the Massachusetts Institute of Technology by Martin Baughman and Dilip Kamat, now at the University of Texas, and Paul Joskow of the Massachusetts Institute of Technology. In the sections that follow we present an overview of the model and the modifications introduced in order to integrate it into our framework. The main thrust of the modification is to allow the demand model to deal with the richness of environmental policy explicitly considered in the coal-supply model.

The demand model begins by estimating the demand for energy. In the residential and commercial sectors, this is done through an econometric equation that expresses the demand for energy per capita as a function of a weighted average of energy prices, income, and population. This equation is estimated from combined time-series and cross-sectional state data. A lagged adjustment structure is estimated to separate long- and short-run effects. These demands for energy are then broken into demand for fuel types by fuel-split equations. The proportion of total energy demand satisfied by a particular fuel is related to the relative price of each type of fuel.

The industrial sector is treated in a similar way. Total energy demand is first estimated as a function of an energy price index and value added in manufacturing. This total demand is allocated to states by a cross-sectionally estimated location equation. This equation estimates the relative share of each state in total energy consumption as a function of prices, income, and populaton relative to a reference state. Finally, the demand per state is divided into fuel types by fuel-split equations.

The output of this procedure is the demand for coal in the industrial sector and the demand for electricity. The former is demand for direct use of coal.

The latter must be further decomposed into the demand for the different fuels used in generating electricity.

Demand for Coal in the Electric Utility Sector. The essence of the electricity model is the economics of plant expansion first worked out by Turvey (1968). The cost-minimization problem is complicated in the electric utility sector by the pattern of electricity demand. The demand level fluctuates during the day as well as from month to month. For example, the highest level of demand typically occurs during the hottest day of the year, when everyone is using air conditioning. The problem, from the standpoint of the utility, is that it must supply such high levels of demand only a few days a year.

The solution to this problem is to use different kinds of plants. Clearly, a utility does not want to build a capital-intensive plant and use it only a few days a year. To satisfy those peak demands, it will build a less capital-intensive plant that uses fuel more intensively. Then the utility will pay for fuel on only those days when plant is in use and thus will carry smaller capital costs. However, there is some amount of electricity that they must generate throughout the year, day in and day out. To satisfy this base-load demand, utilities will choose the more capital-intensive, less fuel-intensive plants. These base-load plants realize high rates of utilization, and thus capital costs per unit of output are reduced. Since the plants operate for much of the year, the variable costs, particularly of fuel, are lower than for peaking units.

The Baughman-Joskow-Kamat model solves for the optimum configuration of capacity, given the pattern of demand fluctuations and the cost characteristics of alternative plants. The model then builds the difference between existing capacity and the optimum configuration. The demand model assumes that utilities use a trend-adjusted moving average to forecast demand levels. With these forecasts, the utility determines how much capacity it needs. In any year, actual demand is determined by the econometric model described above. Over time, the demand forecasts are adjusted as actual demand evolves. The simulation approach attempts to capture actual utility behavior.

In any given year with fixed plant capacity, the problem is to minimize the cost of generating the actual demand. This is done by using most frequently those plants that cost the least. This minimization routine determines actual plant use. Each plant type is ranked in terms of fuel and operating costs. The total demand is met by going down the list of plants from low cost to high cost until the total demand is met.

In the model as we have modified it, the plant types available to the decisionmaker are the following: coal plants without stack-gas scrubbers, coal plants with scrubbers, nuclear plants, oil and gas plants, and hydroelectric plants. The expected costs of generation for each plant are compared in making capacity-expansion decisions. The decision as to type of coal plant depends on the premium paid for low-sulfur coal. The electric-utility model receives from the coal model two coal prices, one for coal satisfying pollution regulations without scrubbers and the other for high-sulfur coal that would require the use of scrubbers in order to meet pollution requirements. The model compares the expected cost of plants with and without scrubbers and assesses whether low-sulfur coal or scrubbers are the least-cost alternative.

A decision is added to the model: the choice of retrofitting existing coal plants. In each period the utility calculates whether it would be cheaper to meet sulfur regulations by burning low-sulfur coal for the life of the plant or by adding a scrubber to an already existing coal-burning plant. The calculation compares the present value of the expected additional costs of low-sulfur coal to the costs of retrofitting an existing plant.

Link between the Supply and Demand Models

The locational aspects of the coal industry are important. Transport costs are relatively large and substantially affect the outcome of the model. Simulation of tax policies requires disaggregation by producing states. Tax policy is intertwined with sulfur regulations, necessitating disaggregation by sulfur content. We have disaggregated further by strip and deep mining, because policy initiatives affect each segment of the supply function differently. Allowing for all this disaggregation yields 360 supply curves, a number that makes a solution to the linear program too expensive.

In order to make the model computationally practical as well as preserve necessary detail, the following procedures were adopted. We sum all strip- and deep-mining supply functions within a state and then sum all state supply functions within a region to get six regional supply functions. We do this for eight different sulfur categories, so that there are forty-eight supply functions. The linear program deals with only these forty-eight supply functions. After the program solves for production by region and by sulfur category, we can determine the split among states and by mining type, using the aggregation process in reverse. With the price in the region known, from the individual state supply functions we can calculate how much is supplied at the calculated price in each individual state.

One other aggregation issue bears discussion. In the demand model, the present value of fuel expenditures is added to capital expenditures to determine the least-cost capacity addition. We introduced into the demand model the choice between high-sulfur coal plus scrubbers and low-sulfur coal to meet pollution regulations. Given eight different sulfur levels, it is computationally infeasible to consider all possible combinations of coal for meeting regulations.

The solution adopted is to create two categories of demand within each region. There is a demand for coal to be used in plants that chose scrubbers. Scrubbers are assumed efficient enough to reduce sulfur emissions of any coal to meet environmental regulations (90 percent removal is assumed). In those plants and in industrial uses where scrubbers are not used, the coal burned must meet the constraint on sulfur emissions. The linear program described in the following section solves for the mixture of coals that satisfy both demand categories at least cost. The average price of each mixture becomes the low- and high-sulfur coal prices that are fed to the demand models. These two prices are, in fact, the marginal costs of the least-cost combination of coals satisfying the respective sulfur constraints. The utility models decide on the basis of these prices whether to build scrubbers. The linearized supply curves and the demand model are linked through a linear programming formulation. The supply and tranposrt models yield the cost of producing and transporting coal from supply region to demand region. The demand model yields coal demanded by the nine U.S. census regions. This is then disaggregated into twelve smaller demand regions and fed to the linear program as constraints. The linear program minimizes the cost of meeting these demands, given mining cost and transport cost, subject to additional constraints on the sulfur content of the coal.

The linear program is solved year by year. However, Mining costs in each year are updated, given the additional cumulative output represented by last year's output. Thus, the costs in any year are a function of the sum of previous years' outputs. In this way, depletion is explicitly accounted for. Each year a new set of step functions is calculated.

Formal Statement of the Linear Program

Following is an algebraic expression of the problem. Discussion of the formulation follows.

$$\text{Minimize} \sum_{i=1}^{i=6} \sum_{j=1}^{j=12} t_{ij} \sum_{s=1}^{s=8} \sum_{k=0}^{k=1} Z_{ijsk} + \sum_{i=1}^{i=6} \sum_{s=1}^{s=8} \sum_{m=1}^{m=8} C_{ism} Y_{ism}$$

such that

$$\sum_{m=1}^{m=6} Y_{ism} - \sum_{j=1}^{j=12} \sum_{k=0}^{k=1} Z_{isk} = 0 \qquad \begin{aligned} & i = 1,\ldots,6; \; s = 1,\ldots,8; \\ & S = D \text{ in tons} \end{aligned}$$

$$\left(\sum_{i=1}^{i=6} B_i\right)\left(\sum_{s=1}^{i=8} Z_{ijsk}\right) > D_{jK} \quad \text{demand constraint}$$

$$\frac{2,000 \times 2 \sum_{i=1}^{i=6} \sum_{s=1}^{s=8} P_s Z_{ijsk}}{\left(\sum_{i=1}^{i=6} B_i\right)\left(\sum_{s-1} Z_{ijsk}\right)} < S_j \quad \text{sulfur constraint}$$

$$Y_{ism} < U_{ism} \quad i = 1,\ldots,6$$
$$s = 1,\ldots,8$$
$$Z_{ijsk}, \; Y_{ism} > 0 \qquad m = 1,\ldots,6$$

where t_{ij} = transport cost (in dollars per ton) from supply region i to demand region j

Z_{ijsk} = shipment of coal (in tons) from supply region i to demand region j of sulfur type s and scrubbing activity k; $k = 1$ indicates scrubbing, $k = 0$ indicates no scrubbing

Y_{ism} = production in region i of coal of sulfur category s from stem m

C_{ism} = cost of coal (i, s) from step m

B_i = Btu value per ton of coal in region i

D_{jk} = demand (in Btu) in region j for coal from scrubbing category k

P_s = sulfur content (in percentage by weight) of coal of sulfur category s

U_{ism} = length (in tons) of step m in region i for sulfur category s

The Objective Function. The objective is simply to minimize the sum of the cost of mining and transporting coal. Transport cost is estimated in the transport model. The variables C_{ism} represent the segments on the linear approximation to the cumulative-cost curves.

Constraints. The first set of constraints and the last two sets of constraints are housekeeping. The first set forces quantities produced Y_{ism} to equal quantities shipped Z_{isk}. In essence, this is the constraint forcing supply to equal demand.

The next-to-last set of contraints simply reflects the stepwise approximation to the cost curves. The U_{ism} are the quantities of coal in each step, and production from each step must be less than that level. The last set of constraints is the set of nonnegativity constraints.

Demand Constraint. The second set of constraints forces the model to satisfy the demand in each region as determined in the demand model. The demand model calculates two sets of demands in each region, as denoted by the subscript k. One set is the demand for coal in plants using scrubbing devices. The second set is for coal demanded in plants not using scrubbers.

The most important feature of the demand constraints is that we calculate a set of *market-sensitive* demands. These are fed as the demands to the linear program to adjust for the fact that we have a long-run equilibrium model and the long-run equilibrium will be approached only gradually.

Thus, subtracted from demand is what was fixed before the model runs begin, that is, before 1976. We assume that the coal shipments fixed by 1975 decline at a rate of 5 percent per year, the historical rate of mine closures. We take the total amount fixed before 1976 as a fixed fraction of 1975 demand (90 percent of 1975 demand is assumed fixed by contracts). In each period, we optimize the production and distribution pattern for incremental demand equal to the net increase in demand over the fixed 1975 levels plus the cumulative amount of mine closures through that year. The model does have the option of assuming that shipments resulting from the optimization routine in years after 1975 are fixed by contracts for a given number of years. However, in this analysis, we opt for the simplest form. We write incremental demand (ID) as

$$\text{ID} = D(t) - \delta D(0) - 0.05[\delta D(0)]t$$

$$\text{ID} = D(t) + \delta D(0)(0.05t - 1) \qquad t \leq 20$$
$$D(t) \qquad\qquad\qquad\qquad\qquad t > 20$$

where ID = incremental demand

δ = fixed fraction of 1975 demand, taken as 0.9

$D(0)$ = 1975 demand

$D(t)$ = demand in year t, $t = 0, 1, 2, \ldots, 25$

Sulfur Constraints. The coal used must satisfy restrictions on sulfur emissions. These contraints reflect the impact of federal and state air-pollution regulations. Actual regulations are extremely complex and are applied on a plant-by-plant basis. It would be computationally infeasible to deal with the complexity of the regulations. The situation is made even more difficult for the analyst by the wide gap between the law and enforcement of the law.

There are four sets of air-pollution regulations to be concerned with. New-source performance standards (NSPS) limit the sulfur content of coal used in plants that come onstream after 1975. These plants have to emit less than 1.2 pounds of sulfur dioxide per million Btu burned. Furthermore, regulations force scrubbing in plants for which construction began after 1978. The third category of constraints represents the state implementation plans (SIPs). These limit emissions, in many cases on a plant basis, in order to meet general ambient-air standards. These standards were supposed to have gone into effect by June 30, 1975, at the latest. Current schedules call for complete enforcement by 1983. In most areas, NSPS are more stringent than SIP standards and are therefore binding on new plants. Last, in areas that are already below the applicable standards, there are nondegradation requirements. These standards mean that plants cannot contribute to increasing air pollution in the area. Again, standards can be met by scrubbing or by burning any mixture of sulfur contents that satisfies the standards.

To reflect these complicated standards, the following procedure was adopted. First, current levels of actual emissions were calculated for each region. Two sets of standards operate on incremental coal demand. Pollution must not increase above the current levels to reflect nondegradation. Second, incremental coal is forced to meet new-source performance standards of 1.2 pounds of SO_2 per million Btu. Only one of these constraints is binding, and that is entered as constraint S_j. In the runs reported here, the standard of 1.2 pounds of SO_2 takes effect in 1985. Between 1975 and 1985 the standard declines linearly from current average emissions to the 1.2-pound level. This reflects the fact that not all incremental coal is subject to NSPS. The bias introduced by this methodology is small. Finally, Best Available Control Technology (BACT) requirements are imposed by forcing scrubbing on plants scheduled to come onstream after 1983.

Base-Case Assumptions

Nuclear Power. The only competition remaining for coal is nuclear power. And, as pointed out at the outset, even this is of questionable viability. To reflect the current difficulties facing nuclear power, we make the following assumptions. Nuclear power is limited to already announced plants through 1988. We assume, in addition, a five-year moratorium on new announcements. Given the ten-year construction lag, this means we constrain nuclear power through 1993. After that, decisions are based on minimizing costs.

Our constraints on new commitments are very loose after 1983. The object is to provide a lower bound on the estimate of the rents available to the Western states. In fact, because of public opposition, it is problematic as to whether any more nuclear plants will be built in the United States. The base case assumes away these difficulties in the long run. However, in order to bracket likely outcomes, we repeat the analysis that follows, assuming the opposite extreme—a complete nuclear moratorium.

Discount Rates. The choice of a rate with which to discount future revenue streams is arbitrary. We conduct the analysis with a 4.5 and an 11.5 percent real rate of discount. The states we analyze can at present borrow at real rates below 4.5 percent. However, if we view the coal deposits as being held by the state residents, we see there is no way for the residents to shed nonsystematic risks associated with coal revenues. This would argue for a higher rate than the 1 to 2 percent real rate that the states currently borrow at. The 11.5 percent rate reflects the private rate of cost capital to the coal industry (Alt 1978). As we see below, the results of our analyses are robust with respect to the interest rate.

Evaluation of Tax Scenarios

Methodology

Three methods for evaluating the effect of taxes are utilized in this chapter. The conventional methods employed by economists analyze how tax regimes affect the allocation of economic resources and resulting losses of economic efficiency and determine who bears the ultimate burden of a tax by evaluating relative tax-induced price changes. The third and unconventional measure utilized stems from the implied preferences of coal state-tax policymakers discussed previously; it is the change in tax revenues and the industry wage bill.

Measuring this latter indicator is straightforward and requires no additional explanation except to mention that no attempt is made to aggregate these two separate components, as was done in chapter 1. Policymakers should be conscious of the final incidence of tax burdens and how this, in turn, affects income and wealth. The model of the copper and coal industries permits us to estimate how much of a tax increase is shifted forward to consumers in the form of price increases and backward to owners of the natural-resource firms. However, both models are somewhat hampered by the assumption of fixed wage rates, which means that either individual rates are not affected at all or labor income is reduced to zero at mine sites which no longer realize a profit after the imposition of a tax and are shut down. If labor mobility were instantaneous and costless, then displaced workers would be immediately hired in the next best mining location or in other jobs with minimal disruption. However, this is not the case. Many workers choose to wait out layoffs because mobility is expensive and unemployment benefits help cushion the burden of waiting for reemployment at their old jobs. We estimate the tax burden on labor in a rough way by estimating changes in employment or the wage bill, rather than by attempting to measure tax-induced changes in wage rates.

Since the nonrenewable natural resources are owned mainly by extractive firms and federal and state governments, we estimate tax burdens on owners by changes in economic rent flowing to private owners and by changes in tax revenues under various tax regimes. The concept of economic rent is widely misunderstood, and model limitations permit us to estimate it for the copper industry only. For this exercise we define *economic rent* to be the difference between marginal cost (to include tax liability per unit) and price at the refinery times the output of copper. In the short run, many costs are invariant with respect to output and are fixed. Thus positive economic rents (producer surplus) are possible under this definition, even though an accountant would conclude that profits were less than economic rent or even negative (for as long as the firm is able to cover its variable costs, it will operate in the short run even through it is making negative accounting profits). In the long run, the firm operates only if total revenues exceed total costs of both the variable and fixed varieties. In this case, economic and accounting profits are closer, the major difference being in the cost of equity capital; and other differences do exist in the copper model becuase of the way in which the long-run marginal-cost functions are derived.

Exporting Tax Burdens

Changes in prices relevant to the consumer, resource exploiter, labor, and state-government constituencies reveal the extent to which state natural-resource taxes are exported. Since most copper and coal (in the raw form or as transformed into electricity) is exported by the producing states, price increases to consumers are one way that taxes are exported. The second way is through reduced earnings to nonresident owners of extractive firms. Since the geographical distribution of resource-firm stockholders is unknown, the precise degree of this exporting is unknown, although it is most likely a large component. State residents benefit by increased tax revenues, because either government expenditures and services are increased or other taxes are reduced. Miners who are forced to relocate and possibly take wage reductions or suffer unemployment are the state residents who suffer most from state tax burdens. The models are incapable of calculating the indirect impact of these employment reductions in income on business throughout the state, although based on various input/output studies, the income multiplier is probably on the order of 1.0 to 2.5.

Efficiency Losses

The final tax effect measured here is the efficiency or welfare loss engendered by taxes. We have chosen to use the standard welfare-economics measure as derived by Harberger (1964). The method accounts for the fact that taxes raise the cost of production (marginal cost) and thereby result in less output being produced in the taxed region. However, higher gross prices engendered in a taxing state may induce additional output in other regions (a shift in demand), which would induce a positive benefit if a tax were in effect in these other locations. The indirect shift in demand may induce these other states to produce at a level closer to that which would have occurred had they not imposed their own taxes. This, in effect, *reduces* an existing efficiency loss. The simplified estimating method assumes that marginal cost in each region is constant, and it measures only the consumer-surplus component of efficiency loss. Thus, if a tax induces the closing of a mine site(s) in the taxing jurisdiction and the opening of a higher-cost site(s) in other regions, this method underestimates welfare losses by the differential in marginal costs net of taxes. However, the resulting underestimate of efficiency loss is relatively small compared to the consumer-surplus component and is ignored in this initial analysis. The Harberger measure linearizes the demand curve and is

$$-\tfrac{1}{2}\Sigma_i T_i \, \Delta X_i$$

where i = *each producing state*

T_i = per-unit tax in each state

ΔX_i = change in production induced by tax change

The alternative tax scenarios in the copper and coal industries are formulated to depict a few plausible situations from an enormous spectrum of possible futures. The current (1977) tax situation versus a zero-tax condition is evaluated for both industries. This provides insight into the existing burden of taxes and establishes a benchmark condition for either case. However, utilizing the existing tax regime as a base case is preferred to a zero-tax situation because this is the initial condition from which changes may be made. Different future tax scenarios were posited for the two industries. For copper, a uniform tax increase in all states and a unilateral increase by Arizona, the largest producer, are hypothesized. For coal, the hypothesis is that the low-cost, low-sulfur coal-producing states of Wyoming and Montana act in concert to maximize the present value of the sum of tax revenues plus the wage bill. This scenario is not appropriate for copper because no single state or case enjoys as large quality or cost advantages, and the more simplified structure of the model, particularly market demand, diminishes one's confidence in the forecasts as large changes are made for any variable.

The scenarios are evaluated in the form of how production and consumption are altered from a base case (predominately the 1977 tax level). A changed tax regime causes a short-run change in the price of output, and levels of production also change but potentially by greater magnitudes because investment may take place in existing sites and in jurisdictions which levy lower taxes. For coal, the model is dynamic, so that depletion (increasing costs of extraction) and substitution among newly constructed coal and nuclear electric-power plants occur. These changes engendered by a permutation of tax regimes are evaluated for their effects on economic efficiency and variables which indicate who bears the burden of the tax. Economic-efficiency losses (a positive number denotes a loss) are measured by forgone consumer surplus. The burden of tax change to consumers is measured by an increase in price of the resource (or electricity), to labor by a change in the wage bill or number of jobs (recall wage rates are assumed constant), to owners of resources and extractive firms by a change in economic rent (pertinent to copper only), and to state residents by tax revenues from the extractive industries. Tax exporting can be measured only from the standpoint of consumption. In all cases, the bulk of production is exported which implies that tax-induced price increases are borne by out-of-state residents and are thereby exported.

Evaluation of Copper Tax Scenarios

The copper model is employed to simulate the effects of various tax regimes. However, a few caveats are in order. The first is that the tax scenarios are sensitive to the estimated demand curve for domestic copper producers. Both the short- and long-run demand curves selected for copper are price-inelastic, whch means that as output is reduced and prices are increased, total industry revenues increase. Demand elasticities were derived from econometric models of copper markets derived by other researchers (Charles River Associates 1970; Fisher, Cootner, and Baily 1972; Hibbard et al. 1977; Hartman 1977). Producers are assumed to operate at capacity as long as profits are greater than zero. Our model of the industry is static, and time is depicted as either the short run (mine sites are fixed) or the long run (mine sites may be opened up or shut down). However, the widespread availability of existing and potential capacity in foreign locations means that the dynamic long-run demand curve for domestic producers is undoubtedly quite elastic, and the results shown in the scenarios are questionable over the long haul. The reader is urged to evaluate these results may making relative comparisons among tax scenarios and not placing too much confidence in the absolute numbers estimated or for periods exceeding ten years or so.

The estimated price, tax revenue, economic rent, and output for the domestic copper industry for the short and long runs under a zero-tax regime and for current (1977) state taxes are shown in table 6–4. The short-run equilibrium price represents only variable costs, while the long-run equilibrium price accounts for all costs. Four tax scenarios are evaluted in both the short and long runs: current state taxes versus zero taxes; a tax increase of 20 percent of marginal costs pursued jointly by all producing states versus zero taxes; a unilateral tax increase by Arizona, the largest producer, of 30 percent of marginal cost versus zero taxes elsewhere; and a similar move by Arizona versus current state taxes. Each scenario is evaluated by the methods mentioned earlier. In order to accomplish this, six pieces of information for each change in tax regimes are presented in table 6–5: the change in market price, tax rate with respect to the value of output, change in tax revenue, change in employment, change in economic rent, and the Harberger measure of efficiency loss.

The imposition of 1977 state taxes from a zero-tax regime results in

Table 6–4
Copper-Scenario Tax Effects

	Price per Pound ($)	Tax Rate (%)	Tax Revenue ($000)	Total Employment	Rent ($000)	Pounds of Output (000)
Short Run						
Zero taxes	0.45					
Arizona				10,355	$187,929	1,933,124
New Mexico				1,283	5,894	257,238
Utah				3,300	48,666	580,576
Current taxes	0.47					
Arizona		2.59	23,464	10,355	172,183	1,933,124
New Mexico		1.40	1,690	1,283	7,050	257,238
Utah		0.81	2,197	3,300	54,034	580,756
Long Run						
Zero taxes	1.10	0	0			
Arizona				8,256	175,404	1,568,402
New Mexico				105	2,363	46,898
Utah		1.07	8,428	3,977	12,700	700,756
Montana		1.80	5,304	1,394	39,838	250,000
Nevada				47	9,927	14,000
Current taxes	1.12					
Arizona		1.9	33,270	8,232	130,784	1,559,767
New Mexico		0.80	419	105	1.977	46.898
Utah	1.07	8,428	3,977	12,700	700,756	
Montana	1.80	5,034	1,394	39,838	250,000	
Nevada		0.01	2	47	10,159	14,000

Table 6–5
Copper-Scenario Tax Evaluation

	ΔSR, Price per Pound ($)	Tax Rate on Value of Output (%)	Tax Revenue ($000)	Change in Employment	Change in Rent ($000)	Efficiency Loss ($000)
Short Run						
1977 Taxes versus zero tax	+0.02					
Arizona		2.59	24,464	0	+7,718	0
New Mexico		1.40	1,690	0	+2,846	0
Utah		0.81	2,179	0	+7,565	0
			28,333		+18,129	
Joint tax increase to 20% of unit cost versus zero tax	+0.11					
Arizona		11.03	106,440	−1,041	+41,051	+5,416
New Mexico		14.05	20,234	0	+588	0
Utah		9.42	30,642	0	+10,169	0
			157,316	−1,041	+51,808	+5,416
Arizona unilateral tax increase to 30% of cost versus zero taxes	+0.10					
Arizona		16.66	109,638	−3,617	−57,699	+33,692
New Mexico		1.17	2,302	+656	+30,988	−318
Utah		0.69	2,198	0	+55,742	−0.3
Montana		2.74	3,178	+1,139	+,831	−1,582
Michigan		0.17	170	+2,102	+3,698	−6.5
			117,486	190	+34,560	+32,055
Arizona unilateral tax increase to 30% of costs versus current taxes	+0.08					
Arizona		16.66	86,174	−3,617	−42,053	+33,962
New Mexico		1.17	612	+656	+29,832	−1,145
Utah		0.69	1	+1,139	+50,374	0
Montana		2.74	3,178	+2,012	+1,839	−1,582
Michigan		0.17	170		+3,698	−86
			90,135	190	+43,690	+31,149
Long Run						
1977 taxes versus zero tax	+0.02					
Arizona		1.9	33,270	−23	−11,350	+91.9
New Mexico		0.8	419	0	+33	0
Utah		1.07	8,428	0	+311	0

Montana	1.80	5,034	0	−1,692	0
Michigan	0.01	2,050	0	+234	0
		49,201	−23	−12,374	+92.0
Joint tax increase to 20% of unit cost versus zero tax	+0.26				
Arizona	13.49	285,738	−24	+8,909	+792.1
New Mexico	8.96	5,711	0	−1,459	0
Utah	14.55	41,737	−3,084	+478	+48,442.0
Montana	12.03	40,849	0	+6,047	0
Nevada	6.07	1,154	0	+2,169	0
		375,189	−3,108	+16,144	+49,234
Arizona unilateral tax increase to 30% of cost versus zero taxes	+0.26				
Arizona	17.79	100,416	−6,688	−135,620	+139,586
New Mexico	1.12	5,825	+2,070	+58,851	−2,545
Utah	0.88	8,430	0	+168,397	0
Montana	1.48	5,038	0	+58,179	0
Michigan	1.60	4,036	+2,012	+19,255	−2,012
Nevada	0.0	2	0	+3,595	0
		123,747	−2,606	+155,357	+135,029
Arizona unilateral tax increase to 30% of costs versus current taxes	+0.24				
Arizona	17.79	67,146	−6,364	−90,990	+138,541
New Mexico	1.12	5,406	+2,070	+59,237	−2,545
Utah	0.88	2	0	+176,508	0
Montana	1.48	4	0	+64,905	0
Michigan	1.60	4,036	+2,012	+19,255	−209
Nevada	0.0	0	0	+3,363	0
		76,594	−2,282	+214,978	+135,787

virtually no dislocation of employment or production. This, in turn, means that there are no efficiency losses in the short run and less than a $100,000 welfare loss in the long run resulting from a small reduction in production in Arizona. Arizona is the major producing state, and therefore copper mining is a seductively attractive tax target, as evidenced by the heaviest effective tax rate among all the producing states.

The current tax rates (1977) produce modest increases in short- and long-run prices ($0.02 per pound) which are borne by consumers and therefore are exported. In the short run, industry benefits through higher rents in all three states. However, in the long run, industry rents fall significantly in Arizona and Montana, but the total decrease in rents is less than one-fourth of the tax revenues raised. Current (1977) taxes seem to achieve both tax goals of not hindering employment or industry growth but producing significant revenues. State tax revenues are substantial in both the long and short runs (except for New Mexico, where tax revenues are minimal in the long run). The strong opposition by industry in Arizona to property taxes, which are invariant with respect to output in the short run, is understandable because of significant loss in long-run rents to the industry. Note that tax revenues are nearly 3 times larger than these rental losses in Arizona and nearly 4 times those for all producing states taken together. However, from the point of view of the efficient use of resources (the economists' sacrosanct Pareto optimality), the current tax regime is nearly without cost to society as a whole. However, this cannot be said of the remaining scenarios, for both long- and short-run losses are large, going to $136 million for a 30 percent tax imposed by Arizona compared to current tax levels.

In the second scenario, it is assumed that all producing states impose a 20 percent tax on marginal costs. The resulting annual tax revenues can be described only as magnificent (totaling $52 million in the short run and $375 million in the long run). This outcome is due primarily to the relatively inelastic estimated demand curve. Arizona experiences a significant employment loss in the short run, and Utah has a far greater one in the long run. The rise in price ($0.11 in the short run, or 22 percent, 5 times the increase due to 1977 taxes; $0.26 in the long run, or 24 percent, 7 times the increase due to 1977 taxes) means that consumers bear most of the tax burden. In fact, industry economic rents increase dramatically in the short run and to a smaller but comfortable extent in the long run, except for a loss in New Mexico. In the short run, efficiency losses are relatively modest ($5.4 million), but are significant over the long haul ($49.2 million in total). Nevertheless, increases in tax revenues swamp efficiency losses and forgone income to unemployed miners.

Should Arizona decide to introduce a tax structure at a 30 percent level of marginal costs, industry will shut down much of its capacity in the short run and leave the state in the long run. The 3,617 workers unemployed in the short run have the opportunity to gain reemployment in New Mexico, Michigan, and Nevada in the short run or New Mexico and Michigan in the long run; however, one-third will have to leave the industry in the latter case. At the political "cost" of this large loss of employment, Arizona maintains short-run tax revenues above those found in the preceding scenario, but takes nearly a two-thirds cut in the long-run take (although the $100 million increase over the existing tax structure might look attractive to some legislators despite an employment loss of 6,688 workers). Since production shifts to other states, their tax revenues increase substantially. This scenario burdens consumers with price increases equal to those in the previous scenarios, and they continue to bear the brunt of the tax burden along with Arizona labor and industry (a $42 million short-run loss in economic rents and $90.9 million long-run loss compared to 1977 taxes). However, gains in economic rent in other states more than compensate for Arizona's losses, so net industry rents increse. As would be expected, both short- and long-run welfare losses are substantial—3 to 6 times higher than a move to a 20 percent joint tax. It must be said that it is rather unlikely that Arizona would pursue such a dramatic course although in the short run, interestingly enough, increased tax revenues are nearly twice industry rent losses but are less than the loss in rents in the long run. Further, the loss of 3,664 jobs represents an enormous loss in private incomes and state taxes, which would appropriate the gain in tax revenues. Also the burden on copper consumers of Arizona pursuing such a dramatic course is no larger than in the joint 20 percent tax case.

Evaluaton of the Coal-Tax Scenarios

The coal-tax scenarios are evaluated similarly to the copper-tax scenarios. However, the coal model itself and the associated alternative tax scenarios are considerably different from those for copper. The demand for coal is derived from a computer simulation model of the demand for electricity and industrial coal, and both demand and supply functions are dynamic (from 1980 through 2000). Electricity demand is sensitive to population, income, and the price of generating power by alternative energy resources (hydroelectric, natural gas, oil, nuclear fission, and coal). Interfuel competition means that the derived demand curve for coal becomes ever more price-elastic as the time horizon becomes long enough for substitution of one technology for another to take place. Recall that the de facto moratorium on nuclear plants assumed in the model expires, so that by 1990 the nuclear role may expand. The supply of coal is determined by the extent and quality of coal reserves in each state, the prices of inputs into coal extraction and transportation, and the tax regime in each state. More important, the supply curve in the coal model reflects the economic effect of rising extraction costs as deeper and lower-quality reserves are extracted. The effects of inflation are portrayed by increasing prices exogenously at plausible rates.

Computer simulations are made for four tax cases:

1. Base case reflecting 1977 severance taxes
2. A no-tax (state and federal) regime
3. 1977 effective taxes as estimated here
4. A consortium of Montana and Wyoming maximizing the sum of tax revenues plus the wage bill to coal miners

These four scenarios are evaluated by comparing the base case (1) versus a no-tax regime (2) and the base case (1) versus the northern Plains consortium maximizing tax (4). The reason for using the base case rather than the 1977 estimated effective tax rates (2) as a benchmark is that the computer simulation model is calibrated to reflect the former. Substituting the 1977 estimated effective taxes into the model produced inaccurate estimates when compared to actual data on production and prices. This result is caused by the effect of taxes other than severance taxes being embedded in the derivation of the model. Thus, adding the total estimated effective tax rate (chapter 5) to marginal costs of production results in double-counting of a portion of the tax burden and consequent inaccuracies. The severance taxes for the base case are shown in table 6–6 and are expressed in ad valorem (percentage-of-value) terms so that the severance-tax burden is assumed to increase with prices over time.

The objective criterion which is used to derive the optimum severance tax rate for a Montana/Wyoming cartel is the objective function employed in analyzing coal-tax policy (chapter 1)—a weighted sum of tax revenues plus the wage bill for the coal industry. We assume that other producing states fail to act strategically by either joining the cartel or unilaterally increasing their tax rates (constant at the rates reported in table 6–6). The linear-programming simulation of coal markets is solved iteratively for increasing tax rates (assumed to be constant from 1979 to 2000). Discount rates of 4.5 and 11.5 percent are employed to compute the discounted present value of tax revenues and the wage bill.

In order to carry out the derivation of the optimum tax, the base-case tax for the two-state consortium was assumed to equal Wyoming's severance-

Table 6–6
Severance-Tax Rates Used in Base-Case Scenario

State	Tax Rate (%)
Ohio	0.2
Pennsylvania	0.0
West Virginia	3.9
Alabama	1.3
Kentucky	4.3
Tennessee	0.8
Virginia	1.0
Illinois	5.0
Indiana	0.0
Montana	25.0
Wyoming	16.8
Colorado	3.0
Utah	0.0
Arizona	2.5
New Mexico	3.8

tax rate (16.8 percent) in 1977. If Montana's 30 percent marginal severance-tax rate were applied, then all greenfield development would occur in Wyoming (as it has from 1975 to 1981) until exhaustion or until the consortium imposed tax equality. Therefore the initial tax rate is assumed to be 16.8 percent for both states.

The results of this exercise are shown in table 6–7. It is readily apparent that the optimum severance tax for the Montana/Wyoming cartel is 62.5 percent except where the weight placed on the wage bill is extraordinarily high. This occurs because of the low labor intensity of Western strip mining. This result was achieved after several modifications were made to an earlier model. Utilizing the earlier version produced an optimum tax of 75 percent. The implications of the 62.5 percent rate are reviewed first, and then the 75 percent rate is employed to measure efficiency effects and tax burdens at the state level. However, the two simulations are not substantially different. Note that the effect of assumptions regarding nuclear-power-plant construction is noted in table 6–7.

The effects of the tax conform to expectations. In the long run, the tax reduces mining openings and expansions in Montana/Wyoming and shifts this output to other states as well as accelerates the substitution of nuclear power after 1993 (see table 6–7). The effect on output disaggregated into the six producing regions is shown in table 6–8. The optimum tax reduces output by the coal cartel from 657 million to 463 million tons annually, and total steam coal production falls by 107 million tons by the year 2000. The

Table 6–7
Present Value of Tax Revenues and Wage Bill for Montana/Wyoming under Alternative Tax Levels
(millions of 1978 dollars)

| | Base-Case Nuclear Assumptions[a] Real Discount Rates | | | |
| | 4.5 Percent | | 11.5 Percent | |
Tax Level (%)	Tax Revenue	Wage Bill	Tax Revenue	Wage Bill
18	8,265	5,814	3,848	2,867
35	11,045	5,086	5,379	2,496
50	14,425	4,663	7,012	2,291
62.5	16,380	4,246	7,717	2,028
75	14,733	3,198	6,834	1,513
100	11,593	1,914	5,496	939
150	11,219	1,260	5,827	685
200	11,317	971	6,208	563

Note: The 1979–1983 period reflects the de facto current moratorium.

[a]The maximum nuclear commitments in the base case (in gigawatts) were: pre-1979, announced as of 1978; 1979—1983, 0; 1984, 4.8; 1985, 5.4; 1990, 10.1; 1995, 15.6; 2000, 30.2.

Table 6–8
Regional Coal Production under Base Case and 62.5 Percent Tax
(millions tons/year)

Year		1	2	3	4	5	6	Total
					Region[a]			
1980	Base	144.7	144.7	122.8	190.0	4.5	60.4	667
	62.5%	130.4	160.0	141.9	62.9	64.3	81.7	641
1985	Base	145.1	140.2	76.3	324.9	15.7	80.0	782.2
	62.5%	151.9	140.2	71.1	265.6	50.8	78.9	758.5
1990	Base	247.3	50.3	141.7	501.9	53.1	85.5	1,079.7
	62.5%	262.4	38.0	143.9	406.5	83.3	108.2	1,042.2
1995	Base	245.1	51.6	146.9	562.8	85.8	81.5	1,174.6
	62.5%	258.3	49.3	158.0	515.0	58.7	107.5	1,147.1
2000	Base	86.5	25.0	337.2	657.1	37.8	92.5	1,236.2
	62.5%	170.5	31.5	312.1	463.4	83	68.5	1,129.0

Note: This excludes metallurgical and export coal. It also assumes that substantial nuclear-power commitments can be made after 1983 for plants to come onstream after 1993.

[a]The regions are: 1, northern Appalachia; 2, southern Appalachia; 3, Midwest; 4, Montana/Wyoming; 5, Colorado/Utah; 6, Arizona/New Mexico.

difference is supplied by the most efficient competitors—the Midwest, the Colorado/Utah region, and Arizona/New Mexico. A detailed breakdown of interregional coal flows is shown in table 6–9. (Note that the twelve consuming regions are defined in table 6–12.) Table 6–10 details the effect of the tax on electrical-power-plant types for the nation and for consuming region 8, which is made up of the Rocky Mountain states. This latter consuming region is more dependent on Powder River Basin low-sulfur coal and consequently bears most of the effects in terms of more expensive coal and electricity, but this is balanced (if we ignore the undesirable effects of coal extraction) by increased coal output. The long-run effect is a substitution of nuclear power generation for coal with scrubbers by the year 2000. This substitution is accompanied by only a modest (1 percent) increase in electricity prices in this region. The effect on electricity prices is shown in table 6–11.

The role that railroads play in pursuing economic rents was mentioned previously. The model can be used to show how successfully they have achieved this objective. Western coal is transport-intensive because of the large distances to major demand centers, and transport cost accounts for a large proportion of the delivered cost of Powder River Basin coal. If we temporarily assume that the rents go exclusively to the single railroad originating coal in the Powder River Basin, we may estimate the increase in railroad tariffs that is comparable to the optimal severance tax. The 62.5 percent tax is equivalent to a charge of roughly $4 per ton in 1980 (net of state and federal taxes and fees). Current tax levels of 16.8 percent account

Table 6–9
Interregional Coal Flows in Base Case and in 62.5 Percent Tax Case
(millions of tons)

From \ To:		1	2	3	4	5	6	7	8	9	10	11	12	Total
							Shipments to Region:							
1		8.76	74.44	14.77	0.02	149.20	0.12	0.00	0.00	0.00	0.00	0.00	0.00	247.32
2		0.08	0.83	5.44	0.04	36.52	6.12	1.26	0.00	0.00	0.00	0.00	0.00	50.28
3		0.00	0.00	41.01	4.30	3.20	92.80	0.00	0.00	0.00	0.34	0.00	0.00	141.65
4		0.00	0.00	207.42	125.31	0.00	0.06	0.00	69.97	0.00	30.75	68.33	0.00	501.85
5		0.00	0.00	0.13	0.09	0.00	0.00	15.68	1.27	35.95	0.00	0.00	0.00	53.13
6		0.00	0.00	0.00	0.00	0.00	0.00	12.42	0.94	24.87	0.00	0.00	47.23	85.45
Total		8.83	75.27	268.77	129.77	188.92	99.09	29.36	72.18	60.82	31.10	68.33	47.23	1,079.68
62.5 Percent Tax Case, Year: 1990														
1		9.08	75.47	14.77	0.02	162.93	0.12	0.00	0.00	0.00	0.00	0.00	0.00	262.4
2		0.08	0.83	5.44	0.04	24.20	6.12	1.26	0.00	0.00	0.00	0.00	0.00	38.0
3		0.00	0.00	41.72	4.30	3.20	94.36	0.00	0.00	0.00	0.35	0.00	0.00	143.9
4		0.00	0.00	194.45	118.64	0.00	0.06	0.00	0.41	0.00	28.80	64.08	0.00	406.5
5		0.00	0.00	0.13	0.09	0.00	0.00	15.61	53.07	14.37	0.00	0.00	0.00	83.3
6		0.00	0.00	0.00	0.00	0.00	0.00	12.36	0.94	49.58	0.00	0.00	45.31	108.2
Total		9.16	76.30	256.50	123.10	190.33	100.65	29.23	54.42	63.95	29.16	64.08	45.32	1,042.7
Base Case, Year: 2000														
1		17.59	68.94	0.00	0.00	0.00	0.00	0.00	0.00	0.00	0.00	0.00	0.00	86.53
2		0.00	0.00	0.00	0.00	22.73	2.30	0.00	0.00	0.00	0.00	0.00	0.00	25.03
3		0.00	0.00	28.96	0.00	204.25	104.02	0.00	0.00	0.00	0.00	0.00	0.00	337.23
4		0.00	0.00	195.58	86.20	0.00	5.80	0.00	135.89	68.09	25.61	116.69	23.22	657.08
5		2.00	4.78	0.00	0.00	6.15	0.00	15.38	0.00	9.48	0.00	0.00	0.00	37.79
6		0.00	0.00	0.00	0.00	0.00	0.00	24.27	0.00	0.00	0.00	0.00	68.20	92.47
Total		19.59	73.72	224.54	86.20	233.13	112.13	39.65	135.89	77.57	25.61	116.69	91.42	1,236.64
62.5 Percent Tax Case, Year: 2000														
1		18.20	74.33	0.00	0.00	77.94	0.00	0.00	0.00	0.00	0.00	0.00	0.00	170.5
2		0.00	0.00	0.00	0.00	27.66	3.82	0.00	0.00	0.00	0.00	0.00	0.00	31.5
3		0.00	0.00	76.56	0.00	119.70	105.11	10.73	0.00	0.00	0.00	0.00	0.00	312.1
4		0.00	0.00	139.49	77.13	0.00	0.00	0.00	108.27	24.94	22.52	91.08	0.00	463.4
5		2.03	0.00	0.00	0.00	6.23	2.87	15.05	0.00	44.17	0.00	0.00	12.65	83.0
6		0.00	0.00	0.00	0.00	0.00	0.00	11.82	0.00	0.00	0.00	0.00	56.72	68.5
Total		20.23	74.33	216.05	77.13	231.52	111.81	37.60	108.27	69.11	22.52	91.08	69.37	1,129.0

Table 6–10
Generation Capacity by Plant Type in the Base Case Compared to the 62.5 Percent Tax Case
(gigawatts)

| | Base Case, Nuclear | | |
| | Coal with | Coal without | |
Year	Scrubbers	Scrubbers	Nuclear
1990			
Region 8			
Base	49.8	22.9	2.0
62.5%	46.9	22.9	2.0
National			
Base	268.9	172.0	124.9
62.5%	260.3	162.0	124.9
1995			
Region 8			
Base	59.0	20.8	1.9
62.5%	55.1	20.8	1.9
National			
Base	347.8	153.9	217.7
62.5%	339.2	154.0	218.4
2000			
Region 8			
Base	84.3	18.4	1.7
62.5%	64.1	18.4	17.7
National			
Base	403.0	133.8	351.4
62.5%	378.9	133.9	369.3

Table 6–11
Electricity Prices
(mills per kilowatthour)

| | 1980 | | 1990 | | 2000 | |
Region	Base Case	62.5 Percent	Base Case	62.5 Percent	Base Case	62.5 Percent
New England	55.7	55.7	101.6	101.7	157.2	156.7
Middle Atlantic	44.3	44.3	93.1	93.3	157.1	157.1
East North Central	39.8	40.2	69.2	70.4	126.0	126.0
West North Central	39.5	40.9	72.6	73.1	124.6	124.5
South Atlantic	39.6	39.6	78.3	78.4	136.3	137.1
East South Central	31.9	32.0	61.2	61.1	98.9	99.0
West South Central	50.3	50.3	69.3	70.2	124.1	127.3
Rocky Mountain	31.8	32.5	68.4	69.8	104.6	107.3
Pacific	34.4	34.4	72.4	72.0	126.9	127.3
National Average	39.9	40.2	74.2	74.7	126.4	127.5

for approximately \$1 of that figure. Thus, the difference between today's level and the "optimum" amounts to about \$3 per ton. That figure represents 16 to 24 percent of the transport cost to Chicago. As of 1980, the Burlington Northern has a monopoly on transporting coal from this region. The expectation of profitable coal hauling induced their management to invest \$1 billion from 1975 to 1980 and another \$1 billion projected by 1985 to upgrade lightweight trackage to withstand the grueling task of supporting 100-ton hopper cars speeding along in 100-car trains. These investments have enabled Burlington Northern to argue successfully for rate increases from the Interstate Commerce Commission (ICC). However, the railroad deregulation act (passed in October 1980) may enable the small Chicago and North Western Transportation Co. to qualify for federal loan guarantees (\$310 million) and hook up to the powerful Union Pacific Railroad. This competition would reduce Burlington Northern's bargaining power to extract economic rents. Another source of potential competition is the proposed coal-slurry pipeline whose promoters are asking the right of eminent domain so that they can compel railroads to share rights-of-way.

The outcomes from the coal-tax scenarios employing the 75 percent optimum tax rate are evaluated for the tax burden and efficiency loss in tables 6–12 and 6–13. The base-case, tax versus no-tax scenario and the 75 percent optimum tax versus the base case are analyzed for tax-induced effects at both consumption and production stages. The percentage change in wage bill and in tax revenue and the welfare loss are calculated for each producing state in 1980, 1990, and 2000 (table 6–12). When taxes are already in place and are increased in one region, as in comparing the maximum tax versus base case, the net welfare cost is calculated as the incremental distortion in the states increasing their taxes (one-half the induced change in output times the change in the tax rate). In states maintaining their tax rates, an increase in production owing to demand shifting from another, higher-taxing region will reduce the efficiency loss caused by their own tax.

The tax increase shifts demand curves for competing coal states, which results in increased output *and* a reduction in the efficiency loss induced by their previously levied own taxes (that is, output is closer to what it would have been in a no-tax situation). The tax-induced percentage change in price of coal and electricity in each consuming region (table 6–13) is calculated as is the net welfare loss (one-half the tax-induced change in price times the change in output). However, the complexities inherent in the electricity demand and supply model and the wide array of preexisting taxes and regulations which create distortions preclude computing the true net welfare costs at the final demand for coal stage. These are approximated as one-half the tax-induced change in price of coal at the consuming region times the change in amount consumed.

Table 6–12
Coal Scenarios for Producing States

Scenario	Total	Ohio	Pennsyl-sylania	West Virginia	Alabama	Kentucky	Tennes-see
1980							
No-Tax versus Base Case							
Percentage change in wage bill							
Percentage change in tax revenue							
Welfare loss	−83.8	−0.4	−0.1	−9.2	−0.1	−22.2	−0.1
Base Case versus 75 Percent Tax							
Percentage change in wage bill		−13.4	−0.1	−3.4	+5.4	+2.8	+7.2
Percentage change in tax revenue		−14.3	0.0	−3.4	+7.4	+3.4	+6.6
Welfare loss	+218.1	+0.1	0.0	+9.0	−0.1	−2.6	−0.1
1990							
No-Tax Versus Base Case							
Percentage change in wage bill							
Percentage change in tax revenue							
Welfare loss	−254.1	−0.1	0.0	−6.7	−0.5	−1.1	−0.3
Base Case versus 75 Percent Tax							
Percentage change in wage bill		−6.0	+7.7	+4.2	−3.2	−6.9	+1.6
Percentage change in tax revenue		+3.9	0.0	+4.3	−3.5	−1.5	0.0
Welfare loss	+1,458.8	−0.1	0.0	−1.6	+0.1	+1.8	−0.2
2000							
No-Tax versus Base Case							
Percentage change in wage bill							
Percentage change in tax revenue							
Welfare loss	−476.6	−3.3	−0.7	−52.5	−0.3	−33.2	−0.3
Base Case versus 75 Percent Tax							
Percentage change in wage bill		+0.9	+1.9	+9.9	+37.1	+14.7	+47.7
Percentage change in tax revenue		+0.6	0.0	+9.9	+40.2	+16.7	+41.8
Welfare loss	+1,238.2	0.1	0.0	−48.2	−2.5	−46.0	−1.0

Virginia	Illinois	Indiana	Mon-tana	Wyo-ming	Colo-rado	Utah	Arizona	New Mexico
−0.3	−42.6	0.0	−5.9		−0.1	0.0	−0.9	−1.6
+2.6	+81.2	+17.8	−66.6		+427.4	+274.0	+108.0	+37.5
+3.1	+77.5	0.0	+13.4		+442.0	0.0	+116.7	+37.7
−0.1	−10.4	−0.2	+235.7		+8.2	−0.5	−0.7	−4.8
−0.2	−5.8	0.0	−103.5	−127.9	−5.6	0.0	−0.3	−2.6
−0.1	+4.4	+6.7	+12.9	−50.0	+94.6	+62.6	+47.4	+26.0
0.0	+3.7	0.0	+142.1	+123.0	+89.6	0.0	+46.4	+26.2
+0.1	−5.5	0.0	−51.7	+1,532.5	−5.2	0.0	−1.6	−9.7
−0.1	−126.5	0.0	−104.8	−132.0	−17.5	−0.1	−2.6	−3.6
+7.0	+11.6	+4.8	−21.1	−50.7	+29.0	+28.3	+22.5	−93.9
+8.0	+13.0	0.0	+69.3	+120.1	+26.8	0.0	−22.3	−38.1
−1.8	−52.5	0.0	+406.5	+946.3	−7.0	0.0	+4.0	+43.2

Table 6-13
Coal-Tax Scenarios for Consumption Regions

Scenario	New England	Middle Atlantic	East North Central	West North Central	Southern Atlantic	East South Central	West South Central	Rocky Mountain	Pacific	Total
1981										
No-Tax versus Base Case										
Percentage change in delivered price										
High-sulfur	0.0	0.0	0.0	−6.9	−5.6	−2.9	−2.1	−8.0	−1.5	
Low-sulfur	−1.9	−2.0	−4.1	−6.7	−2.9	−3.0	−2.1	−8.2	−2.1	
Welfare loss ($)										
High-sulfur	0.0	0.0	0.0	0.0	0.0	0.0	0.0	0.0	0.0	
Low-sulfur	−0.05	−0.6	−1.3	−0.8	−0.5	+0.3	−0.2	−0.7	−0.4	
Total	−0.05	−0.6	−1.3	−0.8	−0.5	+0.3	−0.2	−0.7	−0.4	−4.25
Base Case versus 75 Percent Tax										
Percentage change in delivered price										
High-sulfur	0.0	0.0	0.0	+16.5	+3.6	+1.1	+0.5	+10.5	+0.6	
Low-sulfur	0.0	0.0	+4.8	+14.6	0.0	+0.4	0.0	+11.9	0.0	
Percentage change in price of electricity	0.0	0.0	+0.8	+4.1	0.0	0.0	0.0	+1.9	0.0	
Welfare loss ($)										
High-sulfur	0.0	0.0	0.0	0.0	0.0	0.0	0.0	0.0	0.0	
Low-sulfur	0.01	0.0	+0.5	+0.2	0.0	0.0	0.0	+0.2	0.0	
Total	+0.01	0.0	+0.5	+0.2	0.0	0.0	0.0	+0.2	0.0	+0.9
1991										
No-Tax versus Base Case										
Percentage change in delivered price										
High-sulfur	−1.6	−1.8	−3.2	−6.2	−1.6	−4.4	−1.7	−8.4	−1.7	
Low-sulfur	−2.0	−2.1	−7.3	−9.9	−2.2	−2.9	−1.8	−10.6	−2.9	
Welfare loss ($)										
High-sulfur	−0.2	−0.1	−0.7	−1.6	−0.2	−0.3	−0.1	−1.6	−0.1	
Low-sulfur	−0.1	0.0	−0.7	−0.8	−0.1	−0.2	0.0	−0.8	−0.1	
Total	−0.3	−0.1	−1.4	−2.4	−0.3	−0.5	−0.1	−2.4	−0.2	−7.7

Base Case versus 75 Percent Tax

	1	2	3	4	5	6	7	8	Total
Percentage change in delivered price									
High-sulfur	0.0	0.0	+19.1	0.0	0.0	+6.8	+19.7	+3.3	
Low-sulfur	0.0	+10.2	+14.8	0.0	0.0	+4.8	+16.1	+5.2	
Percentage change in price of electricity	+1.1	+2.0	+4.5	0.0	0.0	+1.4	+1.5	+0.5	
Welfare loss ($)									
High-sulfur	0.0	0.0	+4.9	−0	−0	+0.2	+2.9	+0.1	
Low-sulfur	0.0	+0.9	+1.2	−0	−0	+0.1	+1.2	+0.1	
Total	0.0	+0.9	+6.1	0	0	+0.3	+4.1	+0.2	+12.6

2000

No-Tax versus Base Case

	1	2	3	4	5	6	7	8	Total
Percentage change in delivered price									
High-sulfur	−1.7	−2.3	−7.5	−3.0	−2.3	−2.2	−9.8	−4.3	
Low-sulfur	−2.0	−1.4	−9.5	−7.1	−2.9	−4.8	−13.4	−7.4	
Percentage change in price of electricity	−0.3	−1.6	−1.0	−0.9	−0.2	−1.3	−5.8	−0.3	
Welfare loss ($)									
High-sulfur	−0.1	−0.1	−1.0	−0.9	−0.2	−1.0	−5.8	−0.6	
Low-sulfur	−0.2	−1.7	−5.6	−2.0	−0.1	−0.4	−20.0	−0.3	
Total	−0.3	−1.8	−6.6	−2.9	−0.3	−1.4	−25.8	−0.9	−42.9

Base Case versus 75 Percent Tax

	1	2	3	4	5	6	7	8	Total
Percentage change in delivered price									
High-sulfur	+4.2	+0.3	+15.7	+5.8	+0.3	+3.9	+19.3	+11.4	
Low-sulfur	+1.8	+9.7	+14.1	+2.6	+4.1	+7.2	+11.2	+10.5	
Percentage change in price of electricity	−2.1	+0.4	+0.9	+0.3	−0.2	+1.8	−8.3	+0.3	
Welfare loss ($)									
High-sulfur	−0.1	+0.4	+2.8	+0.6	0.0	+1.7	+96.5	−1.6	
Low-sulfur	0.0	+29.6	+3.4	+0.1	+0.1	+0.4	+10.8	−1.0	
Total	−0.1	+30.0	+6.2	+0.7	+0.1	+2.1	+107.3	−2.6	+149.1

What can be said about the data from the scenarios? For consuming regions (table 6–13) the welfare costs of the base case versus a no-tax situation are relatively small ($4.25 million in 1981 and $42.9 million in 2000, and note that a minus signifies an improvement in welfare). However, the imposition of a 75 percent severance tax in Montana and Wyoming generates significantly larger, but still modest, welfare losses as more costly coal is extracted in other regions (an incremental loss above the base case of $0.9 million in 1981 and $149.1 million in 2000) with the East North Central (Michigan, Ohio, Wisconsin, Illinois, Indiana, and Kentucky), West North Central (Minnesota, Nebraska, Kansas, Iowa, North Dakota, South Dakota, and Missouri), and Rocky Mountain regions (New Mexico, Colorado, Utah, Nevada, Idaho, Wyoming, and Montana) receiving the brunt of these efficiency losses in consumer surplus implied by their demand for coal. These losses are small because the relatively inelastic demand (in the short run) for electricity generated from coal means that although prices increase significantly, demand falls only modestly and thus the loss of consumer surplus is relatively small.

Coal-price increases indicate that most of the tax is passed forward to consumers in these three regions. However, welfare losses measured for the producing states are substantial ($218.1 million in 1980 and peaking at $1458.8 million in 1990) for the 75 percent tax increase above the base case (table 6–12). The welfare costs of the 1977 tax regime are considerably less but still substantial compared to the zero-tax condition (the −$83.8 million in 1980 signifies the gain if all taxes were rescinded, growing to −$476.6 million in 2000). These losses exceed the losses measured at consuming regions because of the differences in regional elasticities of demand for coal at the point of production and consumption, and the delivered price of coal includes shipping costs. These two sets of estimates may be interpreted as upper and lower bounds to welfare losses.

The linked electricity-demand, coal-supply model produces an estimated effect on the price of electricity (busbar price or price at the base-load power plant). The increased price due to the 75 percent tax versus the base case is shown in table 6–13. Because of the regulatory regime of utilities as well as the assumption in the model that all costs are passed on to consumers, this estimate provides insight into both the welfare losses associated with the 75 percent tax (the direct measure of welfare loss was not calculated for a variety of reasons having to do with the complexity of the model) and the *burden* placed on consumers. It is clear that the price increase borne by consumers in the Midwest (East North Central and West North Central) and Rocky Mountain regions are relatively modest. Furthermore, by 2000 the price increases have been mitigated by conversion to nuclear power and conservation, as detailed in the coal model. Although this option is technologically viable, the future of nuclear power is highly uncertain because of political and social constraints, as evidenced by the occurrence and response to Three-Mile Island. Given the assumptions built into this model, the modest effects on electricity prices imply that welfare losses to consumers are commensurately modest. However, the relatively large welfare losses measured for producing states and the uncertainty regarding the economic and political viability of expanded nuclear capacity mean that these estimates of real price increases for electricity must be considered a lower bound on the costs which consumers may bear. This portion of the tax burden is exported almost in toto by the Montana/Wyoming consortium to the Midwest and Rocky Mountain states.

The estimated changes in the wage bill for the coal industry and in tax revenues indicate how the tax burden is incurred within each producing state. The method for deriving the optimum 75 percent tax rate and the data in table 6–2 clearly illustrate that Wyoming and Montana taxpayers are beneficiaries. These additional revenues would be used either to expand government-supplied services or to reduce other taxes. The 50 percent fall of mining employment in Wyoming indicates that miners would have to move to states taking up the slack. These states can be identified by observing where tax revenues and wage bills increase because of shifted production. Wyoming and Montana, Colorado, Utah, Arizona, and Illinois are clear beneficiaries, although Arizona and New Mexico experience more rapid depletion and a subsequent absolute fall in production, tax revenue, and employment by 2000. Presumably miners are able to shift location from Montana and Wyoming to Colorado, Utah, and Arizona with relative ease and reasonable moving costs. The Wyoming/Montana Powder River Basin boomtowns will decline in employment, economic activity, and population, but the resulting tax revenues will make up these losses many times over, although the distribution of income will be affected. The non-coal miner nonboomtown resident will benefit whereas those individuals more directly affected by coal extraction (property owners, business managers, and so on in coal areas) may suffer net losses compared to the base-case scenario. Shifting output means that states in which more is produced are net beneficiaries insofar as tax revenues and the coal-industry wage bill increases are concerned. Without actively participating in the consortium, these states become beneficiaries. It is an empirical question of what their respective benefits might be from active participation in the coal cartel. The tradeoff between tax revenue and industry employment for the Montana/Wyoming consortium was shown to be insensitive to the relative weight placed on these conflicting goals. An implausibly high wage-bill weight is required before the optimum tax is changed, which implies that coal industry representatives would be hard-pressed to alter this tax level.

Conclusions

The conclusions which can be drawn from this investigation of nonrenewable natural-resource tax policy are as follows.

1. Resource tax policy is the result of a wide array of conflicting special interest groups which confront one another in the marketplace and the political environment. We hypothesize that the primary divergent positions are revealed by positing the conflicting tax-policy goals of maximizing tax revenues (the motive being to slow development and export the tax burden) and the industry wage bill (the motive being to encourage resource and economic development). Analysis of actual coal-tax policy implies that Western states, where the mining sector is less deeply entrenched, place a heavier weight on tax revenues and are willing to sacrifice a portion of potential development compared to Midwestern and Eastern producing states.

2. The economics literature elegantly describes how extractive firms operate in a theoretical context and how both the efficiency losses (or gains) from tax policy and who bears the final tax burden may be measured. However, there appears to be no agreed-on "appropriate" tax policy save one of tax neutrality (taxes disturbing the market allocation of resources minimally). The "market failures" identified in the literature are ambiguous and unmeasured to date. Consequently, there is no widely accepted tax policy which is "optimal" in the sense that it will correct these quantitatively unknown distortions. On the other hand, there is ample evidence of distortions created by federal and state taxes owing to differential rates and special provisions which create disparities in effective tax rates. On the grounds of neutrality, these conflicting tax treatments should be eliminated. On the grounds of extracting economic rents and exporting tax burdens, there is little likelihood that state governments will voluntarily limit their authority.

3. There is empirical evidence to indicate that the federal corporate income tax (the effective rate) is discriminatory and is shifted by extracting firms (owners of capital) to consumers and suppliers of complementary inputs (labor and owners of natural resources).

4. Taxes imposed by states on coal and copper extraction are levied on inputs to the production process as well as output in a complex fashion. This means that descriptions of the technology of resource extraction and resource markets must be location-sensitive before it is possible to form meaningful estimates of effective tax rates.

5. The domestic-copper-industry simulation model on the supply side represents new research. However, the demand estimate is derived from others' research, and its inelastic nature is most likely not borne out in long-run copper markets where foreign producers are playing an increasingly important role. The static simulations indicate that existing taxes are

forward-shifted to copper consumers in the short run, although a significant portion is borne by producers' economic rents. This implies that welfare losses are small in the short run, but are substantial in the long run for all tax scenarios. Tax increases made jointly by producing states are primarily forward-shifted. A unilateral tax increase undertaken in Arizona results in heavy employment losses in the long run as production is shifted to other states, but the tax appears to shift roughly two-thirds of potential economic rents from producers to the state. However, economic rents, tax revenues, and employment all increase in other producing states. Both short- and long-run efficiency losses aree substantial.

6. Results from coal-tax scenarios reveal that existing state and federal taxes (1977) create only small net welfare losses and in all likelihood are primarily shifted forward to consuming areas. Forward shifting is more predominant in the long run. A consortium of Montana and Wyoming would be able to successfully raise taxes if a goal of maximizing tax revenue were pursued. Maximizing the sum of the industry wage bill plus tax revenue (equal policy weights) would leave these outcomes unaltered except in extreme cases. A tax rate of 62.5 percent was found to be an optimum one, and it is insensitive to variation in the discount rate and to the relative weight placed on the coal-industry wage bill and tax revenue collected. The result is shifting production from those states to other states, primarily Western and Midwestern, who benefit from increased employment and tax revenues. Prices to consumers of Western coal increase somewhat, implying that tax exporting is successful; however, long-run substitution to nuclear power reduces the effect on electricity prices from a modest regional one to near zero. A 75 percent Montana/Wyoming severance tax imposes surprisingly low net welfare losses when measured at consuming regions ($149.1 million per year in 2000). However, when measured at producing states, welfare losses are substantial ($1.2 billion per year in 2000) and should be of concern to the federal government.

Bibliography

Aaron, H. 1974. "A New View of Property Tax Incidence." *American Economic Review* 64, no. 2 (May).

Adams, Gerard F., and Behrman, Jere R., eds. 1978. *Econometric Modeling of World Commodity Policy*. Lexington, Mass.: Lexington Books, D. C. Heath.

Adelman, M.A. 1962. "The Supply and Price of Natural Gas." *Journal of Industrial Economics* 1.

———. 1964. "Efficiency of Resource Use in Crude Petroleum." *Southern Economic Journal* 31, no. 2 (October).

———. 1970. "Economics of Exploration for Petroleum and Other Minerals." *Geoexploration* 8.

———. 1972. *The World Petroleum Market*. Baltimore, Md.: Johns Hopkins Press.

Agria, Susan R. 1969. "Special Tax Treatment of Mineral Industries." In *Taxation of Income from Capital*, edited by A.C. Harberger and M.J. Bailey. Washington: The Brookings Institution.

Allais, M. 1957. "Method of Appraising Economic Prospects of Mining Exploration over Large Territories: Algerian Sahara Case Study." *Management Science* 3 (July).

Alt, Christopher. 1978. "The Cost of Capital to U.S. Coal Companies." Mimeographed. Cambridge: Massachusetts Institute of Technology.

Anders, Gerhard; Gorman, W. Philip; and Maurice, S. Charles. 1978. *Does Resource Conservation Pay?* Los Angeles: International Institute for Economic Resources, July.

Anderson, K.P. 1972. "Optimal Growth When the Stock of Resources Is Finite and Depletable." *Journal of Economic Theory* 4 (April).

Anderson, Robert C. 1976. "Federal Mineral Policy: The General Mining Law of 1872." *Natural Resources Journal* 16 (July).

———. 1977. "Taxes, Exploration and Reserves." In *Non-Renewable Resource Taxation in the Western States*. Monograph 77-2. Cambridge, Mass.: Lincoln Institute of Land Policy.

Anderson, Robert C., and Spiegelman, R.D. 1977. *The Impact of the Federal Tax Code on Resource Recovery*. Washington: Environmental Protection Agency.

Ansoff, H.I., and Selvin, D.P. 1968. "An Appreciation of Oil Dynamics." *Management Science* (March).

Arrow, K.J. 1965. "The Role of Securities in the Allocation of Risk-Bearing." *Review of Economic Studies* 31 (April).

———. 1975. *Essays on the Theory of Risk-Bearing*. Chicago: Markam.

Arrow, K.J.; Karlin, S.; and Scarf, H. 1958. *Studies in the Mathematical*

Theory of Inventory and Production. Stanford, Calif.: Stanford University Press.

Attanasi, Emil D. 1978. "Firm Size and Petroleum Exploration Behavior." *Proceedings of the Council of Economics, American Institute of Mining Engineers*, Denver, Colorado.

Attanasi, Emil D., and Johnson, S.R. 1976. "Leasing Policies for the Extractive Resources." *Annals of Regional Science*, July.

Atwood, J. 1973. "Energy Resources in New Mexico." *New Mexico Business* 26, no. 5 (May).

Ayres, R.U., and Kneese, A.V. 1969. "Production, Consumption, and Externalities." *American Economic Review* 59, no. 3 (June).

Bailly, Paul. 1976. "Converting Resources into Reserves." *Mining Engineering*, January 28.

Banks, F.E. 1971. "The Economics of Exhaustible Resources: A Note." *Economic Sanfundets Tijskr* 24. (4).

———. 1974. "A Note on Some Theoretical Issues of Resource Depletion." *Journal of Economic Theory* 9, no. 2 (October).

Barger, H., and Schurr, S.H. 1944. *The Mining Industries, 1899–1939*. New York: National Bureau of Economic Research.

Barnett, H.J. 1950. *Energy Use and Supplies, 1939, 1947, 1965*. Bureau of Mines, Information Circular 7582. Washington: U.S. Department of the Interior, October.

Barnett, H.J., and Morse, C. 1963. *Scarcity and Growth: The Economics of Natural Resource Scarcity*. Baltimore, Md.: Johns Hopkins Press.

Baughman, M.L.; Joskow, Paul L.; and Kamat, Dilip. 1979. *Electric Power in the United States: Models and Policy Analysis*. Cambridge, Mass.: M.I.T. Press.

Baumol, William J., and Oates, Wallace E. 1975. *The Theory of Environmental Policy, Externalities, Public Outlays and the Quality of Life*. Englewood Cliffs, N.J.: Prentice-Hall.

Baxter, Charles H., and Parks, Roland D. 1949. *Examination and Valuation of Mineral Property*, 3d ed. Cambridge, Mass.: Addison-Wesley.

Beasley, Charles A.; Tatum, Charles R.; and Lawrence, Brian W. 1974. "A Program for the Determination of the Technical and Economic Feasibility of Mining Operations." In *Twelfth Annual International Symposium on Industry*, edited by Thys B. Johnson and Donald W. Gentry. Golden: Colorado School of Mines.

Beazley, Ronald, 1967. "Conservation Decision-Making, A Rationalization." *Natural Resources Journal* 7.

Beckerman, W. 1972. "Economists, Scientists, and Environmental Catastrophe." *Oxford Economic Papers* 24, no. 3 (November).

Beckmann, M.J. 1974. "A Note on the Optimal Rates of Resource

Exhaustion." *Review of Economic Studies, Symposium on the Economics of Exhaustible Resources.*

Benelli, G.C. 1967. "Forecasting Profitability of Oil-Exploration Projects." *American Association of Petroleum Geologists Bulletin,* November.

Bennett, Harold J.; Moore, Lyman; Welborn, Lawrence E.; and Toland, Joseph E. 1973. *An Economic Appraisal of the Supply of Copper from Primary Domestic Sources.* Bureau of Mines, Information Circular 8598. Washington: U.S. Department of the Interior.

Bennett, Harold J.; Thompson, Jerrold G.; Quirinn, Herbert J.; and Toland, Joseph E. 1970. *Financial Evaluation of Mineral Deposits Using Sensitivity and Probabilistic Analysis Methods.* Bureau of Mines, Information Circular 8495. Washington: U.S. Department of the Interior.

Bergman, Karl G., and McLean, James H. 1971. "Risk Analysis for Evaluating Capital Investment Proposals." In *Decision Making in the Mineral Industry,* edited by J.I. McGerrigle. 9th International Symposium on Techniques for Decision Making in the Mineral Industry.

Bertold, Michael J. 1977. *Preliminary Economics of Mining a Thick Coal Seam by Dragline, Shovel Truck and Scraper Mining Systems.* Bureau of Mines, Information Circular 8761. Washington: U.S. Department of the Interior.

Bieniewski, Carl L.; Persse, Franklin H.; and Brauch, Earl F. 1971. *Denver: Availability of Uranium at Various Prices from Resources in the United States.* Bureau of Mines, Information Circular 8501. Washington: U.S. Department of the Interior.

Bingaman, Anne K. 1970. "New Mexico's Effort at Rational Taxation of Hard Minerals Extraction." *Natural Resources Journal* 10, no. 3 (July).

Borovskikh, B. 1973. "Urgent Problems in Planning the Reproduction of Natural Resources." *Problems in Economics* 16, no. 7 (November).

Bosselman, Fred P. 1969. "The Control of Surface Mining: An Exercise in Creative Federalism." *Natural Resources Journal* 49, no. 2 (April).

Boulding, K.E. 1966. "The Economics of the Coming Spaceship Earth." In *Environmental Quality in a Growing Economy,* edited by H. Barell. Baltimore: Johns Hopkins Press.

———. 1973. "The Economics of Energy." *Annals of the American Academy of Political Social Science* 410 (November).

Boyle, Gerald J. 1977. "Taxation of Uranium and Steam Coal in the Western States." In *Non-Renewable Resource Taxation in the Western States.* Monograph 77-2. Cambridge, Mass.: Lincoln Institute of Land Policy.

Bradley, Paul G. 1967. *The Economics of Crude Petroleum Production.* Amsterdam: North-Holland Publishing Co.

———. 1973. "Increasing Scarcity: The Case of Energy Resources." *American Economic Review* 63 (May).

————. 1979. *Modelling Mining: Open Pit Copper Production in British Columbia*. Resource Paper no. 31. Department of Economics, University of British Columbia, January.

Brannon, Gerard M. 1974. *Energy Taxes and Subsidies*. Report to the Energy Policy Project of the Ford Foundation. Cambridge , Mass.: Ballinger Publishing Company.

————. 1975*a*. "Existing Tax Differentials and Subsidies Relating to the Energy Industries." In *Studies in Energy Tax Policy*, edited by Gerard M. Brannon. Cambridge, Mass.: Ballinger Publishing Company.

————. 1975*b*. "U.S. Taxes on Energy Resources." *American Economic Review* 65 (May).

Brannon, Gerard M., ed. 1975. *Studies in Energy Tax Policy*. Papers prepared for the Energy Policy Project of the Ford Foundation. Cambridge, Mass.: Ballinger Publishing Company.

Breyer, Stephen. 1979. "Taxes as a Substitute for Regulation." *Growth and Change* 10 (January).

Broadbent, Lorin J. 1961. "Eminent Domain Valuation of Land Containing Minerals." *The Appraisal Journal* 29 (January).

Brobst, Donald A., and Prall, Walden P. 1973. *United States Mineral Resources*. Washington: U.S. Department of the Interior, 1973.

Brobst, Donald A.; Pratt, W.E.; and McKelvey, V.E. 1971. *Summary of United States Mineral Resources*. Geological Survey Circular 687. Washington: GPO.

Brock, B.M. 1969. "Benefit-Cost Analysis of Surface Coal Mining." *Mining Engineering*, May.

Bronder, Leonard D. 1976*a*. "Taxation of Coal Mining: Review with Recommendations." Denver, Colo.: Western Governors' Regional Energy Policy Office.

————. 1976*b*. "Taxation of Surface and Underground Coal Mining in Western States." Special Report to the Governors. Denver, Colo.: Western Governor's Regional Energy Policy Office, August.

Bronder, Leonard D.; Carlisle, D.; and Savage, M.D. 1977. "Financing Strategies for Alleviation of Socioeconomic Impacts in Seven Western States." Denver, Colo.: Western Governors' Regional Energy Policy Office.

Brookes, L.G. 1972. "More on the Output Elasticity of Energy Conservation." *Journal of Industrial Economics* 1 (November).

Brooks, David B. 1964. "The Supply of Individually Mined Minor Metals and Its Implications for Subsidy Programs." *Land Economics* 60 (February).

————. 1965. "Goals and Standards of Performance—The Conservation of Minerals: A Comment." *Natural Resources Journal* 5.

————. 1966. "Strip Mine Reclamation and Economic Analysis." *Natural Resources Journal* 6.

Brooks, David B., and Andrews, P.N. 1974. "Mineral Resources, Economic Growth and World Population." *Science*, July 5.

Brown, G.A. 1970. "The Evaluation of Risk in Mining Ventures." *Canadian Mining and Metallurgical Bulletin* 60 (October).

Brown, Gardner, Jr., and Field, Barry. 1975. *Alternative Measures of Resource Scarcity*. Discussion Paper 75-16, Institute for Social Research. Seattle: University of Washington, December.

Brown, K.C. 1967. "The Distribution of Louisiana Outer Continental Shelf Lease Bids." *Land Economics* 43 (August).

Brown, Martin S., and Butler, John. 1968. *The Production, Marketing and Consumption of Copper and Aluminum*. New York: Praeger.

Brown, Robert Douglas, 1974. "British Columbia's New Mining Royalties." *Canadian Mining and Metallurgical Bulletin* 67, no. 745 (May).

———. 1975. "At Long Last—1974 Ontario Mining Tax Changes Finally Enacted." *Canadian Mining and Metallurgical Bulletin* 68, no. 757 (May).

———. 1978. "The Impact of Canadian Federal and Provincial Mining Taxation." *Proceedings of the Council of Economics, American Institute of Mining and Engineering*, Denver, Colorado.

Bucovetsky, Meyer W. 1967. "The Taxation of Mineral Extraction." *Studies of the Royal Commission on Taxation* 8 (July). Ottawa: Queen's Printer.

———. 1972. "Tax Reform in Canada: The Case of the Mining Industry." Ph.D. thesis, University of Toronto.

Burgin, Lorraine B. 1976. *Time Required in Developing Selected Arizona Copper Mines*. Bureau of Mines, Circular 8702. Washington: U.S. Department of the Interior.

Burness, Stuart H. 1976. "On the Taxation of Nonreplenishable Resources." *Journal of Environmental Economics and Management* 3 (December).

Burt, O.R. 1964. "Optimal Use of Resources over Time." *Management Science* 11 (September).

Bush, James M. 1972. "The Effect of Taxation on Mine Management." In *The Valuation of Oil and Mineral Rights*. Chicago: International Association of Assessing Officers.

Byrne, R.F., and Sparvero, L.J. 1969. "A Study and Model of the Exploration Process in the Non-Fuel Minerals Industry." Report to U.S. Bureau of Mines through Clearinghouse for Federal and Technical Information. Springfield, Va.: U.S. Department of the Interior.

Calhoun, D.A. 1973. "Oil and Gas Taxation." *Management and Accounting Journal* 55, no. 5 (November).

Capen, E.C.; Clapp, R.V.; and Campbell, W.M. 1971. "Competitive Bidding in High-Risk Situations." *Journal of Petroleum Technology* 23 (June).

Carlisle, D. 1954. "The Economics of a Fund Resource with Particular Reference to Mining." *American Economic Review* 44 (September).

Carter, Kenneth LeM. (Chairman), and others. 1967. *Report of the (Canadian) Royal Commission on Taxation*. Vol. 4: *Taxation of Income*, chap. 23, appendix K. Ottawa: Queen's Printer.

Charles River Associates. 1970. "An Econometric Model of the Copper Industry." Prepared for Government Services Administration, September.

―――. 1977. *The Feasibility of Copper Price Stabilization Using a Buffer Stock and Supply Restrictions from 1953 to 1976*. Report no. 379. Prepared for United Nations Conference on Trade and Development, October.

Chung, Pham; Church, Albert M.; and Kurry, Channing. 1980. "Taxation of Electricity Generation: The Economic Efficiency and Equity Bases for Regionalism within the Federal System." *Natural Resources Journal* 20 (December).

Church, Albert M. 1977. "Market and Non-Market Evaluation of the Mining Firm." In *Non-Renewable Resource Taxation in the Western States*. Monograph 77-2. Cambridge, Mass.: Lincoln Institute of Land Policy.

―――. 1978a. "Pollution Abatement Incentives in the Mineral Industries with Specific Application to Copper and Coal." *Proceedings of the American Institute of Mining and Metallurgical and Petroleum Engineers*, Denver, Colorado.

―――. 1978b. "Trends in State and Local Energy Taxation—Copper and Coal—A Case in Point." *National Tax Journal* 31 (September).

―――. 1980. "State Taxation of Coal—Revealed Preferences of Economic Development versus Tax Exploitation of a Natural Resource." *Western Tax Review* 1 (May).

Church, Albert M., and Foley, Patricia. 1979. "The Short-Run Supply Curve for the Domestic Copper Industry." *Proceedings of the American Institute of Chemical Engineers*.

Ciriacy-Wantrup, S.V. 1944. "Taxation and the Conservation of Resources." *Quarterly Journal of Economics*, February.

―――. 1952. *Resource Conservation: Economics and Politics*. Berkeley: University of California Press.

Colby, Donald S., and Brooks, David B. 1969. "Mineral Resource Valuation for Public Policy." Washington: U.S. Department of the Interior, Bureau of Mines.

Commoner, Barry. 1976. *The Poverty of Power*. New York: Alfred A. Knopf, Inc.

Conaway, O.B. 1972. "Coal Mining: New Efforts in an Old Field." *Annals of the American Academy of Political Social Sciences* 62, no. 1 (March).

Conley, L.A., and Sabatini, J. 1977. *Bibliography of Investment Costs,*

Operating Costs and Related Economic Information for the Mineral Industries, January–December 1976. Bureau of Mines, Information Circular 8748. Washington: U.S. Department of the Interior.

Conrad, Robert F. 1978*a*. "Royalties, Cyclical Prices and the Theory of the Mine." *Resources and Energy* 1 (October).

———. 1978*b*. "Taxation and the Theory of the Mine." Ph.D. thesis, University of Wisconsin.

———. 1979. "State Taxation of Mineral Resources." Unpublished, Duke University, 1979.

Conrad, Robert F., and Hool, Bryce. 1978*a*. "Taxation of Mineral Resources." Unpublished manuscript.

———. 1978*b*. "Theory of Mineral Resource Extraction." Unpublished manuscript.

———. 1979*a*. "Reserve Determination, Output Effects and Mining Taxation." Unpublished, Duke University.

———. 1979*b*. "A Theory of the Mine." Unpublished, Duke University.

CONSAD Research Corporation. 1969. "The Economic Factors Affecting the Level of Domestic Petroleum Reserves." Part 4 of U.S. Congress, House Committee on Ways and Means, and Senate Committee of Finance, *Tax Reform Studies and Proposals.*

———. 1970. "A Study and Model of Exploration Process in the Non-Fuel Minerals Industry." No. S0180874, Open File. Washington: U.S. Bureau of Mines.

Cooper, D.O.; Davidson, L.B.; and Reim, K.M. 1973. "Simplified Financial and Risk Analysis for Minerals Exploration." In *Eleventh International Symposium on Computer Applications in the Mineral Industry*, edited by John R. Sturgueld. Tucson: University of Arizona.

Corcoran, P.J. 1974. *Will Today's Future Price be Tomorrow's Spot Price?* Research Paper no. 7411. New York: Federal Reserve Bank of New York, June.

Cotner, M.L. 1969. "A Policy for Public Investments in Natural Resources." *American Journal of Agriculture Economics* 51, no. 1 (February).

Cox, D.P. et al. 1973. "Copper." U.S. Geological Survey, Professional Paper 820. In *United States Mineral Resources*, edited by D.A. Brobst and W.P. Pratt. Washington: GPO.

Cox, J.C., and Wright, A.W. 1975. "The Cost-Effectiveness of Federal Tax Subsidies for Petroleum Reserves." In *Studies in Energy Tax Policy*, edited by G.M. Brannon. Cambridge, Mass.: Ballinger Publishing Company.

———. 1976. "The Determination of Investment in Petroleum Reserves and Their Implication for Public Policy." *American Economic Review* 66, no. 1 (March).

Cragg, J.G., Harberger, A.C.; and Mieszkowski, P. 1967. "Empirical

Evidence in the Incidence of the Corporate Income Tax." *Journal of Political Economy* 75 (December).

Creamer, Daniel; Dobrovolsky, Serge P.; and Borenstein, Israel. 1960. *Capital in Manufacturing and Mining—Its Formation and Financing.* Princeton, N.J.: Princeton University Press.

Crommelin, M., and Thompson, A.B., eds. 1977. *Mineral Leasing as an Instrument of Public Policy.* Vancouver: University of British Columbia Press.

Cummings, R.G. 1968. "Some Extensions of the Economic Theory of Exhaustible Resources." *Western Economic Journal* 7, no. 3 (September).

Cummings, R.G., and Burt, O.R. 1969. "The Economics of Production from Natural Resources." *American Economic Review* 59.

Cummings, R.G.; Grigalunas, T.A.; McFarland, J.W.; and Kuller, R.G. 1975. "Energy Commodities and Natural Resource Exploitation.' *Southern Economic Journal* 41, no. 3 (January).

Cummings, R.G., and Mehr, Arthur F. 1977. "Investments for Urban Infrastructure in Boomtowns." *Natural Resources Journal* 17 (April).

Cummings, R.G., and Schulze, William D. 1978. "Optimal Investment Strategy for Boomtowns: A Theoretical Analysis." *American Economic Review* 68 no. 3 (June).

Curry, A.V., and Kiessling, O.F. 1938. *Grade of Ore.* Report no. E-6, Mineral Technology and Output per Man Studies. Philadelphia: Works Progress Administration, National Research Project, August.

Dapen, E.C.; Clapp, R.U.; and Campbell, W.M. 1971. "Competitive Bidding in High Risk Situations." *Journal of Petroleum Technology,* June.

d'Arge, R.C., and Kogiku, P. 1973. "Economic Growth and the Environment." *Review of Economic Studies* 40, no. 2 (January).

Dasgupta, P. 1974. "On Some Alternative Criteria for Justice between Generations." *Journal of Public Economics* 3, no. 4 (November).

Dasgupta, P., and Heal, G. 1974. "The Optimal Depletion of Exhaustible Resources." *Review of Economic Studies, Symposium on the Economics of Exhaustible Resources.*

Dasgupta, P., and Stiglitz, J.I. 1975. "Uncertainty and the Rate of Extraction under Alternative Institutional Arrangements." Mimeographed. Palo Alto, Calif.: Stanford University.

Davidson, Paul. 1963. "Public Policy Problems of the Domestic Crude Oil Industry." *American Economic Review* 53 (March).

———. 1970. "The Depletion Allowance Revisited." *Natural Resources Journal* 10 (January).

Davidson, Paul; Falk, Laurence; and Lee, Hoesong. 1975. "The Relation-

ships of Economic Rents and Price Incentives to Oil and Gas Supplies."
In *Studies in Energy Tax Policy*, edited by Gerard M. Brannon.
Cambridge, Mass.: Ballinger Publishing Company.

Davis, Ronald R. 1967. "Income Approach to Value of Quarry Land."
Assessors Jorunal 2, no. 3 (October).

Davis, Ronald R.; Kisiel, Chester C.; and Duckstein, Lucien. 1973. "Basic
Methods for Decision Making in Mineral Exploration and Exploitation."
In *Eleventh International Symposium on Computer Applications in the
Mineral Industry*, edited by John R. Sturgueld. Tucson: University of
Arizona.

Day, Richard H. 1978. "Adaptive Economics and Natural Resource
Policy." *Materials and Society* 2(3).

DeCarvalho, M. 1970. "The Taxation of Income from Land." *Arquivo
Institute* 5 (1).

Denver Research Institute. 1979. *Socioeconomic Impacts of Western
Energy Resource Development*. Washington: Council of Environmental
Quality.

DeYoung, John H., Jr. 1978. "Measuring the Economic Effects of Tax
Laws on Mineral Exploration." *Proceedings of the Council of
Economics, American Institute of Mining and Engineering*, Denver,
Colorado.

Dolman, Anthony J., ed. 1976. *Rio: Reshaping the International Order*.
New York: New American Library.

Dran, John J., Jr., and McCarl, Henry N. 1974. "A Critical Examination of
Mineral Valuation Methods in Current Use." *Mining Engineering* 26,
no. 7 (July).

Duchesneau, Thomas D. 1975. *Competition in the U.S. Industry*.
Cambridge, Mass.: Ballinger Publishing Company.

Due, John F. 1970. "The Developing Economies, Tax and Royalty
Payments by the Petroleum Industry, and the United States Income
Tax." *Natural Resources Journal* 10, no. 1 (January).

Dumars, Charles T.; Brown, F. Lee; and Browde, Michael B. 1979. "Legal
Issues on State Taxation of Energy Development." Report made to the
New Mexico Energy Institute.

Dunn, J.R.; Wallace, W.A.; and Brooks, D.B. 1971. *Mineral Resource
Valuation in the Public Interest*. American Institute of Mining and
Engineering, Society of Mining Engineers, preprint 71-H-87, February.

Dusansky, Richard. 1972. "The Short-Run Shifting of the Corporation
Income Tax in the United States." *Oxford Economic Papers* 24
(November).

Eldridge, Douglas H. 1950. "Tax Incentives for Mineral Enterprise."
Journal of Political Economy 58 (June).

———. 1962. "Rate of Return, Resource Allocation, and Percentage Depletion." *National Tax Journal* 15 (June).

Epple, D. 1976. *Petroleum Discoveries and Government Policy.* Cambridge, Mass.: Ballinger Publishing Company.

Erickson, E.W. 1968. "Economic Incentives, Industrial Structure and the Supply of Crude Oil Discoveries in the U.S., 1946–1958/9." Ph.D. dissertation, Vanderbilt University.

———. 1970. "Crude Oil Prices, Drilling Incentives and the Supply of New Discoveries." *Natural Resources Journal* 10, no. 1 (January).

Erickson, E.W.; Millsaps, S.W.; and Spaun, R.M. 1974. "Oil Supply and Tax Incentives." *Brookings Paper on Economic Activity* 2.

"Exxon's High Roller in Oil Lease Sales." *Business Week*, September 20, 1976.

Falternayer, E. 1972. "The Energy 'Joyride' is over; Technology and Good Sense Can Stretch Our Resources, but Only a Big Breakthrough Can Bring Back Cheap Fuel and Power." *Fortune Magazine*, September.

Feldstein, Martin. 1975. "The Surprising Incidence of a Tax on Pure Rent." Harvard Institute of Economic Research Discussion Paper no. 424, July.

Fisher, A.C., and Krutilla, J.V. 1975. "Resource Conservation, Environmental Preservation and the Rate of Discount." *Quarterly Journal of Economics* 89, no. 3 (August).

Fisher, A.C., and Peterson, F.M. 1976. "The Environment in Economics." *Journal of Economic Literature* 14, no. 1 (March).

Fisher, Franklin M. 1964. *Supply and Costs in the U.S. Petroleum Industry: Two Econometric Studies.* Baltimore, Md.: Johns Hopkins Press.

Fisher, Franklin M.; Cootner, Paul H.; and Baily, Martin Niel. 1972. "An Econometric Model of the World Copper Industry." *Bell Journal of Economic and Management Science* 3, no. 2 (Autumn).

Foley, Patricia T. 1979. "A Supply Curve for the Domestic Primary Copper Industry." Master's thesis, Massachusetts Institute of Technology, January.

Foley, Patricia T., and Clark, Joel P. 1980. "U.S. Copper Supply: An Economic/Engineering Analysis of Cost-Supply Relationships." Submitted to *Resources Policy* in October.

Francis, A.A. 1971. "Actual Resource Absorption and Taxation: A Micro Theoretic Approach." *Society and Economic Studies* 7, no. 3 (November).

Frank, H.J., and Schory, J.J., Jr. 1973. "The Future of American Oil and Natural Gas." *Annals of the American Academy of Political Social Sciences* 410 (November).

Gaffney, Mason. 1977a. "Intergovernmental Competition for Energy Reserves: The Public Interest." *Proceedings of the 76th Annual Conference of the National Tax Association*, St. Louis, Mo.

————. 1977b. "Oil and Gas Leasing Policy: Alternatives for Alaska in 1977." Report to the Alaska Department of Mineral Resources and the Alaska Legislature, February.

Gaffney, Mason, ed. 1967. *Extractive Resources and Taxation*. Madison: University of Wisconsin Press.

Gardner, B. Delworth. 1967. "Toward a Disposal Policy for Federally Owned Oil Shales." In *Extractive Resources and Taxation*, edited by Mason Gaffney. Madison: University of Wisconsin Press.

Garnaut, R., and Clienies, Ross A. 1975. "Uncertainty, Risk Aversion and the Taxing of Natural Resource Projects." *Economic Journal* 85, no. 338 (June).

Gaskins, D.W., and Tiesberg, T. 1976. "An Economic Analysis of Pre-Sale Exploration in Oil and Gas Lease Sales." In *Essays in Industrial Organization in Honor of Joe S. Bain*, edited by R. Masson and P.D. Qualls. Cambridge, Mass.: Ballinger Publishing Company.

Gaventa, John. 1975. "Property Taxation of Coal in Central Appalachia." A Report for the Senate Subcommittee on Intergovernmental Relations from Save Our Cumberland Mountains, Inc. Washington.

Gentry, Donald W. 1971. "Two Decision Tools for Mining Investment—And How to Make the Most of Them." *Mining Engineering*, November.

————. 1977. "Financial Modeling of Mining Ventures—The Effects of State Mine Taxation." In *Non-Renewable Resource Taxation in the Western States*. Monograph 77-2. Cambridge, Mass.: Lincoln Institute of Land Policy.

Georgescu-Roegen, Nicholas. 1971. *The Entropy Law and the Economic Process*. Cambridge, Mass.: Harvard University Press.

————. 1975. "Energy and Economic Myths." *Southern Economic Journal* 41, no. 3 (January).

Gilbert, R. 1975a. "Decentralized Exploration Strategies for Nonrenewable Resource Deposits." Mimeographed. Palo Alto, Calif.: Stanford University.

————. 1975b. "Resource Depletion under Uncertainty." Mimeographed. Palo Alto, Calif.: Stanford University.

Gillis, S. Malcolm. 1977. "Taxation, Mining and Public Ownership." In *Non-Renewable Resource Taxation in the Western States*. Monograph 77-2. Cambridge, Mass.: Lincoln Institute of Land Policy.

————. 1979. "Severance Taxes on North American Energy Reserves: A Tale of Two Minerals." *Growth and Change* 10(1).

Gillis, S. Malcolm, and Bucovetsky, Meyer. 1978. "The Design of Mineral Tax Policy." In *Taxation and Mining: Non-Fuel Minerals in Bolivia and Other Countries*. Cambridge, Mass.: Ballinger Publishing Company.

Gillis, S. Malcolm, and McLure, Charles E., Jr. 1975a. "Taxation of Natural Resources—Incidence of World Taxes on Natural Resources with Special Reference to Bauxite." *American Economic Review* 65 (May).

———. 1975*b*. "The Distributional Implications of the Taxation of Natural Resources." *Rice Studies*, Fall.

Gilmore, John S. et al. 1976. *Analysis of Financing Problems in Coal and Oil Shale Boomtowns*. Washington: Federal Energy Administration.

Goldsmith, O.S. 1974. "Market Allocation of Exhaustive Resources." *Journal of Political Economy* 82, no. 5 (September/October).

Goldstein, Morris, and Smith, Robert S. 1975. "Land Reclamation Requirements and Their Estimated Effects on the Coal Industry." *Journal of Environmental Economics and Management* 2.

Gordon, R.J. 1967. "Incidence of Corporation Income Tax." *American Economic Review* 62 (September).

Gordon, Richard L. 1966. "Conservation and the Theory of Exhaustible Resources." *Canadian Journal of Economics and Political Science* 32, no. 3 (August).

———. 1967. "A Reinterpretation of the Pure Theory of Exhaustion." *Quarterly Journal of Economics* 3 (June).

———. 1975*a*. *Economic Analysis of Coal Supply: An Assessment of Existing Studies*. University Park: Pennsylvania State University.

———. 1975*b*. *U.S. coal and the Electric Power Industry*. Baltimore, Md.: Johns Hopkins Press.

———. 1978. "Price and Output Behavior of Exhaustible Resource Industries." *Materials and Society* 2(3).

Gordon, Scott. 1973. "Today's Apocalypses and Yesterday's." *American Economic Review* 63 (May).

Gray, J.L. et al. 1977. *Socioeconomic Impacts of Coal Mining on Communities in Northwestern New Mexico*. Bulletin no. 652. Las Cruces: New Mexico Agricultural Experiment Station.

Gray, Lewis C. 1914. "Rent under the Assumption of Exhaustibility." *Quarterly Journal of Economics*, May.

Grayson, C.J. 1960. *Decisions under Uncertainty*. Boston: Harvard Business School.

Gregory, Richard B. 1974. "Tax Consequences and Distinctions Involved in the Sale or Lease of Oil and Gas Interests." *Natural Resources Journal* 14 (April).

Griffin, James M. 1972. "The Process Analysis Alternative to Statistical Cost Functions: An Application to Petroleum Refining." *American Economic Review*, March.

———. 1977. "The Econometrics of Joint Production: Another Approach." *Review of Economics and Statistics* 59, no. 4 (November).

Griffin, Keynon N., and Shelton, Robert B. 1978. "Coal Severance Tax in the Rocky Mountain States." *Policy Studies Journal* 7, no. 1 (Autumn).

Griffin, Keynon N., and Steele, Henry B. 1980. *Energy Economics and Policy*. New York: Academic Press.

Grossman, Philip W. 1935. "The Valuation of Land with Underlying Natural Resources." *Journal of American Institute of Real Estate Appraisers*, April. Reprinted in *Selected Readings in Real Estate Appraisal*. Chicago: R.R. Donnelley and Sons, 1953.

Groundwater, T.D. 1967. "Role of Discounted Cash Flow Methods in the Appraisal of Capital Projects." *Institute of Mining and Metallurgy Transactions*, April.

Guerin, J. 1960. "Excise Taxation and Quality of Product." *Public Finance* 1.

Guither, H.D. 1974. "Illinois Lands Affected by Strip Mining." *Illinois Agriculture Economics* 14, no. 2 (July).

Halls, J.L.; Bellum, D.P.; and Lewis, C.K. 1969. "The Determination of Optimum Ore Reserve and Plant Size by Incremental Financial Analysis." *Proceedings of the Council of Economics of American Institute of Mining and Engineering*. New York.

Hamilton, H.D. 1954. "Taxes and Taconite: Iron Ore Tax Legislation in the Lake Superior Region." *National Tax Journal* 7 (December).

Hansen, C.J. 1977. "If Cabbages Were Kings... A Practical Approach to the Taxation of Mining Properties." In *Non-Renewable Resource Taxation in the Western States*. Monograph 77-2. Cambridge, Mass.: Lincoln Institute of Land Policy.

Harberger, Arnold C. 1955. "The Taxation of Mineral Industries." *Federal Tax Policy for Economic Growth and Stability*. Washington: Joint Committee on the Economic Report.

———. 1957. "Some Evidence on the International Price Mechanism." *Journal of Political Economy* 65 (December).

———. 1961. "The Tax Treatment of Oil Exploration." In *Proceedings of the Second Energy Institute*. Washington: American University.

———. 1962. "The Incidence of the Corporation Income Tax." *Journal of Political Economy* 70 (June).

———. 1964. "Taxation, Resource Allocation and Welfare." In *The Role of Direct and Indirect Taxes in the Federal Reserve System*. Princeton, N.J.: Priceton University Press.

Harris, D.P. 1965. "An Application of Multivariate Statistical Analysis to Mineral Exploration." Ph.D. dissertation, Pennsylvania State University.

———. 1970. Risk Analysis in Mineral Investment Decisions." *American Institute of Mining and Engineering, Society of Mining Engineers Transactions*, September.

Hartman, Raymond Steve. 1977. "An Oligopolistic Pricing Model of the U.S. Copper Industry." Ph.D. dissertation, Massachusetts Institute of Technology, January.

Hartman, R.; Bozdogan, Z.; and Nadkarni, R. 1979. "The Economic Impacts of Environmental Regulations to the U.S. Copper Industry." *Bell Journal of Economics* 10, no. 2 (Autumn).

Hartwick, John M. 1975. *Man's Exhaustible Resources in a Dynamic Aggregate Model.* Discussion Paper no. 183. Cambridge, Mass.: Massachusetts Institute of Technology, July.

Hausman, Jerry A. 1975. "Project Independence Report: An Appraisal of U.S. Energy Needs up the 1985." *Bell Journal of Economics* 6 (Autumn).

Haveman, R.H. 1973. "Efficiency and Equity in Natural Resource and Environmental Policy." *American Journal of Agriculture Economics* 55, no. 5 (December).

Hayen, R.L., and Watt, G.L. 1975. "A Description of Potential Socio-economic Impacts from Energy-Related Development on Campbell County, Wyoming." Washington: U.S. Department of Interior.

Hazlett, John M. 1972. "Effect of Minerals on the Tax Base." In *Valuation of Oil and Mineral Rights.* Chicago: The International Association of Assessing Officers.

Headington, Robert C. 1974. "Assessment of a Mineral Property." *Assessors Journal* 9, no. 2 (July).

Heal, G. 1976. "The Relationship between Price and Extraction Cost for a Resource with a Backstop Technology." *Bell Journal of Economics* 7, no. 2 (Autumn).

Hellerstein, Walter. 1978. "Constitutional Constraints on State Taxation of Energy Resources." *National Tax Journal* 31, no. 3 (September).

Helliwell, J.F. 1978. "Effects of Taxes and Royalties on Copper Mining Investment in British Columbia." *Resource Policy* 4 (March).

Herfindahl, Orris C. 1955. "Some Fundamentals of Mineral Economics." *Land Economics* 31 (May).

———. 1959. *Copper Costs and Prices: 1870–1957.* Baltimore, Md.: Johns Hopkins Press.

———. 1961. "The Long Run Cost of Minerals." In *Three Studies in Minerals Economics.* Washington: Resources for the Future.

———. 1963. "Goals and Standards of Performance for the Conservation of Minerals." *Natural Resources Journal* 3, no. 1 (May).

———. 1967. "Depletion and Economic Theory." In *Extractive Resources and Taxation,* edited by Mason Gaffney. Madison: University of Wisconsin Press.

Hespos, R.F., and Strassman, P.A. 1965. "Stochastic Decision Trees for the Analysis of Investment Decisions." *Management Science,* August.

Hibbard, W.R.; Kelly, M.A.; Rapoport, L.A.; Rohrer, M.W.; and Soysler, A.L. 1977. "Midas Materials Interchangeability Development and Supplies: Phase 1 Model of the Copper Industry from Mine to Conductor

Wire." Prepared for U.S. Bureau of Mines, U.S. Department of Interior. Blacksburg: Virginia Polytechnic Institute and State University, December 15.

Hileman, D.H.; Collins, Bruce A.; and Wilson, S.R. 1970. *Coal Production From the Vintu Region, Colorado and Utah.* Bureau of Mines, Information Circular 8497. Washington: U.S. Department of the Interior.

Hoel, M. 1978*a.* "Resource Extraction and Recycling with Environmental Costs." *Journal of Environmental Economics and Management* 5 (September).

———. 1978*b.* "Resource Extraction, Uncertainty and Learning." *Bell Journal of Economics* 9 (Autumn).

Hogan, John D. 1967. "Resource Exploitation and Optimum Tax Policies: A Control Model Approach." In *Extractive Resources and Taxation*, edited by Mason Gaffney. Madison: University of Wisconsin Press.

Hogan, Timothy D., and Shelton, Robert B. 1973. "Interstate Tax Exportation and States' Fiscal Structures." *National Tax Journal* 26, no. 4 (December).

Holland, Charles T. 1972. "The Value of Coal in the Ground." In *Valuation of Oil and Mineral Rights*. Chicago: International Association of Assessing Officers.

Hoskold, H.D. 1877. *The Engineers' Valuing Assistant.* London: Longmans, Green and Co.

Hotchkiss, W.O., and Parles, R.D. 1939. *Total Profits vs. Present Value in Mining.* Technical Publication no. 708, American Institute of Mining and Metallurgical Engineers, February.

Hotelling, H. 1931. "The Economics of Exhaustible Resources." *Journal of Political Economy* 39 (April).

Howard, H.A. 1971*a.* "A Measurement of the External Diseconomies Associated with Bituminous Coal Surface Mining, Eastern Kentucky 1962–67." *Natural Resources Journal* 11, no. 1 (January).

———. 1971*b.* "Are Landowners Underpaid of Overpaid for Surface Mining Mineral Rights?" *American Journal of the Economic Society* 30, no. 4 (October).

Hubbert, M. King. 1969. "Energy Resources." In *Committee on Resources and Man, National Academy of Sciences, Resources and Man.* San Francisco: W.H. Freeman.

Hudson, E.A., and Jorgenson, D.W. 1974. "U.S. Energy Policy and Economic Growth, 1975–2000." *Journal of Public Economics* 5, no. 2 (Autumn).

Hughart, D. 1975. "Informational Asymmetry, Bidding Strategies, and the Marketing of Offshore Petroleum Leases." *Journal of Political Economy* 83, no. 5 (October).

Hughes, Helen. 1974. "The Distribution of Gains from Foreign Direct Investment in Mineral Development." SEADAG paper for the United Nations. New York.

Humphrey, D.B., and Moroney, J.R. 1975. "Substitution among Capital, Labor, and Natural Resource Products in American Manufacturing." *Journal of Political Economy* 83, no. 1 (February).

Ingham, A., and Simmons, P. 1975. "Natural Resources and Growing Population." *Review of Economic Studies* 42 (April).

Ise, J. 1925. "The Theory of Value as Applied to Natural Resources." *American Economic Review* 15, no. 2 (June).

Jacob, H. Myles. 1977. "Taxes—Some Philosophical and Financial Musings." In *Non-Renewable Resource Taxation in the Western States*. Monograph 77–2. Cambridge, Mass.: Lincoln Institute of Land Policy.

Jenkins, Glenn, and Wright, Brian. 1975. "Taxation of Income of Multinational Corporations: The Case of the U.S. Petroleum Industry." *Review of Economics and Statistics*, February.

Johnson, Edward E., and Bennett, Harold J. 1969. "An Economic Evaluation of an Ore Body." *Quarterly of Colorado School of Mines* 64, no. 3 (July).

Johnson, Lowell C. 1953. "Valuation of Petroleum Mineral Rights for Ad Valorem Taxation." *Appraisal Journal* 21, no. 4 (October).

Johnston, J. 1970. *Statistical Cost Analysis*. New York: McGraw-Hill.

Kahn, A.E. 1960. "Economic Issues in Regulating the Field Price of Natural Gas." *American Economic Review* 50, no. 2 (May).

Kalter, R.J.; Stevens, P.H.; and Bloom, O.H. 1975. "The Economics of Outer Continental Shelf Leasing." *American Journal of Agricultural Economics* 5, no. 2 (May).

Katell, Sidney; Hemingway, E.L.; and Berkshire, L.H. 1976. *Basic Estimated Capital Investment and Operating Costs for Coal Strip Mines*. U. S. Bureau of Mines. Washington; GPO.

Kaufman, G.M. 1963. *Statistical Decision and Related Techniques in Oil and Gas Explorations*. Englewood Cliffs, N.J.: Prentice-Hall.

Kay, John, and Mirrless, James. 1975. "The Desirability of Natural Resource Depletion." In *Economics of Natural Resource Depletion*, edited by C.W. Pearce and J. Rose. London: MacMillan Press.

Kearns, Desmond P. 1970. "Property Taxation of the Mining Industry in Arizona." *Arizona Law Review*, Winter.

Keith, C.K. 1938. *Mineral Valuation of the Future*. American Institute of Mining and Metallurgical Engineers. York, Pa.: Maple Press.

Kentucky Council of State Governments. 1976. "Coal State Commercial Severance Taxes and Distribution of Revenues." Lexington, Ky., September.

Kneese, Allen V., and Herfindahl, Orris C. 1974. *Economic Theory of Natural Resources*. New York: Charles Merrill Co.

Kneese, Allen V., and Schultz, C.L. 1975. *Pollution, Prices and Public Policy*. Washington: The Brookings Institution.

Knight, Frank H. 1971. *Risk, Uncertainty and Profit*. Boston: Houghton Mifflin.

Koopmans, T.C. 1973. "Some Observations on 'Optimal' Economic Growth and Exhaustible Resources." In *Economic Structure and Development: Essays in Honor of Jan Tinbergen*, edited by H.C. Bos, H. Linnemann, and P. de Wolff. Amsterdam: North-Holland Publishing.

———. 1974. "Proof for a Case where Discounting Advances Doomsday." *Review of Economic Studies, Symposium on the Economics of Exhaustible Resources*.

Krauss, Melvyn. 1972. "Differential Tax Incidence: Large versus Small Tax Changes." *Journal of Political Economy* 80 (January/February).

Krauss, Melvyn, and Johnson, H.G. 1972. "The Theory of Tax Incidence: A Diagramatic Analysis." *Economica* 11 (November).

Krige, D.G. 1964. "Recent Developments in South Africa in the Application of Trend Surface and Multiple Regression Techniques to Gold Ore Valuation." *Quarterly of Colorado School of Mines* 59, no. 4 (October).

Krutilla, John V.; Fisher, Anthony U.; with Rice, Richard E. 1978. *Economic and Fiscal Impacts of Coal Development: Northern Great Plains*. Baltimore, Md.: Johns Hopkins Press.

Krzyzaniak, Marian, and Musgrave, Richard D. 1963. *The Shifting of the Corporation Income Tax*. Baltimore,Md.: Johns Hopkins Press.

Kuer, James E., and Montgomery, W. David. 1973. "Resource Allocation, Information Cost and the Law of Government Intervention." *Natural Resources Journal* 13.

Kuller, Robert G., and Cummings, Ronald G. 1974. "An Economic Model of Production and Investment for Petroleum Reservoirs." *American Economic Review* 64, no. 1 (March).

Lago, A.M. 1968. "A Quantitative Analysis of Mining Industry Finance." Available through Clearing House for Federal Science and Technological Information. Springfield, Va.: Department of the Interior.

Laing, Glen J.S. 1976. "An Analysis of the Effects of State Taxation on the Mining Industry in the Rocky Mountain States." M.S. thesis, Colorado School of Mines.

Lamont, William et al. 1978. "Tax Lead Time Study." Governor's Committe on Oil Shale Environmental Problems. Denver: Colorado Geological Survey.

Landesberg, H.H.; Fischman, L.L.; and Fisher, J.L. 1963. *Resources in America's Future*. Baltimore, Md.: Johns Hopkins Press.

Lane, K.F. 1964. "Choosing the Optimum Cutoff Grade." *Quarterly of the Colorado School of Mines* 59, no. 4 (October).

Lasky, S.G. 1950. "How Tonnage and Grade Relations Help Predict Ore Reserves." *Engineering and Mining Journal* 151 (April).

Leholm, Arnold G. et al. 1976. "Fiscal Impact of a New Industry in a Rural Area: A Coal Gasification Plant in Western North Dakota." *Regional Science Perspectives* 6.

Leith, J.C. 1975. "The Ontario Mining Profit Tax: An Evaluation." In *Natural Resource Revenues*, edited by A. Scott. Vancouver: University of British Columbia Press.

Lemon, James D. 1971. "Problems in Mine Evaluation." *Management and Accounting Journal* 53, no. 3 (September).

Leontief, Wassily W. 1941. *The Structure of the American Economy, 1919–1929*. Cambridge, Mass.: Harvard University Press.

Leontief, Wassily et al. 1977. *The Future of the World Economy*. New York: Oxford University Press.

Lewis, C.K. 1969. "An Economic Life for Property Valuation." *Engineering and Mining Journal* 170 (October).

Lewis, F. Milton, and Bhappu, Roshan B. 1975. "Evaluating Mining Ventures via Feasibility Studies." *Mining Engineering* 27 (October).

Lewis, Tracy R. 1975. *A Note on Theoretical Issues of Resource Depletion*. Social Science Working Paper no. 91. Pasadena, Calif.: Environmental Quality Laboratory, August.

———. 1976. "Monopoly Exploitation of an Exhaustive Resource." *Journal of Environmental Economics and Management* 3 (October).

Lewis, Tracy R.; Matthews, Steven A.; and Burness, H.S. 1979. "Monopoly and the Rate of Exhaustible Resources: Comment." *American Economic Review* 68 (March).

Lichtenberg, A.J., and Nogaard, R.B. 1974. "Energy Policy and the Taxation of Oil and Gas Income." *Natural Resources Journal* 14, no. 4 (October).

Lindholm, R.W., ed. 1967. "Property Taxation in U.S.A." *Proceedings of a Symposium Sponsored by the Committee on Taxation, Resources, and Economic Development* TRED. Madison: University of Wisconsin Press.

Link, Arthur A. 1978. "Political Constraint and North Dakota's Coal Severance Tax." *National Tax Journal* 31, no. 3 (September).

Lockner, Allyn O. 1962a. "A Proposed Mineral Tax for Colorado." *National Tax Journal* 15, no. 3 (September).

———. 1962b. "The Economic Effect of a Progressive Net Profit Tax on Decision-Making by the Mining Firm." *Land Economics* 38 (November).

Long, Millord. 1962. "The Marginal-Cost Price of Coal 1956–57." *Economica* 29, no. 116 (November).

Long, N.V. 1975. "Resource Extraction under Uncertainty about Possible Nationalization." *Journal of Economic Theory*, February.

Long, Stephen C.M. 1976. "Coal Taxation in Western States: The Need for a Regional Tax Policy." *Natural Resources Journal* 16 (April).

MacAvoy, Paul W., and Pindyck, Robert S. 1973. "Alternative Regulatory Policies for Dealing with the Natural Gas Shortage." *Bell Journal of Economics* 4, no. 2 (Autumn).

———. 1975. *The Economics of the Natural Gas Shortage 1960–1980.* New York: North-Holland–American Elsevier.

MacDonnell, L.J. 1976. "Public Policy for Hard Rock Minerals Accession Federal Lands." *Quarterly of the Colorado School of Mines* 17, no. 1 (April).

Mackenzie, B.W. 1969. "Economic Evaluation Techniques Applied to the Mine Development Decision." SME Preprint 69K326, Society of Mining Engineers Fall Meeting, Salt Lake City, Utah, September.

———. 1970, 1971. "Evaluating the Economics of Mine Development." *Canadian Mining Journal* 91 (December) and (March).

Mackenzie, B.W., and Bilodeau, M.L. 1977. "Assessing the Direct Effects of Mining Taxation: The Case of Base Metal Investment in Canada." Paper presented at Workshop on Rate of Return and Taxation of Minerals, Queens University, December.

Mackenzie, B.W.; Bilodeau, M.; and Mascall, G.E. 1974. "The Effect of Uncertainty on the Optimization of Mine Development." In *Twelfth Annual International Symposium on the Applications of Computers and Mathematics in the Minerals Industry*, edited by Thys B. Johnson and Donald W. Gentry. Golden: Colorado School of Mines.

Manne, A.S. 1974. "Waiting for the Breeder." *Review of Economic Studies, Symposium on the Economics of Exhaustible Resources.*

Manvel, Allen D. 1975. "A Survey of the Extent of Unneutrality toward Energy under State Excise, Property and Severance Taxation." In *Studies in Energy Tax Policy*, edited by Gerard M. Brannon. Cambridge, Mass.: Ballinger Publishing Company.

Marcke, R.B. 1970. "The Long-Run Supply Curve of Crude Oil Produced in the United States." *Antitrust Bulletin*, Winter.

Marglin, S.A. 1963. "The Social Rate of Discount and the Optimal Rate of Investment." *Quarterly Journal of Economics* 77 (February).

Martin, John D., and Scott, David F., Jr. 1974. "A Discriminant Analysis of the Corporate Debt-Equity Decision." *Financial Management* 3 (Winter).

Martin, Susan. 1979. "Evaluation of Alternative New Source Performance Standards." *Coal Review*. Data Resources Incorporated, December.

Matson, R.E., and Blumer, J.W. 1973. *Quality and Reserves of Strippable Coal, Selected Deposits, Southeastern Montana*. Montana Bureau of Mines and Geology Bulletin no. 91. December.

McDivitt, David. 1974. *Minerals and Men*. Baltimore, Md.: Johns Hopkins Press.

McDonald, Stephen L. 1962. "Percentage Depletion and Tax Neutrality: A Reply to Messrs. Musgrave and Eldridge." *National Tax Journal* 15 (September).

———. 1963. *Federal Tax Treatment of Oil and Gas*. Washington: The Brookings Institution.

———. 1966. "The Efects of Severance vs. Property Taxes on Petroleum Conservation." In *1965 Proceedings of the Fifty-Eighth Annual Conference on Taxation*. Harrisburg, Penn.: National Tax Association.

———. 1967. "Percentage Depletion, Expensing of Intangibles, and Petroleum Conservation." In *Extractive Resources and Taxation*, edited by Mason Gaffney. Madison: University of Wisconsin Press.

———. 1970. "Distinctive Tax Treatment of Income from Oil and Gas Production." *Natural Resources Journal* 10, no. 1 (January).

———. 1971. *Petroleum Conservation in the United States: An Economic Analysis*. Baltimore, Md.: Resources for the Future, Johns Hopkins Press.

———. 1974. "Incentive Policy and Supplies of Energy Sources." *American Journal of Agriculture Economics* 56, no. 2 (May).

———. 1976. "Taxation System and Market Distortion." In *Energy Supply and Government Policy*, edited by William A. Vogely and Robert J. Kalter. Ithaca, N.Y.: Cornell University Press.

McGeorge, Robert L. 1970. "Approaches to State Taxation of the Mining Industry." *Natural Resources Journal* 10, no. 1 (January).

McKie, James W. 1960. "Market Structure and Uncertainty in Oil and Gas Exploration." *Quarterly Journal of Economics* 74, no. 4 (November).

McKie, J.W., and McDonald, S.L. 1962. "Petroleum Conservation in Theory and Practice." *Quarterly Journal of Economics* 1 (February).

McLure, Charles E., Jr. 1964. "Commodity Tax Incidence in Open Economies." *National Tax Journal* 14 (June).

———. 1967. "The Interstate Exporting of State and Local Taxes: Estimates for 1962." *National Tax Journal* 20 (March).

———. 1969. "The Inter-Regional Incidence of General Regional Taxes." *Public Finance* 24.

———. 1970a. "Taxation, Substitution and Industrial Location." *Journal of Political Economy* 78, no. 1 (January/February).

———. 1970b. "Tax Incidence, Macroeconomic Policy, and Absolute Prices." *Quarterly Journal of Economics* 74 (May).

———. 1974. "A Diagrammatic Exposition of the Harberger Model with One Immobile Factor."*Journal of Political Economy* 82 (January/February).

———. 1975. "The Economic Effects of a Texas Tax on the Processing of Petroleum Products." Report to the Texas Mid-Continent Oil and Gas Association, February 4.

————. 1978. "Economic Constraints on State and Local Taxation of Energy Resources." *National Tax Journal* 31, no. 3 (September).

Mead, Richard W., and Bonem, Gilbert W. 1976. "The Southwestern Copper Industry, 1976–1985." Bureau of Business and Economic Research, University of New Mexico, June 25.

Mead, Walter J. 1967. "Natural Resource Disposal Policy: Oral Auctions versus Sealed Bids." *Natural Resources Journal* 7 (April).

————. 1970. "The System of Government Subsidies to the Oil Industry." *Natural Resources Journal* 1 (June).

————. 1978. "The Use of Taxes, Regulation and Price Controls in the Energy Sector." *National Tax Journal*, September.

Meadows, D.H.; Meadows, D.L.: Randers, J.; and Behrens, W.W.II. 1972. *The Limits to Growth*. New York: University Books.

Meissner, C.R.; Dedil, C.B.; and Stricker, G.D., eds. 1976. *Coal Geology and the Future—Symposium Abstracts and Selected References*. U.S. Geologic Survey Circular no. 756. Washington: GPO.

Melkus, Roy A. 1974. "Toward Rational Future Energy Policy." *Natural Resources Journal* 14.

Michelson, R.W.; Poltannd, H.J.; and Peterson, O. 1970. *Evaluating the Economic Availability of Mesabi Range Taconite Iron Ores with Computerized Models*. Bureau of Mines. Washington: U.S. Department of the Interior.

Mid-Continent Oil and Gas Association. 1969. "Analysis and Comment Relating to the CONSAD Report on the Influence of U.S. Petroleum Taxation on the Level of Reserves." Washington, April 25. Appears in Tax Reform Act of 1969, Senate Committee on Finance, Part 5.

Mieszkowski, Peter. 1966. "The Comparative Efficiency of Tariffs and Other Tax-Subsidy Schemes as a Means of Obtaining Revenue or Protecting Domestic Production." *Journal of Political Economy* 74.

————. 1969. "Tax Incidence Theory: The Effects of Taxes on the Distribution of Income." *Journal of Economic Literature* 7 (December).

————. 1972. "The Property Tax: An Excise or a Profits Tax?" *Journal of Public Economics* 1.

Mikesell, Raymond F. 1974. "Financial Considerations in Negotiating Mine Development Agreements." *Mining Magazine*, April.

————. 1975. *Foreign Investment in Copper Mining, Case Studies of Mines in Peru and Papua, New Guinea*. Baltimore, Md.: Johns Hopkins Press.

————. 1979. *The World Copper Industry—Structure and Economic Analysis*. Baltimore, Md.: Johns Hopkins Press.

Mikesell, Raymond F., ed. 1971. *Foreign Investment in the Petroleum and Mineral Industries*. Baltimore, Md.: Johns Hopkins Press.

Miller, Edward. 1973. "Some Implications of Land Ownership Patterns for Petroleum Policy." *Land Economics* 69 (November).

———. 1975. "Percentage Depletion and the Level of Domestic Mineral Production." *Natural Resources Journal* 15, no. 2 (April).

Milliman, J.W. 1962. "Can People Be Trusted with Natural Resources?" *Land Economics* 3 (August).

"Mining Industry Views Proposed Colorado Severance Tax as a Disaster." *Engineering and Mining Journal* 76 (May).

"Mining and Taxation of Metallic Minerals in Wisconsin." Madison: Wisconsin Legislative Council Staff, September 1976.

Mocrakis, M.K., ed. 1976. *Energy: Demand, Conservation and Institutional Problems*. Cambridge, Mass.: M.I.T. Press.

Morgan, William E., and Shelton, Robert B. 1976. "Natural Resource Taxation, Tax Exportation and the Stability of Fiscal Federalism." Paper no. 114. Laramie: University of Wyoming, August.

Morissett, Irving. 1967. "Economic Theory and Risk Policy." In *Extractive Resources and Taxation*, edited by Mason Gaffney. Madison: University of Wisconsin Press.

Morse, C. 1973. "Natural Resources as a Constraint on Economic Growth: Discussion." *American Economic Review* 63, no. 2 (May).

Murdock, S.H., and Leistritz, F.L. 1979. *Energy Development in the Western United States: Impact on Rural Areas*. New York: Praeger Publishers.

Murphy, J.J., ed. 1972. "Energy and Public Policy." Report No. 575. New York: The Conference Board.

Musgrove, P. 1976. "The Distribution of Metal Resources (Tests and Implications of the Exponential Grade-Size Relation)." *Proceedings of the Council of Economics of the American Institute of Mining, Metallurgical and Petroleum Engineers*. New York: AIME.

Myers, John. 1977. "Optimal Extraction and Reserves of Depletable Resources under Uncertain Discovery." Unpublished, University of New Mexico.

Nehrig, Richard, and Zycher, Robert. 1976. "Coal Development and Government Regulation in the Northern Great Plains: A Preliminary Report." Santa Monica, Calif.: The Rand Corporation.

Nelson, C.W. 1967. "An Ex Ante Profitability Analysis of Capital Investment in Taconite Pellet Production Facilities in Minnesota circa 1950–1951." Ph.D. dissertation, University of Minnesota, July.

Netschert, B.C. 1953. *The Future Supply of Oil and Gas*. Baltimore, Md.: Johns Hopkins Press.

Nordhaus, William D. 1973a. "The Allocation of Energy Resources." *Brookings Papers* 3.

———. 1973b. "World Dynamics—Measurement without Data." *Economic Journal* 82, no. 332 (December).

————. 1974. "Resources as a Constraint on Growth." *American Economic Review* 64 (May).

Noren, Nils-Erik. 1971. "Mine Development—Some Decision Problems and Optimization Models." In *Decision Making in the Mineral Industry*, edited by J.I. McGerrigle. 9th International Symposium on Techniques for Decision Making in the Mineral Industry, 1971.

Norgaard, R.B. 1975. "Resource Scarcity and New Technology in U.S. Petroleum Development." *Natural Resources Journal* 15, no. 2 (April).

Oakland, William H. 1972. "Corporate Earnings and Tax Shifting in U.S. Manufacturing, 1930–1968." *Review of Economics and Statistics* 56 (August).

Oberbillig, Ernest. 1964."Appraisal of Mineral Land." *Appraisal Journal* 32, no. 4 (October).

O'Brian, D.T., and Weiss, A. 1968. "Practical Aspects of Computer Methods in Ore Reserve Analysis." *CIMM*, special vol. 9.

O'Calloghan, Terry A. 1967. "The Mining Law and Multiple Use." *New Era of Conservation* 7.

O'Neil, Thomas J. 1974. "The Minerals Depletion Allowance: Its Importance in Nonferrous Metal Mining." *Mining Engineering* 26, no. 10 (October).

Oren, M.E., and Williams, A.C. 1975. "On Competitive Bidding." *Operations Research* 23, no. 6 (November/December).

Page, Talbot. 1977. *Conservation and Economic Efficiency*. Baltimore, Md.: Johns Hopkins Press.

Paish, F.W. 1938. "Causes of Changes in Gold Supply." *Economica*, New Series 5 (November).

Parks, Ronald D. 1972. "Valuation of Mineral Property." In *The Valuation of Oil and Mineral Rights*. Chicago: International Association of Assessing Officers.

Paschall, Robert H. 1972. "Valuation of an Oil and Gas Producing Property." In *The Valuation of Oil and Mineral Rights*. Chicago: The International Association of Assessing Officers.

————. 1974. "Stock Market Derivation of Discount Rates." *Appraisal Journal*, April.

Pearce, D.W., and Rose, J., eds. 1975. *The Economics of Natural Resource Depletion*. London: Macmillan Press Ltd.

Peck, Ann E. 1976. "Futures Markets, Supply Response and Price Stability." *Quarterly Journal of Economics*, August.

Pepper, H.W.T. 1968. "The Fiscal Treatment of Mineral Operations." *Bulletin for International Fiscal Documentation*. Part I: 22, no. 9 (September); Part II: 22, no. 10 (October).

Peterson, Frederick M. 1972. "The Theory of Exhaustible Natural Resources: A Classical Variational Approach." Ph.D. dissertation, Princeton University.

————. 1975a. "A Theory of Mining and Exploring for Exhaustible Resources." Mimeographed. College Park: University of Maryland.

————. 1975b. "Two Externalities in Petroleum Exploration." In *Studies in Energy Tax Policy*, edited by G.M. Brannon. Cambridge, Mass.: Ballinger Publishing Company.

————. 1976. "An Economic Theory of Mineral Leasing." In *Mineral Leasing as an Instrument of Public Policy*, edited by M. Crommelin and A.R. Thompson. Vancouver: University of British Columbia Press.

————. 1978. "A Model of Mining and Exploring for Exhaustible Resources." *Journal of Environmental Economic and Management* 5 (September).

Peterson, Frederick M., and Fisher, A.C. "The Exploitation of Extractive Resources: A Survey." *Economic Journal* 87 (December).

Petry, Glenn H. 1975. "Empirical Evidence on Cost of Capital Weights." *Financial Management* 4 (Winter).

Pierce, W.S. 1974. "Factors Affecting Responsiveness to Technological Innovations in Coal Mining." Working Paper no. 54, Research Program in Industrial Economics. Cleveland, Ohio: Case Western Reserve University.

Pindyck, Robert S. 1979. "Advances in the Economics of Energy Resources." In *The Structure of Energy Markets, The Production and Pricing of Energy Resources*. Greenwich, Conn.: JAI Press.

Posner, M.V. 1972. "The Rate of Depletion of Gas." *Economic Journal* 82, no. 325 (March).

Potter, N., and Christy, F.T., Jr. 1962. *Trends in Natural Resource Commodities: Statistics of Prices, Output, Consumption, Foreign Trade and Employment in the United States, 1870–1957*. Baltimore, Md.: Johns Hopkins Press.

Preston, Lee E. 1960. *Exploration for Non-Ferrous Metals*. Washington: Resources for the Future.

Pugh-Robert Associates, Inc. 1976. "Materials Policy Analysis: A Case Study of Copper." Draft of Final Report to the Office of Science and Technology Policy, October.

Puo. T. 1977. "On the Profitability of Exhausting Natural Resources." *Journal of Environmental Economics and Management* 4.

Radawsky, O. 1970, 1971. "Economic Evaluation Techniques for Mining Investment Projects." *Mineral Industries Bulletin*, Colorado School of Mines, November and January.

Ramsey, William. 1979. *Unpaid Costs of Electrical Energy*. Baltimore, Md.: Johns Hopkins Press.

Rawls, J. 1971. *A Theory of Justice*. Cambridge, Mass.: Harvard University Press.

Reece, D. 1979. "An Analysis of Alternative Bidding Systems for Leasing Offshore Oil." *Bell Journal of Economics* 10, no. 2 (Autumn).

"Report to the Federal Trade Commission on Federal Energy Land Policy: Efficiency, Review, and Competition." Washington: GPO, 1975.

Reul, R.I. 1968. "Which Investment Appraisal Technique Should You Use?" *Chemical Engineering*, April 22.

Rist, K.A.; Raney, L.C.; and Wiebe, H.A. 1965. "The Equivalent Annual Amount System—A New Approach to Investment Analysis." In *Proceedings of the Symposium on Computers and Computer Applications in Mining Exploration*. Tucson: University of Arizona.

Roberts, Warren Aldrich. 1944. *State Taxation of Metallic Deposits*. Cambridge, Mass.: Harvard University Press.

Robinson, Colin. 1975. "The Depletion of Energy Resources." In *The Economics of Natural Resource Depletion*, edited by D.W. Pearce and J. Rose. London: Macmillan Press, Ltd.

Roff, Arthur W., and Franklin, James C. 1964. "A Statistical Mine Model for Cost Analysis, Planning and Decision Making." *Quarterly of Colorado School of Mines* 59, no. 4 (October).

Rosenberg, Nathan. 1973. "Innovative Responses to Materials Shortages." *American Economic Review* 63 (May).

Ross, Marion. 1965. "British Tax Treatment of Mineral Assets." *National Tax Journal* 18, no. 4 (December).

Ross, R.S. and Garnaut, P. 1975 "Uncertainty, Risk Aversion and the Taxing of Natural Resource Projects." *Economic Studies* 85, no. 338 (June).

Rottlowski, F.E., and Biederman, R.A. 1971. "Detailed Report on Mining in New Mexico in 1970." *New Mexico Business* 24, no. 5 (May).

Rumfelt, H. 1961. "Computer Method for Estimating Proper Machinery Mass for Stripping Overburden." *Mining Engineering* 13, no. 5 (May).

Ryan, J.M., and Gardner, A.O. 1965. "The Measurement of Dynamic Risk." *Southern Economic Journal* 31 (January).

Sandler, Todd M., and Shelton, Robert B. 1972. "Fiscal Federalism, Spillovers and the Export of Taxes." *Kyklos* 25.

Schenk, G.H.K. 1966. "Selected Bibliography Serves as Guide to Modern Mine Valuation Methods." *Engineering and Mining Journal* 167 (September).

Schlottman, Alan, and Abrams, Lawrence. 1977. "Sulfur Emissions Taxes and Coal Resources." *Review of Economics and Statistics* 59, no. 1 (February).

Schultze, C.L. 1977. *The Public Use of Private Interest*. Washington: The Brookings Institution.

Schulze, William D. 1974. "The Optimal Use of Non-Renewable Resources: The Theory of Extraction." *Journal of Environmental Economics and Management* 15.

Scott, Anthony D. 1953, "Notes on User Cost." *Economic Journal* 63 (June).

———. 1955. *Natural Resources: The Economics of Conservation*. Toronto: University of Toronto Press.

———. 1967. "The Theory of the Mine under Conditions of Certainty." In *Extractive Resources and Taxation*, edited by Mason Gaffney. Madison: University of Wisconsin Press.

Scott, Anthony D., ed. 1976. *Natural Resource Revenues: A Test of Federalism.* British Columbia Institute for Economic Policy Analysis Series. Vancouver: University of British Columbia Press.

Seetz, W.D. 1972. "An Analysis of Strip Mining and Local Taxation Practices." *Illinois Agriculture Economics* 15, no. 2 (January).

Shelton, Robert B., and Morgan, William E. 1977. "Resource Taxation, Tax Exportation and Regional Energy Policies." *Natural Resources Journal* 12, no. 2 (Fall).

Shkatov, V. 1969. "Prices on Natural Resources and the Problem of Improving Planned Price Formation." *Problems in Economics* 2, no. 2 (June).

Shoven, John B., and Whaley, John. 1972. "A General Equilibrium Calculation of the Effects of Differential Taxation of Income from Capital in the U.S." *Journal of Public Economics* 1.

Siegfried, John J. 1974. "Effective Average U.S. Corporation Income Tax Rates." *National Tax Journal* 27 (June).

Simon, J.A., and Smith, W.H. 1968. "An Evaluation of Illinois Coal Reserve Estimates." *Proceedings of the Illinois Mining Institute*, Springfield, Illinois.

Singer, D.A. 1976. "Mineral Resource Models and the Alaskan Mineral Resource Assessment Program." In *Non-Fuel Mineral Models: A State of the Art Review*, edited by W. Vogely. Baltimore, Md.: Johns Hopkins Press.

Singer, D.A.; Cox, D.P.; and Drew, L.J. 1975. "Grade and Tonnage Relationships among Cooper Deposits." U.S. Geological Survey, Professional Paper 907-A. Washington: GPO.

Slater, Lloyd E. 1974. "Interstate Allocation of Business Income: U.S. Policy and Experience." *1973 Proceedings of the Sixty-Sixth Annual Conference on Taxation Held Under the Auspices of the National Tax Association—Tax Institute of America*. Edited by Stanley J. Bowers. Columbus, Ohio.

Smith, D., and Wells, L. 1975. "Mineral Agreements in Developing Countries: Structures and Substance." *American Journal of International Law* 67.

————. 1976. *Reconcessions Process: Mineral Agreements in Developing Countries*. Cambridge, Mass.: Ballinger Publishing Company.

Smith, V. Kerry. 1974. "Re-Examination of the Trends in the Prices of Natural Resource Commodities, 1970–1972." Working Paper no. 44, Economic Growth Institute, State University of New York. Binghamton: November.

————. 1978. "Scarcity and Growth Reconsidered." *Materials and Society* 2 (3).

Smith, V. Kerry, ed. 1979. *Scarcity and Growth Reconsidered*. Baltimore, Md.: Johns Hopkins Press.

Smith V.L. 1963. "Tax Depreciation Policy and Investment Theory." *International Economic Review* 4, no. 1 (January).

————. 1968. "The Economics of Production from Natural Resources." *American Economic Review*, June.

————. 1974. "An Optimistic Theory of Exhaustible Resources." *Journal of Economic Theory* 9, no. 4 (December).

Solow, Robert M. 1974*a*. "The Economics of Resources and the Resources of Economics." *American Economic Review* 64, no. 2 (May).

————. 1974*b*. "Intergenerational Equity and Exhaustible Resources." *Review of Economic Studies, Symposium on the Economics of Exhaustible Resources*.

Solow, Robert M., and Wan, F.Y. 1976. "Extraction Costs in the Theory of Exhaustible Resources." *Bell Journal of Economics* 7 (Autumn).

Sophy, Gerald. 1858. "Appraisal of Gravel Pit Sites." *Appraisal Journal* 26, no. 3 (July).

Sorenson, J.B., and Greenfield, R. 1977. "New Mexican Nationalism and the Evolution of Energy Policy in New Mexico." *Natural Resources Journal* 17, no. 2 (Fall).

"South Africa in Explosive Mineral Potency: Leasing and Taxation—The Investment Ground Rules." *Engineering and Mining Journal* 173, no. 11 (November 1972).

Sprague, J.W., and Julian, B. 1970. "An Analysis of the Impact of an All Competitive Leasing System On Off-Shore Oil and Gas Leasing Revenue." *Natural Resources Journal* 10, no. 3 (July).

Stahl, J.E., and Lajzeroqicz, J. 1974. "Mining Investment Evaluation Techniques—A Comparative Analysis." *Canadian Mining and Metallurgical Bulletin* 67, no. 748 (August).

State Board of Equalization. 1966. "Appraisal of Oil and Gas Producing Properties." Sacramento, California.

Steele, Henry. 1967. "Natural Resource Taxation: Resource Allocation and Distribution Implications." In *Extractive Resources and Taxation*, edited by Mason Gaffney. Madison: University of Wisconsin Press.

———. 1974. "Cost Trends and the Supply of Crude Oil in the United States: Analysis and 1973–1985 Supply Schedule Projections." In *Energy: Demand, Conservation, and Institutional Problems*, edited by M.S. Macrakis. Cambridge, Mass.: M.I.T. Press.

Steiner, Peter O. 1959. "Percentage Depletion and Resource Allocation." In *Tax Revision Compendium*, by U.S. Congress, House Committee on Ways and Means. Washington: GPO.

———. 1963. "The Non-Neutrality of Corporate Income Taxation—With and without Depletion." *National Tax Journal* 16 (September).

———. 1964. "Rejoinder to McDonald." *National Tax Journal* 12 (March).

Stermole, F.J. 1971. *Engineering Economy and Investment Decision Methods*. Golden: Colorado School of Mines.

Stiglitz, Joseph E. 1974. "Growth with Exhaustible Natural Resources: Efficient and Optimal Growth Paths." *Review of Economic Studies, Symposium on the Economics of Exhaustible Resources*.

———. 1975a. "The Efficiency of Market Prices in Long Run Allocations in the Oil Industry." In *Studies in Energy Tax Policy*, edited by Gerard M. Brannon. Cambridge, Mass.: Ballinger Publishing Company.

———. 1975b. "Monopoly and the Rate of Extraction of Exhaustible Resources." Mimeographed. Palo Alto, Calif.: Stanford University.

Stinson, Thomas F. 1977, 1978. *State Taxation of Mineral Deposits and Production*. Washington: Environmental Protection Agency (1977), U.S. Department of Agriculture (1978).

Stinson, Thomas F., and Voelker, Stanley W. 1978. *Coal Development in the Northern Great Plains*. Washington: Department of Agriculture.

Stobaugh, Robert, and Yergin, Daniel, eds. 1979. *Energy Future—Report of the Energy Project at the Harvard Business School*. New York: Random House.

Stober, W.T., and Falk, L.H. 1967. "Property Tax Exemption: An Inefficient Subsidy to Industry." *National Tax Journal* 4 (December).

———. 1970. "Poorly Conceived Financial Inducements: A Study of Louisiana's Gas Severance Rebate." *Social Science Quarterly* 51, no. 1 (June).

Sulerud, Allen C. 1972. "Mineral Valuation: Analysis of Taxpayer Reporting." In *Valuation of Mineral Resources*. Chicago: International Association of Assessing Officers.

Surrey, A.J., and Page, William. 1975. "Some Issues in the Current Debate about Energy and Natural Resources." In *Economics of Natural Resource Depletion*, edited by D.W. Pearce and J. Rose. London: Macmillan Press.

Sweeney, James L. 1977. "Economics of Depletable Resources: Market Forces and Intertemporal Bias." *Review of Economic Studies* 64 (February).

Symposium on Mine Taxation. 1969. Department of Mining and Geological Engineering, College of Mines, University of Arizona, March 12–13.

Tambini, Luigi. 1969. "Financial Policy and the Corporation Income Tax." In *The Taxation of Income from Capital*, edited by Arnold C. Harberger and Martin J. Bailey. Washington: The Brookings Institution.

"Taxation with Representation." *The Petroleum Industry's Tax Burden.* Compendium prepared for the House Committee on Ways and Means and the Senate Committee on Finance, Washington, 1973.

Timbrell, D.Y., assisted by Anson-Cartwright, H. 1967. "Taxation of the Mining Industry in Canada." In *Studies of the Royal Commission on Taxation*, no. 9. Ottawa: Queen's Printer.

Toman, N.E. et al. 1977. "A Fiscal Impact Model for Rural Industrialization." *Western Journal of Agricultural Economics* 1(1).

Tomas, L.J. 1973. *An Introduction to Mining.* Sidney: Hicks Smith & Sons.

Turek, Joan L. 1970. "Short-Run Shifting of the Corporate Income Tax in Manufacturing, 1935–1965." *Yale Economic Essays* 10 (Spring).

Turvey, R. 1968. *Optimal Pricing and Investment in Electricity Supply.* Cambridge, Mass.: M.I.T. Press.

Uhler, R.S., and Bradley, P.G. 1970. "A Stochastic Model for Determining the Economic Prospects of Petroleum Exploration over Large Regions." *Journal of the American Statistical Association* 65, no. 339 (June).

United Nations Secretariat. 1972. *Analysis of Taxation of Mineral Resources in Developing Countries.* No. ST/SG/AC.8/R.36.New York: July.

U.S. Bureau of Mines. 1971. *Strippable Reserves of Bituminous Coal and Lignite in the United States.* Information Circular 8531. Washington: GPO.

———. 1972. *Cost Analyses of Model Mines for Strip Mining in the United States.* Information Circular 2535. Washington: GPO.

———. 1974a. *Basic Estimated Capital Investment and Operating Costs for Underground Bituminous Coal Mines.* Information Circulars 8632 and 8641. Washington: GPO.

———. 1974b. *The Reserve Base of U.S. Coals by Sulfur Content.* Information Circulars 8632 and 8641. Washington: GPO.

———. 1976. "Coal—Bituminous and Lignite in 1967–1976." In *Mineral Industry Surveys.* Washington.

U.S. Congress,Committee on Ways and Means. 1973. *General Tax Reform.* Panel Discussions, Part 9, February 26.

U.S. Department of Commerce, Bureau of Census. 1978. *County Business Patterns: 1967–1976.* Washington.

———. 1979. *1977 Census of Mineral Industries.* Washington.

U.S. Department of Energy. 1979. *National Energy Plan II, May 1979.* Washington.

U.S. Department of the Interior, Bureau of Land Management. 1970. *The Role of Petroleum and Natural Gas from the Outer Countinental Shelf in the National Supply of Petroleum and Natural Gas*. Technical Bulletin no. 5. Washington.

U.S. Department of the Treasury. 1978. *Effective Income Tax Rates Paid by United States Corporations in 1972*. Washington: GPO.

U.S. General Accounting Office. 1979. "A Review of the Department of Energy Tax Policy Analysis." Comptroller General's Report to the Congress of the United States, EMD-79-26. Washington.

U.S. Geological Survey. 1973. *United States Mineral Resources*, edited by D.A. Brobst and W.P. Pratt. U.S. Geological Survey, Professional Paper no. 820. Washington: GPO.

U.S. Government, Secretary of the Interior. 1976. *Mining and Minerals Policy—1976 Bicentennial Edition*. Washington: GPO.

United States Mineral Resources. Geological Survey, Professional Paper no. 820. Washington: 1973.

U.S. Office of Management and Budget. 1978. "Energy Impact Assistance Needs." Mimeographed. Washington: March.

"Utilities Will Sue Montana over Coal Tax." *High Country News*, March 21, 1979.

Vernor, William J., and Shurtz, Robert F. 1966. "For Mine Evaluation—A Fresh Model." *Mining Engineering* 8 (November).

Vickrey, William. 1967. "Economic Criteria for Optimum Rates of Depletion." In *Extractive Resources and Taxation*, edited by Mason Gaffney. Madison: University of Wisconsin Press.

Voelker, S.W., and Stinson, T.F. 1979. *Taxation of Lignite Mining in Montana and North Dakota*. Washington: U.S. Department of Agriculture.

Vogely, William, ed. 1975. *Mineral Materials Modeling*. Washington: Resources for the Future.

Vousden, N. 1973. "Basic Theoretical Issues in Resource Depletion." *Journal of Economic Theory* 6, no. 2 (April).

Walters, A.A. 1963. "Production and Cost Functions: An Econometric Survey." *Econometrica*, January–April.

Weinaug, C.F. 1964. "Proposed Ad Valorem Tax Property Valuation Method for Oil." *Computers in the Mineral Industry*. Palo Alto, Calif.: Stanford University Publications, Geological Sciences.

Weinstein, Milton C., and Zeckhauser, Richard J. 1974. "Use Patterns for Depletable and Recyclable Resources." *Review of Economic Studies, Symposium on the Economics of Exhaustible Resources*. 1975.

———. 1975. "The Optimal Consumption of Depletable Natural Resources." *Quarterly Journal of Economics* 89, no. 3 (August).

Wright, Arthur W. 1975. "Taxation of Natural Resources,Discussion." *American Economic Review* 65, no. 2 (May).

Wright, Brian D. 1976. "The Taxation of Petroleum Production." Ph.D. thesis, Harvard University, January.

Yasnowsky, Phillip N., and Graham, Annette P. 1976. "State Severence Taxes on Mineral Production." In *American Institute of Mining, Metallurgical and Petroleum Engineers Council of Economics—Proceedings*, March.

Zimmerman, Martin B. 1977. "Model in Depletion in a Mineral Industry: The Case of Coal." *Bell Journal of Economics* 8, no. 1 (Spring).

————. 1979a. "Estimating a Policy Model of U.S. Coal Supply." In *Advances in the Economics of Energy and Natural Resources*, edited by R.C. Pindyck. Greenwich, Conn.: JAI Press.

————. 1979b. "Rent and Regulation in Unit-Train Rate Determination." *Bell Journal of Economics* 10, no. 1 (Spring).

————. 1980a. *The U.S. Coal Industry: The Economics of Policy Choice*. Cambridge, Mass.: M.I.T. Press.

————. 1980b. "Western Coal Producing Cartel." Paper presented to the International Association of Energy Economists, Denver, September, and Sloan School, Massachusetts Institute of Technology.

Zimmerman, Martin B., and Ellis, R.P. 1980. "What Hapened to Nuclear Power: A Discrete Choice Model of Technology Adoption." Energy Lab Working Paper no. 80-002. Cambridge, Mass.: Massachusetts Institute of Technology.

Zwartendly, K. 1974. "The Life Index of Mineral Resources—A Statistical Mirage." *Canadian Mining and Metallurgical Bulletin* 67, no. 750 (October).

Index

About the Author

Albert M. Church received the B.A. degree from Colorado College in 1963 and the Ph.D. from Claremont Graduate School in 1971. He has taught at Middlebury College and is currently an associate professor at the University of New Mexico. He is the author of numerous publications in professional journals and of *Computers and Statistics in the Appraisal Process* (with R. Gustafson) and *The Sophisticated Investor*.